Tribulation

God's Plan for the End of the Age

Other Books By R. H. Vargo…

Eschatology & Prophecy:

2028 (God's Plan for Mankind): First printing December 2007

Tribulation

God's Plan for the End of the Age

R. H. Vargo

authorHOUSE®

AuthorHouse™
1663 Liberty Drive
Bloomington, IN 47403
www.authorhouse.com
Phone: 1-800-839-8640

Tribulation (God's Plan for the End of The Age)
Prophecy; Christian Eschatology
© 2009 R. H. Vargo. All rights reserved.

First published by AuthorHouse 12/7/2009

ISBN: 978-1-4490-3656-0 (e)
ISBN: 978-1-4490-3654-6 (sc)
ISBN: 978-1-4490-3655-3 (hc)

Library of Congress Control Number: 2009912137

This book is printed on acid-free paper.

Printed in the United States of America
Bloomington, Indiana

Contents

Dedicated to the LORD of the Heavens and Earth…

I said to him, "Sir, you know." And he said to me, "These are they who have come out of the great tribulation; they have washed their robes and made them white in the blood of the Lamb.

Revelation 7:14 (RSV)

Introduction

AFTER WRITING MY FIRST BOOK "2028 (God's Plan for Mankind)" I was sure it would be my last. I'm not a writer and really had no intentions of writing any books in my lifetime. But the Lord has a way of using us as He wishes whether we want Him to or not. After finishing the book, I realized there were a few things that troubled me, which I had not adequately addressed. The first was the heavy use of mathematics made it hard for many to read and comprehend the revelation that God has a plan for mankind.

My daughter said it was like reading a college textbook. I thought by using mathematics no one would be able to refute what I had written. But I was mistaken. Not in that they could refute it, but that they would even try. Some still did not believe it was possible to know when Christ would return because of their preconceived beliefs and their inability to grasp the mathematical concepts. In other words, they argued against what was written without really understanding the Bible prophecies and mathematics that provided the foundation of the conclusions.

Another point that was lacking in the first book was a clearer understanding of the rapture. Not what the rapture is, but whether there is going to be a rapture of the church and when this event would take place if there is one. This was not something the Holy Spirit had revealed to me nor did I do a thorough study of scripture to find this truth. This was not the focal point of that book. That focus was to explain that God had a plan for mankind from the very beginning and He was following His plan.

I do not want to misrepresent that I have not studied the theology of the rapture idea. On the contrary, I have spent many years studying what others have written on this topic, but I have been unable to determine which of the various viewpoints is genuine because I see some truth in all the positions put forth by biblical scholars on this matter. I have not done a lot of self-study with just the Bible as my foundation to see where this path might lead. Understanding more and more of God's purpose for mankind opens one's eyes to seeing how other ideas fit and I believe I now have a better understanding of where this piece belongs.

Lastly, I began to realize just because the Holy Spirit had revealed the year and season of Christ's return, that this might not be enough. Scripture reveals many details about future history that will aid Christians during those terrible times… what to expect and what is expected of us.

My hope is to prepare you for what is to come by telling you as plainly as I can in terms that all can understand. Some of my thoughts may be open for interpretation and in these cases I will give you my best explanation in the hopes it may help. Please know these things may never come to pass because I was not shown any visions for these future events. If I have ever had prophetic dreams, which is possible, I have never recognized them as such and clearly I am unable to interpret them. These are spiritual insights that the Lord has not blessed me with at this time.

All my prophecy interpretations are based on other scripture interpretations, inspiration from the Holy Spirit, projections of current earthly trends, common sense, and my understanding of God's plan, His use of numbers, and Satan's plan of deceiving the world by mimicking Christ and the real plan. Perhaps as I write this book, the Lord will unveil the truth to me with greater conviction and power just as He did in the first book and if I write things that are untrue I will pray for His forgiveness should my words ever lead any astray. My only intent is to prepare Christians for the coming tribulations, inform them of their responsibilities during this time, and to provide advanced training so they will be better equipped to handle what is expected of them.

But before we get into the details of this upcoming period of time—the "tribulation" period, I would like to briefly review the basics of God's overall plan for those who never read my first book. It is important to realize and believe there is a plan God is working from so that these details of the end-times can be seen more clearly. My wish is that, by knowing what the future holds, you may find the inner peace necessary to persevere during the cataclysmic times ahead.

Chapter 1

God's Plan for Mankind

GOD HAS A PLAN AND HE is working His plan… even now… at this very moment. This plan was predetermined from the beginning of creation. A plan for mankind that has been hidden in the words of the Bible and its prophecies for centuries.

A simplified version or understanding of this plan is shown below and reveals many things to those who study it. In my first book these secrets were unlocked and they revealed unimaginable things that the Lord wants you to know before his return.

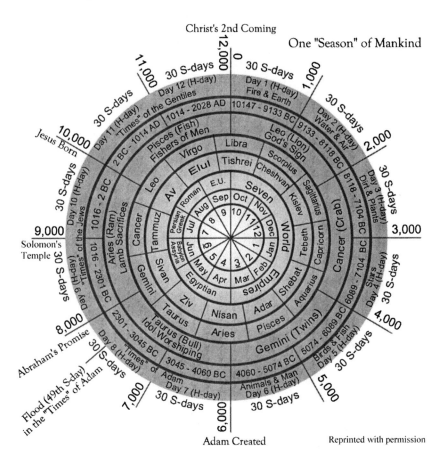

1

This version of the plan shows the beginning of creation through Christ's second coming in 2028 A.D.—after shocking natural disasters, the War of Gog (Russia and an Arab coalition attacking Israel) as prophesied by Ezekiel, World War III's nuclear war, the rise of the antichrist as the dictator (president) of the European Union, the false prophet (president of the European central bank), a global cashless economy requiring a world ID (the "mark of the beast"), and the decline and replacement of the United States as the only super power currently on earth. These are just a few of the things that will take place in our remaining years... the last years of this age.

But before we get into those particulars and many more details later in the book, I would like for you to see the picture of the overall plan God has in store for mankind. By grasping the "big picture," you will be able to see smaller pieces of the plan with greater clarity. I will not go into the particulars on how this was developed, but just provide a quick synopsis of the plan since the focus of this book is to look at the remaining time we have left until the end of the age when Christ returns.

After Creation

Adam was created at the end of the sixth day of creation. This was the six-thousandth "prophetic" year and this event started the clock of human history—of his story: the story of Christ which all of mankind has been part of. This half of God's plan is divided into three equal parts: three trimesters of two thousand prophetic years (P-yrs.) each. Prophetic years are years from God's perspective that have 360 prophetic days in them rather than the present 365.25 days in a year that man measures. Therefore, these two thousand prophetic year increments equate to two thousand twenty-nine years and two months[1] exactly of recorded history when they are converted to time as measured on earth.

The "Times of Adam"

The "Times of Adam" are the first block of time and they cover the events from the creation of Adam until the first promise from God given to Abraham at the age of seventy-five[a] that he would be the father of a great nation who would become God's people... the Israelites... the Jews. This block of time was just over two thousand and twenty-nine

2

years long and is also recognized as the Time of Lawlessness, the Beginning Times, and the first trimester of history, which occurred during the Age of Taurus.

The "Times of the Jews"

The "Times of the Jews" are the second trimester of God's plan and they encompass the events from Abraham's promise to the birth of Christ. This block of time is just over two thousand and twenty-nine years long as well and is also identified as the age under God's Law… the Middle Times and it coincided with the Age of Aries.

The "Times of the Gentiles"

The "Times of the Gentiles" are the third portion of God's plan and they include the events from Christ's birth until Christ's return. We are currently living at the end of the "Times of the Gentiles." This block of time is just over two thousand and twenty-nine years long, by symmetry, and is also acknowledged as the Time of Grace, the third trimester, and the End Times. This period of history is the Age of Pisces.

The Millennial Kingdom

The "Millennial Kingdom" is one thousand, fourteen years and seven months long (one thousand prophetic years)[2] and will follow the three trimesters of man's rule with this half a trimester of Christ's rule before heaven and earth are remade.[b] This is the future age that is knocking at the doorstep of time.[c] This enlightened period will take place during the Age of Aquarius on God's celestial wheel of time.

7,000 Prophetic Years

We see by adding these blocks of time together that the total time for mankind is seven thousand prophetic years long (from God's perspective), but only one week on God's calendar in heaven. Six "days" of work (man's rebellion) for God followed by one day of rest during Christ's reign. Or as God says concisely, "Time (two thousand P-years), Times (four thousand P-years) and half a Time (one thousand P-years)

3

for His whole plan. This truism, "time, times, and half a time," is an accounting method of actual historical time. God uses it for smaller pieces on His timeline as well as the overall time.

Let's now look at a simple linear timeline of God's overall plan for mankind as we have discussed so far.

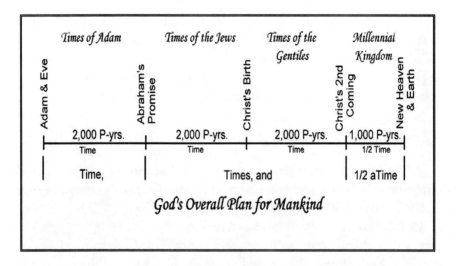

God's Overall Plan for Mankind

We can easily see the symmetry of the plan from this timeline. Next, we are going to add dating information that was developed in the first book, as well as a few other interesting details from that book, to see what this additional information reveals.

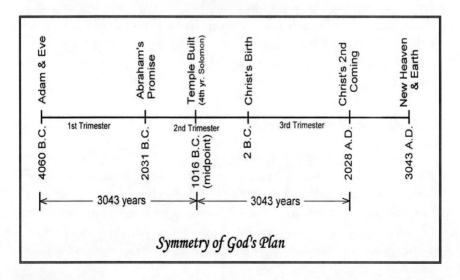

Symmetry of God's Plan

Notice that on a yearly analysis, there are three thousand and forty-three years (and some months) between when Adam and Eve were created and the construction of first temple to God was started. This was the fourth year of Solomon's reign. This event was also the mid-point between Adam and Christ's second coming. One can also see on the second timeline that there are another three thousand and forty-three years (and some months) from the first Jewish temple until Christ returns.

Furthermore, the time from Christ's birth to the end of human history when heaven and earth will be remade is also three thousand and forty-three years long. This observation is not shown on the timeline. What is most surprising is when this information was laid out on the timeline the date 3043 A.D., for the end of the Millennial Kingdom, turned out to be the exact same number as the years between major events on the timeline! Is it by chance that these numbers are exactly the same or evidence of God's planning?

These timelines are nothing more than taking the time from six thousand P-years to twelve thousand P-years on the overall plan (one season of mankind—one S-year) shown on page one and straightening it out; just another way to represent that portion of God's plan that is of further interest to us. Some cannot easily see the things hidden within the plan, but these timelines presented in this fashion help them to understand and grasp God's overall plan.

Understanding the overall picture, the "macro" version of God's calendar allows us to understand how God operates. The Lord of the Universe is the master of master mathematicians and uses patterns over and over. Now the times shown for the overall plan were proved mathematically and with scripture to be true in my first book. It was also shown that God uses smaller chunks of time based on this plan as well. Blocks of time that are reduced by factor(s) of ten depending on whether the timeframe deals with an individual, a nation, or mankind as a whole. Just so you know, God has used seven, seventy, and seven hundred P-years for specific pieces within His overall plan in the past and future.

You will learn that seven hundred P-yrs. is the answer to one of Daniel's prophecies that unlocks the hidden timing of Christ's return. That seventy prophetic years is the answer for how long the "beast" will be in existence before Christ's return.

And lastly, it will be shown that the classical understanding of the tribulation period and the antichrist's rule will be for seven P-years. All these times fit together into God's plan and further support each other as being totally true. Now have any eschatology experts or writers ever provided dates and timing for events? Dates and times that are undeniable when you will see how they were derived.

All the timeframes mentioned pertain to the end-time prophecies given by the prophets to inform man ahead of time of God's plans so that we might not be unaware of their timing or the existence of a supreme being controlling everything. Who else can predict the future with one hundred percent certainty except God? No one! Now most believe the Bible does not provide dates for future events, but you will soon learn that it does! Those readers who know scripture will know the truth of this assertion after reading this book, as long as they seek with open eyes and open hearts. Jesus said,

> *Woe to you experts in the law, because you have taken away the key to knowledge. You yourselves have not entered [kingdom of God], and you have hindered those who were entering [from knowing the truth].*
>
> Luke 11:52 (NIV)

Basically, be careful not to let the religious experts tell you what can't be done and limit the Holy Spirit's power of showing you the truth. Open your eyes like a Berean[(d)] searching for true understanding instead of believing that you already know the truth!

Let's zero in on God's calendar to the most important event in human history—Christ's return. This is the one mystery you want answered is it not? The date Jesus comes back to rule the world and punish all unbelievers and lover's of every evil thing detestable to God. This is the good news! The bad news is those damned are our friends, neighbors, and family members who do not have Christ as the center of their lives and, without us sharing this news, they will most certainly spend eternity in the Lake of Fire.

Chapter 2

"2028" The Condensed Analysis

IN JUNE 2007, I WAS STUDYING the Bible and I was shown the timing of Noah's Flood. This understanding was so incredible that the next morning I began writing my first book "2028." As I wrote that book, God continued to show me His entire plan for mankind in various ways. Although I never intended to write about Christ's return when I began, the Lord had different ideas. As a result of seeing the overall plan, I was shown the timing of Christ's second coming. Not only was this timing uncovered, but I was also shown the timing in many different ways hidden within the Bible that all came to the same time—the autumn of the year 2028 A.D.!

It is impossible to tell you everything that was unveiled to me by the Holy Spirit without you reading my first book. Moreover, I don't believe everything I wrote will stand up as the absolute truth. The Bible says a true prophet of God can be identified from a false prophet because they proclaim only the word, they warn of future events, and one hundred percent of everything they predict comes true. Let me make this clear: I did not receive any visions or dreams that led me to these truths. Many of the "eurekas" and "breakthroughs" I had were while I was wide awake. Others came to me after prayer, while reading the Word, or sleeping. Some of these inspirations and insights really are hard to believe.

The purpose of this chapter is to try and condense many of these proofs of the timing of Christ's return into as small a space as possible so that you too can see there is no doubt as to the year. How many times does one need to be told or shown something before they can comprehend or believe it? Some people take but one time, while others need a few more. Some people no matter how you show them, explain it to them, or teach them, never understand. Let's pray you are not one of those for then I fear your heart may be truly hardened. It took Pharaoh ten plagues to believe (because his heart was hardened) before he would let the Israelites leave Egypt.

I will show you ten different ways that indicate Christ will come in the year 2028 A.D. and there are even a few more. The number ten is the Lord's number of completion and if I can't convince you after showing you ten different ways, then you may need to do some soul searching or studying with the Word. I have been unable to determine the exact day or hour and I find comfort in this knowledge because it confirms the Lord's word; "no one knows the day or the hour except the Father in heaven." But the season and the year are undeniable.

Method 1 (Daniel 10-12: The Kings of the South & the North Prophecy)

"Time, Times, and half a Time"... this is a phrase you will find at the end of Daniel 10-12[a] that is the answer given for the amount of time from the Persian Empire to the return of Christ when he will set up his earthly kingdom. This is also an algebraic problem where:

$$\text{Time} = t,$$
$$\text{Times} = 2t,$$
$$\&\ \tfrac{1}{2}\ \text{Time} = \tfrac{1}{2}t$$

and so we now have 3.5t. If we knew how much time "t" represented, then wouldn't we know the time of Christ's return? All of Daniel's prophecies start with the decree issued to Ezra on March 21, 458 B.C. from Artaxerxes, who was the Persian king at the time. This was the third of four decrees from Persian kings mentioned in the Bible.

If we test every number for "t" (brute-force technique) we find the number seven hundred is the right number. How do we know that? Well that is a long story, but the quick answer is this number is the number of perfection (seven) multiplied by the "intermediate" number of completion (one hundred).[b] So three and a half multiplied by seven hundred equals twenty-four hundred and fifty P-years of history from the seventh year of the Persian king Artaxerxes until Jesus returns.

We know that prophetic years (P-yrs.) are 360-day years from God's perspective and are not normal 365.25-day years (which are from man's perspective) so we need to make the usual adjustment. Making this perspective conversion to our last result gets us to our final answer of 2,485.68 years[3] for the total time.

By traveling forward this span of years from our starting point in 458 B.C. we arrive 2,027.898 years[4] into the future from the beginning year of our Lord (A.D.). Making a one-year correction because we crossed the B.C./A.D. threshold, and there is no "zero" year using this dating system, gives us an actual date. This date is 2028.898 A.D. and the date of Christ's appointed return. The 0.898 decimal portion of the date when converted further gives us the day... November 24, 2028 A.D.

Method 2 (Daniel 9: Seventy 'Weeks' Prophecy)

Daniel chapter nine implies there will be seventy 'weeks' (sevens: NIV) before the final remaking of heaven and earth when everything is finished. Daniel then separates out seven 'weeks' and sixty-two 'weeks'; then another 'week' and informs us of some things that will take place at these break points. For instance, after the sixty-nine 'weeks' the Messiah will be "cut-off (die)." Let's make that calculation so you can see the proof that all times given in Bible prophecies are "prophetic" and need to be converted: even though this prophecy pertains to the Messiah's first coming.

The Messiah is "Cut-Off" Prophecy

The first of Nisan in 458 B.C. (3303 A.M. on the Jewish calendar) was when Ezra officially left to go to Jerusalem.[c] This was March 9, 458 B.C. according to the Internet calendar website www.abdicate.net/ cal.aspx. However, Ezra really did not leave until the twelfth of Nisan after gathering things together for the long journey.[d] This would have been March 21st (vernal equinox and the first day of spring) and the day the Persian king Artaxerxes issued him the decree mentioned in Daniel's Seventy 'Weeks' prophecy to aid him in his travels through the Persian Empire.

We need to start counting sixty-nine 'weeks of years' which are 483 P-years,[5] from this date to determine when the Messiah was "cut-off (crucified)." Since this prophecy was given from God's perspective of 360-day years we need to make our conversion to normal years so we can compare this prophecy with actual history. Therefore, four hundred and eighty-three P-years are really four hundred and ninety years and thirteen days.[6]

By making this adjustment, we gained one additional 'week' (seven more years) using Daniel's accounting method! Adding this amount of time to the twelfth of Nisan (March 21, 458 B.C.) gives us Friday, April 3, 33 A.D. as the exact date of Jesus' crucifixion! Those who use the Word of God as being completely true and completely reliable will recognize this to be the day of our Lord's death!

This is an example of using prophetic years that comes to an exact date. Jesus said, "no one knows the day or the hour (of his return)." He didn't say this applied to everything foretold in the Bible.

This prophecy not only predicts the day of Jesus' death, but if the hour of the issuing of the decree were known, it would almost certainly predict the hour of his death (which the Bible indicates was around 3 pm). Those who teach, "no one can know the time" twist the truth of Christ's words and apply them to other parts of scripture. It's clear in this case the mathematics prove that this practice is in error. Let's return to Daniel's prophecy.

Daniel's prophecy of the seventy 'sevens' was one I prayed to the Lord for understanding and He showed me that Christ would return after five periods of seventy 'weeks' which are really five periods of four hundred and ninety P-years. This translates into twenty-four hundred and fifty P-years.[7] The same number we arrived at in the first method! Finishing the rest of the calculations would yield the same exact end date in 2028 A.D.

I searched the Internet for substantiation of this time and found a site (http//www.returntogod.com/jubilee/jubileehypothesis.htm) that had written a book which determined our Lord's return would be in 2027 A.D. I, however, believe their starting point was in error or they would have calculated 2028 A.D. as I had. It stated the authors had a revelation from God in 1997 that told them there would be seven 'weeks' of "Jubilee years" before Christ's return. Jubilee years are fifty-year periods and so, seven 'weeks' is forty-nine Jubilees and forty-nine Jubilees are twenty-four hundred and fifty normal years.[8] A similar answer to the one I was shown, but a different revelation than the one I received when I prayed to God for understanding.

They realized these were regular years when counting from the first Jubilee year after Ezra's decree (423/422 B.C.) and this method works too, but by using the wrong starting year an incorrect ending year was determined. It's also possible adding a year when crossing the B.C./A.D. date threshold was missed which would cause the same error. Nevertheless, they received from the Holy Spirit the same answer and the same number of years (2,450) as I had.

Method 3 (Luke 3: The 15th year of Tiberius)

The fifteenth year of Tiberius (Tiberius Caesar officially took the throne in September 14 A.D. after the death of Augustus in August) was September 28 A.D. to September 29 A.D. John the Baptist began his preaching in the spring of the fifteenth year. This could only mean that he began to preach in the spring of 29 A.D. because that was the only spring in the fifteenth year of Tiberius.

Since John was a Levite, he would have begun his service to the Lord on his thirtieth birthday and finished when he was fifty according to the Mosaic Laws. Jesus was John's cousin and six months younger. Thus, Jesus would have begun his ministry in the fall of 29 A.D. just short of his thirtieth birthday ("about thirty" as scripture records) since we now realize John was thirty.

When you add two thousand years (two days in heaven) to this date you reach the fall of 2029 A.D. However, Christ said "the appointed time would be cut short,"[e] which I believe this shortened time will be the amount of time necessary to get to the Day of Trumpets, the Feast of the Tabernacles, or the Day of Atonement in 2028 A.D. This is a repeat performance of the resurrection, but on a larger scale. The grand scale of God's overall plan. Jesus was in the tomb two days and rose on the morning of the third day.[f] We see here Jesus is gone "two days (in heaven)" and will return at the beginning of the third day.

Method 4 (God's Overall Plan… His Calendar)

Creation was seven days long, but these are seven days in heaven and not earth days! A day in heaven is like a "thousand years on earth"[g] and so we have seven thousand years. But these are prophetic years too. Man was made at the end of the sixth day in heaven after all the animals were created and on the seventh day God rested. This means the Lord rested for the first one thousand P-years of mankind's history.

This was like a "week" to God... a week in heaven. Let's call this the first week. The next week, the second week, is God's plan for mankind. It covered two thousand P-years (two H-days; days in heaven) from Adam to Abraham's promise (the "Times of Adam"). Then there were another two thousand P-years from Abraham to Christ's birth (the "Times of the Jews"). Next, there will be two thousand P-years from Christ's birth to his return (the "Times of the Gentiles").

Lastly, there will be one thousand P-years of Christ's rule on earth (the Millennial Kingdom; a Sabbath of rest for God when He can relax as Christ reigns). This is seven thousand P-years total and another week in heaven from God's viewpoint. A few individuals can calculate the time from Adam to Christ's ministry as four thousand years and then they add two thousand years to get to 2028 A.D. (www.cephasministry.com/water/bible_believers_6000_years.html). But there are minor errors with their math like Christ starting his ministry in 28 A.D. when in fact it was 29 A.D. However, these are really prophetic years and using their chronology of biblical events renders some passages from Genesis false.

We know that two thousand P-years are actually 2,029 years and two months exactly.[3] Thus, there were 2,029 yrs. from Adam to Abraham, 2,029 yrs. from Abraham to Jesus' birth in 2 B.C. and 2,029 yrs. from the birth of Jesus to his return in 2028 A.D.[9]

Now some modern day scholars want us to believe that Christ was born around 6 to 5 B.C. because of a William Whiston's footnote pertaining to the death of Herod in his 1737 A.D. translation of Josephus' "Antiquities of the Jews." This is a false teaching!

Ancient historians did not agree. Isaac Newton quotes many "old" historians who believed the year was 2 B.C. in a book published posthumously in 1733 A.D. dealing with Daniel and Revelation. He further agrees Christ began his ministry in 29 A.D. Who do you want to believe—the Bible account and Isaac Newton (who was also a theologian and historian) to do your math, or Mr. Whiston and present day scholar's who do Satan's work to change the "set times" as Daniel prophesied?[h] We will explore this claim in detail later.

Those who want you to believe 30 A.D. was the date of Christ's death, want you to believe the fifteenth year of Tiberius was around 25/26 A.D. Three years earlier than it really was. Check any history book that covers this period of history and I do mean any! If you are to believe them, then you have to believe the Bible is in error on this point. In other words, a lie! Using this idea projects the end of the tribulation to be in 2025 A.D.—two thousand years after this false teaching.

Those who believe this lie, will trust the lie the antichrist will tell them as well—that he is the messiah because he came back two thousand years exactly after his first coming (claiming he is the Christ returned). Stating this differently, the antichrist may use "Method 3" to justify (falsely assert) he is the legitimate messiah.

Summarizing, God's overall plan is "Time, Times, and half a Time" long where…

Time equals	2,000 P-yrs.,
Times equals	4,000 P-yrs., and
half a Time	<u>1,000 P-yrs</u>. for a total of 7,000 P-yrs.

from God's perspective. One week long on God's calendar and Christ's return is marked on that calendar for 2028 A.D.

Method 5 (Daniel 8: 2,300 Days Prophecy)

Daniel's twenty-three hundred "evenings and mornings"… "days" are prophetic years. The sanctuary will be reconsecrated (Christ will return because he is our only sanctuary) at the end of these "days." Now I believe this number to be an error in most Bibles. Why? It is reported that the eight copies of the book of Daniel recovered from the Dead Sea scrolls all had this part of the text damaged. Other writers claim the Vatican's copy of the Bible has twenty-four hundred days (a number I support) and still other Bibles twenty-two hundred days. I'll let you decide the truth of this matter since I believe this to be an example of Satan trying to change the "set times." If all the Bibles had the same number, then there would be no question as to the correct number. But with these different numbers it is clear other forces have been at work.

First, twenty-three hundred isn't a number that God would typically use because of its numerical factors (23×10^2), but twenty-four hundred (24×10^2) is! Ten is the number of completion and twenty-four is used in prophecy.[1] Next, let's go back to Daniel's seventy 'weeks' prophecy to

13

get the starting date (Ezra's decree). This prophecy first indicates there will be seven 'weeks' without any explanation as to why those seven 'weeks' were separated out in the text or what will occur after those seven 'weeks' are up.

I believe this timing was signaled out in the text to give us the starting date for the twenty-four hundred days because this prophecy was given to Daniel twelve to thirteen years before the "Seventy Sevens" prophecy with no starting date.[j] If you count ahead in time seven 'weeks' of years, which are forty-nine P-years from Ezra's decree, you get to the very beginning of 408 B.C.[10]

Now this date is indirectly verified as part of this prophecy in Daniel 8 because it began with conflict between the Ram (Persia) and the Goat (Greece). This new starting year is exactly at the midpoint between when the greatest king of Persia (Xerxes) attacks Greece for his first time (480 B.C.) and when the greatest king of Greece (Alexander) attacks Persia for his first time (335 B.C.). If we count twenty-four hundred P-yrs. from Ezra's adjusted starting time, which are twenty-four hundred and thirty-five years exactly,[11] we arrive at the year 2028 A.D. again.[12]

Now you may not want to accept this, but if you count twenty-three hundred prophetic years from the birth of the first antichrist (Alexander the Great)[k] you get to his rebirth in February 1979.[13] This year was the sixty-ninth Jubilee from the Egyptian Exodus.[l]

Yes, the antichrist came once before[m] just as Christ came once before. Satan the deceiver is always mimicking (copying) God's plan for humanity. That is one of his ways of deceiving both the Jews and Christians when he assumes power and is revealed at the end of this age. This is the ultimate deception he will employ claiming he is the prophesied messiah foretold in the Bible.

Method 6 (Jubilee Years)

Daniel's seventy 'weeks'... or seventy 'sevens' are really seventy weeks of 'Sabbath year cycles... heptades... Jubilees (fifty years).[n] This deeper understanding translates into seventy times fifty or thirty-five hundred actual years. Now these are normal years because Jubilee years are not prophetic years, but real years that are easily measured. If we start counting from Ezra's decree in 458 B.C. and move thirty-five hundred years into the future we arrive in the spring of the year 3043 A.D. This

14

prophecy confirms what we've already learned about God's Plan by other means! The description in Daniel 9 of the seventy 'sevens' is really a description of utopia... paradise... the time when heaven and earth will be remade[o]... not a description of any period of time in past history or even a description of the Millennial Kingdom.

The Millennial Kingdom on earth will be wonderful, but there will still be sin. However, that sin will be controlled with the iron scepter of Christ's rule and will not be influenced by the Devil who will be chained up for that time period.[p] In other words, there will be a lot less sinning going on; a new beginning with Christians only and all those things and people who caused sin in the past ages will be "weeded out" by angels and thrown in the Lake of Fire at his second coming.[q]

So we know the date of the remaking of heaven and earth and need only to subtract the length of time for the Millennial Kingdom to get to the date of our Lord's return. Most evangelical scholars would subtract one thousand years, but we know these are P-years from God's perspective and are really one thousand fourteen years and seven months exactly.[2] Making this last subtraction we arrive in the autumn of 2028 A.D.[14] as the year of Christ's return!

Method 7 (The Microcosm of my Life)

I prayed to the Lord and asked Him why He kept showing me all these things as I wrote my first book and on September 1, 2007 (a Saturday... a Sabbath) He finally gave me the answer I sought after falling asleep. I awoke with an explanation that troubled me, but we do as the Lord commands if we are His.[r] The answer given was my life was a microcosm of His plan... a living blueprint.

The simple version is I was born in 1958 A.D. on a Sabbath day and I will be 'seventy' years old in 2028 A.D. This revelation was seven years to the very day from when I retired at forty-two years of age in the year 2000 A.D. I began writing my first book when I was forty-nine years old.

Do you notice anything unusual about these numbers? They all match numbers in prophecy. What an awesome responsibility for an average person like me to tell everyone of this revelation; a responsibility that I am currently flunking right now. I am a voice in the wilderness crying out, but most refuse to listen. I have tried telling people the time is short with my first book and I am trying again with this one, but most can't hear my whispers on the noisy winds of everyday life.

Method 8 (The Year of Jubilee)

The Jews believe the Messiah will come in a year of Jubilee, but over history they lost track of the true years these were because they were not keeping the Lord's commands. I was shown these years of Jubilee and calculated all of them from the beginning of Adam all the way to the remaking of heaven and earth using a computer spreadsheet. Counting from the decree issued to Ezra (which by the way is the midpoint of history both past and future; God is awesome in His use of numbers), Christ started his ministry on the tenth Jubilee in the year 28/29 A.D. and will come again after forty Jubilees on the fiftieth Jubilee counting from Ezra's declaration in the Jubilee year 2028/2029 A.D.

Jesus was born on the third Sabbath year 2/1 B.C. counting from the Jubilee year 23/22 B.C. as mentioned in Whiston's Josephus' translation. Sabbath and Jubilee years are not measured by Gregorian dating, but by using the Jewish calendar and these years listed are translated from that calendar.

The great flood was the year before the third Sabbath year after the thirty-third Jubilee as measured from Adam. This reckoning is another form of God's number (333.333), but that subject was covered in the first book. The first antichrist died on the third Jubilee from Ezra's decree in 323 B.C. and he will die again on his forty-ninth birthday in the fiftieth Jubilee year from that same decree when Christ returns. Again, this is another topic we will not rehash and I only highlight these events to show how important Sabbath and Jubilee years are to God's plan. So realize the year 2028/2029 A.D. is a year of Jubilee and the Jews are correct in their belief of the Messiah's coming.

Method 9 (The "Good Samaritan" Prophecy)

The Good Samaritan is a parable with a deeper understanding of Christ's return. It is really a prophecy, as it was revealed to a brother in Christ a few weeks after I had begun writing my first book. Clearly it was another sign from God for me to spread the news of Christ's return.

Jesus represents the Samaritan... a half breed... half Jew (man) and half God. He comes to the aid of an injured stranger when no one else could or would save him. Then the Samaritan leaves the man with an innkeeper and pays him two denarii (a denary was one day's wage at Christ's time) to take care of him until he returns. Realizing He will be away for two days and return on the third day... just like the resurrection account. Hence, two days from God's perspective (a day is like a thousand years) is two thousand P-years or 2,029 years.[1]

If we start counting from when Jesus first came (at his birth in 2 B.C.) to symbolize when he first meets the stranger, we get to the year 2028 A.D.[15] Starting from when Christ began his ministry and adding just two thousand "normal" years, yields the year 2029. However, we know that there will not be two thousand years precisely because Christ said the "appointed time" would be cut short or else all life would perish. If we shorten this time by roughly a year, we get to the autumn of 2028 A.D.

Method 10 (The "Fig Tree" Prophecy)

When the nation of Israel is "reborn" (the fig tree in Matthew chapter twenty-four), the generation who witnesses this key sign will see Christ return to earth. This messianic event occurred on May 14, 1948 A.D. and if you add eighty years for the length of a generation,[5] you get to the year 2028 A.D. However, I like to believe the computation goes more like this...

> *"Thus there were fourteen generations in all from Abraham to David, fourteen from David to the exile to Babylon, and fourteen from the exile to Christ."*
>
> Matthew 1:17 (TNIV)

... forty-two generations in total. Three trimesters of fourteen generations. You have witnessed this kind of accounting technique used by God in His overall plan! I explained that from "Abraham to Christ," these were the "Times of the Jews" and the second trimester of God's overall plan.

The first trimester was the "Times of Adam" and the third trimester is the "Times of the Gentiles" (forty-two generations... forty-two months the gentiles will trample on the holy city).[t] Three trimesters just like childbirth[u] and each of these blocks of time are 2,029 years and two months long as you have learned.

The first trimester had twenty generations, but after Noah's flood, man's life expectancy was limited[v] and stabilized out. The forty-two generations mentioned in Matthew are the "Times of the Jews" and therefore are 2,029+ years long or 48.313 years[16] per generation. This math also applies to the "Times of the Gentiles."

Now many who have looked for the timing of Christ's return, started with the year 1948 A.D. because Israel became a country in that year and they added forty years for a generation because they believed there are biblical passages that support this belief.[w] However, there are other passages that don't support this idea as well.[x] When they did this they came to the year 1988 A.D. as our Lord's return date and that time has come and gone. These people were only doing what Jesus told them to do: to keep watch.[y] But they did not realize all the prophecies of the Bible that speak to the Lord's return need to converge to the same point in time and they didn't know the timing of the other prophecies, which you now do.

Next, they reasoned using the starting year of 1967 A.D. and added forty years again, but that date passed as well so now they were stumped. My first "eureka" inspired by the Holy Spirit came on the fortieth anniversary of the Six-Day War of 1967 and actually passed just as I wrote about this subject in my first book. A very strange coincidence indeed. However, there are no coincidences in God's plan. It is precise, contrary to what others might tell you.

Both of these starting dates were obviously in error or the length of a generation was incorrect. Regardless of the Six-Day War (when the Jews finally captured Jerusalem), the Jews didn't officially control all of Jerusalem until July 30, 1980 A.D. when the Jerusalem Law was passed (a decree by modern standards). Adding the 48.313 years we calculated earlier for a generation to the July 1980 A.D. date, gets us to the autumn of 2028 A.D.

There are a few more methods that will confirm this timing which were covered in my first book, but if ten is not enough for you to believe, then you will not believe this revelation regardless of how many ways I show you. You will believe what you wish and do what you want because you are blind to the truth or you are unfamiliar with God's Word. I have done the best I could to show you the condensed version of my understanding of Christ's return. I know it is still a lot to absorb, but it is much shorter than the version in "2028."

For those of you who are unfamiliar with scripture and the prophecies I used to derive the date of Christ's return, I apologize for going too fast. Most people who buy eschatology books already have a working knowledge of many of these prophecies I used and they were covered in detail in the first book. If you are unfamiliar with any of them it may be a good time to put this book aside and read the complete prophecies for yourself. The major prophecies of Daniel I used here are covered in Daniel chapters seven through twelve.

The Olivet Discourse, discussed in three of the gospels (Matthew 24, Mark 13, Luke 21), is worth studying as it covers end-time events and the return of Jesus. Most Christians do not view Christ as a prophet, but indeed he was and made many prophecies (in the form of parables) about the events of the last days.

Let's move ahead and deal with some of the heresies of Christianity that affect people's abilities to see the truth. For example, I mentioned that Christ was not born in 6 or 5 B.C., which many modern scholars believe and teach. Those who believe this and other heresies will find it hard to come to grips with a few calculations I provided. They might argue I started with the wrong year of Christ's birth and so the return date should be something besides 2028 A.D. or my analyses calculated the wrong year of Christ's death because the interpretations I used were flawed, but they cannot dispute the mathematical techniques.

I find those who believe so strongly in these wrong ideas, are blinded to seeing the real truth. I have conversed with a few of them and have even shown them the errors of their evidence and logic and yet they hold fast to them. I am amazed at some of the wild beliefs that are being taught to unsuspecting Christians. My analogy to this problem would be showing someone that one plus one equals two and yet they still would argue it equals three.

You will see what I mean in the next chapter. If you are one of those Christians who believe in some of these heresies, I ask you to personally investigate what I am going to tell you. Don't just believe blindly what others have taught you in the past, but weigh my words and let the Holy Spirit speak to the truth of these words. Believing some of these heresies may keep you from seeing God's plan and understanding that everything fits together and has purpose. Everything I have shown you and will show you fits together from every angle imaginable and if the things you believe don't agree, then in all probability they are false and don't fit within the Lord's plan.

Chapter 3

Heresies of Christ's Life

The coming of the lawless one will be in accordance with how Satan works. He will use all sorts of displays of power through signs and wonders that serve the lie, and all the ways that wickedness deceives those who are perishing. They perish because they refused to love the truth and so be saved. For this reason God sends them a powerful delusion so that they will believe the lie and so that all will be condemned who have not believed the truth but have delighted in wickedness.

2nd Thessalonians 2:9-12 (TNIV)

WE ARE LIVING AT THE "END" of the end-times. As each day passes by and we get closer to that end, more and more lies are being spread and believed by the unsuspecting masses. After writing my first book, and while waiting to find a publisher, I began the task of setting up a website to spread this knowledge. I began searching for those Christians who the Holy Spirit had been sharing the same thoughts and ideas that I had been receiving. As I surfed the net, I began to see many websites and find many people who promote false ideas about Christ, which surprised me. Things seemed so clear to me, and yet, they could not see the truth. Most of these professed Christians believe these false ideas so blindly that, even when confronted with the truth, they refuse to change their beliefs. They even presume I am the one who is blinded.

This is troublesome and hard to understand. What is clear is there are unseen forces at work clouding their judgment, but I cannot determine if they are God's doing or Satan's. From the verses above we see Satan is deceiving many and because they refuse to see the truth, God is sending a powerful delusion to those who reject the Holy Spirit. Many of these false ideas are a hindrance to the knowledge I am sharing. I do not know exactly when these lies began to take root, but I do know they are gaining traction. These heresies are gaining footholds in some Christian minds and if left unchecked, might make it impossible for them to see the truth of Christ's imminent return. Most of these heresies have to do with Jesus' birth, death, when he started his ministry, when he will come again, and much more.

For example, the well known television evangelist Jack Van Impe preaches about the end-time tribulation, Christ's second coming, and Bible prophecies weekly on television. He routinely chastises pastors and preachers for not informing their congregations about the coming Day of the Lord and for preaching wrong doctrines. I agree with him on these points.

He further says the rapture of the church can occur at anytime before the seven-year tribulation period begins, a period of time when God will unleash His wrath on the inhabitants of the earth because their wickedness has reached the heights of heaven. So it logically follows from his teachings that everyday a Christian wakes up in the morning, Christ's second coming must be at least seven years later. Since the current year is 2009 A.D. we can infer that Christ cannot return before the year 2016 A.D. if Dr. Van Impe's doctrine is correct, and yet he promotes and sells a video to raise money for his ministry that predicts doom and gloom is coming by the end of 2012 A.D. … December 21, 2012 to be exact. Now I have not viewed his video, but based on the sales pitch these ideas appear to be based on Mayan calendar hype and not God's Word. Should you buy this video if you really believed the rapture is at least seven years before Christ's second coming?

I wrote him a short letter when I sent him a copy of my book "2028" and asked him why he promotes these inconsistencies. I got the book returned about a month later with a written reply from someone in his ministry indicating Dr. Van Impe was too busy to read my book due to his demanding schedule. More importantly there was no answer to my simple question. Now Dr. Van Impe claims he is bringing more people to Christ than ever before and so I can only take for granted these are non-believers since you can't bring people to Christ who already belong to the Lord.

Clearly he is doing God's work because he leads more people to the faith in one week than I will in my entire lifetime. But what will these new believers think when the year 2012 A.D. rolls by just like other dates he and others have gotten worked up about (2000 A.D. e.g.)? Predicted dates of biblical events with little or no scriptural foundation. Will they become disillusioned and fall away from the faith because their faith began with the seeds of half-truths? Or will they have time to grow in the faith so that when they realize they were misled, their faith will be strong enough to overcome any misunderstandings? I have seen both cases. Some will be won for Christ and some lost forever.

At one time, I had a pastor who would tell "little white" lies to motivate the congregation to do the "right" things. I wasn't aware of it for many years until I got more involved with day-to-day church activities. This practice irritated me when I heard him stretch the truth. Is this how the Holy Spirit wants us to spread the Word of God? I don't believe so. It is one thing to preach untruths based on misunderstandings of the Word and quite another to knowingly twist the truth for some alleged greater purpose. The Holy Spirit doesn't need the help of deceit to spread the good news of Christ.

Christ's Birth

I would like to begin at what I believe is the root of most of these lies about Christ's life. When William Whiston translated Josephus' "Antiquities of the Jews" in 1737 A.D. he added some of his own comments in the footnotes. One of these comments appears to be the source of these falsehoods. Many scholars and Christians who study these matters quote Josephus as the source of this information, but it was not Josephus who wrote these footnotes. It was Whiston's commentary on things Josephus wrote.

Josephus stated that Herod the Great died shortly after a lunar eclipse just before the Passover. Whiston, after consultation with eighteenth century English astronomers, decided Herod must have died in early 4 B.C. because there was an eclipse in March of that year. Modern science, using computers instead of manual calculations, shows there was indeed an eclipse during this month. This eclipse was a partial lunar eclipse. Furthermore, computer programs that make these kinds of calculations have calculated other eclipses around this time that could also have been the right one. We will learn shortly that the eclipse of 4 B.C. was not the correct one.

I find many of the theologians who study these things are very poor with mathematics, science, and logical thinking. They come up with explanations, which for most people seem reasonable on the surface, but under scientific and logical inspection fall far short of the target. If these ideas are left unchecked, they begin to propagate out of control. If these theologians are well respected or have a lot of influence, then the problem is compounded even further.

Isaac Newton in his book "Observations Upon the Prophecies of Daniel and the Apocalypse of St. John" published in 1733 A.D. after his death, relates the birth of Jesus as follows:

> "Now Nehemiah came to Jerusalem in the 20th year of this same Artaxerxes, while Ezra still continued there, Nehemiah 12:36, and found the city lying waste, and the houses and wall unbuilt, Nehemiah 2:17, 7:4, and finished the wall the twenty-fifth day of the month Elul, Nehemiah 6:15, in the twenty-eighth year of the King, that is, in September in the year of the Julian Period 4278. Count now from this year threescore and two weeks of years, that is 434 years, and the reckoning will end in September in the year of the Julian Period 4712 which is the year in which Christ was born, according to Clemens Alexandrinus, Irenaeua, Eusebius, Epiphanius, Jerome, Orosius, Cassiodorus, and other ancients; and this was the general opinion, till Dionysius Exiguus invented the vulgar account, in which Christ's birth is placed two years later."

We notice from Newton's account pertaining to the birth of Jesus that we can figure out what year this event occurred using our current dating system. Isaac Newton believed the seventh year of Artaxerxes' reign was in 457 B.C. Therefore, the twenty-eighth year (4278 J.P.) was twenty-one years later in 436 B.C. Subtracting the four hundred and thirty-four years (sixty-two weeks of Daniel's prophecy), we get to September 2 B.C. as the time of Christ's birth according to those ancient historians who lived closest in history to Christ's life.

Therefore, Christ was born in the year of the Julian Period 4712 (2 B.C.) and not 1 A.D. as Dionysius contended. Newton refuses to rewrite history and scripture when he defers to the ancient historians, unlike many scholars today who imagine they know more than those historians who lived closer to the actual events. Newton further rips Dionysius for his poor calculations, assumptions, and research in resetting history's timeline incorrectly. So too must those who do this shoddy work in today's age, regardless of their intentions, be reprimanded. Let's look further at what I wrote on this subject in my book "2028,"

> "According to the translation of Josephus' The Antiquities of the Jews by William Whiston in 1737 A.D., "…the thirteenth mid fourteenth years of Herod are the twenty-third and twenty-fourth years before the Christian era." Simply restated; at the end of 23 B.C., Herod the Great had completed approximately 13½ years of his reign in Judea. Josephus further records that Herod reigned 34 or 37 years before his death depending on the starting point. Combining these two accounts makes Herod's death in mid 1 B.C. using the 37 year account and mid 4 B.C. using the 34 year account."

So the question is this: When Whiston recorded this footnote was his date calculations based on the 4 B.C. lunar eclipse? I believe they were. His faulty reasoning was based on eighteenth century astronomy that was done by manual calculations and he didn't have the fortune of consulting computer programs that do this work for us today. Most theologians recite this reference to substantiate their premise (belief) that Christ was born one to two years earlier in 6/5 B.C.

The computer programs that make calculations on lunar and solar eclipses show that a total lunar eclipse occurred on January 10, 1 B.C. This eclipse is a closer match to what Josephus wrote of Herod's death and what the ancient historians recorded, than was Whiston's deductive reasoning of the year 4 B.C.

This would place Herod's death just before the Passover of 1 B.C. about six to seven months after the real birth of Christ. Matthew records the Magi visited Herod and as a result of this visit Herod was able to learn of the Messiah's birth. When the Magi did not return, Herod sent orders to kill all boys two years old and younger. Theologians use this scripture, along with Whiston's footnote, to derive the 6/5 B.C. date they claim is the true time of Jesus' birth. I guess these theologians should be unhappier with Dionysius Exiguus than Newton was. Actually, Newton would be just as stern with modern day "experts" as he was with Dionysius for trying to change history. Knowing that Herod never hesitated to kill any of his own family, one should not be surprised when he decided to kill male children a little older than what may have been reported so he could make sure he had killed the messiah.

The most logical explanation is that Christ was born in the autumn of 2 B.C. Clearly God would send heavenly signs to announce the Messiah's advent ahead of time. The Lord always foretells His intentions before He takes action just as I am telling you now. He told Noah to build an ark before the flood came didn't he? He instructed Noah seven days before the actual flood to get the food and animals aboard. He has given many prophecies on the coming end-times beforehand so that you too can be prepared in advance. You can bet the celestial signs of Jesus' birth were provided ahead of time as well. This is how God works so that no one should be caught off guard of His intentions: that is, no one who follows the Word of God and listens to the Holy Spirit.

Modern astronomers have confirmed there were astonishing movements of the planets beginning in 3 B.C. and continuing on over the next two years. Since the Magi studied the stars and obviously the prophecies of Daniel, these signs would have caught their interest. After months of observations, consultation with distant colleagues, and study to determine the signs relevance, they must have determined they predicted the birth of the Jewish Messiah.

Packing and travel after this revelation would have taken at least four months if they were coming from Babylon[a] and even longer if the Magi were journeying from Persia or Arabia. Providing the details of their trip to Herod doesn't mean Jesus was about two years old when they arrived in Jerusalem. It only means the signs were in place ahead of time and that Herod was extra conservative with his two-year-old estimate on whom to kill!

Sign of Jonah

There are some on the Internet who teach a literal "three days and three nights" that Jesus was in the tomb. This doctrine even goes as far as to make sure the timing is exactly seventy-two hours. They condemn most mainstream Christians who believe Christ died on Friday. When I first heard this idea, I was caught off guard without an answer. However, after careful biblical study of the entire Gospel accounts, I easily determined the truth and concluded that mainstream Christians had it right all along!

I'm not sure where or when this idea originated, but those who teach and believe it are blind to the truth. One Christian wrote me and asked why I thought Christ died on a Friday in response to a short message I had posted on a Christian forum. His tone implied he had never heard this idea before. In retrospect, I think he was testing me for surely all Christians have heard of Good Friday. I gave him an answer and asked what he believed. He wrote back and said Christ died on a Thursday and rose Sunday morning and that is how you get three days and three nights.

That does seem to be the most logical answer (but there are other answers which we will see) and maybe he was taught this and never looked into it any deeper. I wrote him back giving him a few more reasons why his belief could not be true. The main reason being the Jewish Passover never occurs on a Friday and for his belief to be true, this Jewish festival day had to be observed on a Friday. I invited him to research calendar websites, history, or to even ask any Jew to provide evidence of a year the Passover fell on a Friday. His reply was the same as many of these heretics… what I wrote him was untrue (with no explanation given) and the reasons I gave were too hard to comprehend. He indicated it was just easier for him to believe Jesus was literally in the tomb three days and three nights. Hmm… Basically it was easier for him to continue believing the lie than to actually look for the facts or to believe someone who was telling him the truth. I was in shock.

In first Thessalonians Paul writes about the return of Jesus. In these verses we learn something of interest about the end-times.

> Be joyful always; pray continually; give thanks in all circum-
> stances, for this is God's will for you in Christ Jesus. Do not put
> out the Spirit's fire; do not treat prophecies with contempt. Test
> everything. Hold to the good. Avoid every kind of evil. May God
> himself, the God of peace, sanctify you through and through. May
> your whole spirit, soul and body be kept blameless at the coming
> of our Lord Jesus Christ. The one who calls you is faithful and he
> will do it.
>
> 1st Thessalonians 5:16-23 (NIV)

We read here we are to "test everything" and "hold to the good." Good advice for the times we live in. There are many false prophets and false teachers in the world today and from what I am seeing, they are multiplying. If you are an unbeliever or a Christian who was taught or believes Christ died on a Thursday, then "test" what I have written. See if you can find even one time in history when the Passover was on a Friday using an actual calendar. You see, the Jews have a calendar rule (lo badu Pesah) that effectively prevents two Sabbath days from falling in a row and the Passover is a Sabbath day just like Saturday. They adjust their calendar to make sure the first day of the Feast of Unleavened Bread (Passover) never occurs on Monday, Wednesday, or Friday.

As a side note, Paul said not to treat prophecies with contempt. I find this instruction interesting because most pastors and priests treat prophecies of the Bible with contempt. How? By rarely studying them or preaching to their congregations about them. They would never openly say anything bad, but their actions say otherwise. I also find some Christians who study prophecy on their own, or are lucky to have a Christian shepherd teach them these things, unknowingly show contempt for some prophecies as well. They chastise those who look for the timing of Jesus' return saying "no one can know the time" and those who predict a time are "tying God's hands." If God wanted His followers to be totally clueless as they profess, then He wouldn't have given mankind prophecies or provide actual times in Daniel and Revelation that, if understood, do predict the Messiah's first and second coming as I have already shown you. Are these attitudes considered contempt or are they given in the spirit of teaching, but just misunderstood?

There are still others who believe Christ died on a Wednesday who support "three days and three nights" in the tomb. I was surprised to find people who actually believe this version of the Sign of Jonah. I could not imagine why anyone would come up with this false doctrine until I realized if one believed Christ died in a year other than the real year, then they would need to come up with some kind of rationale to support this belief.

Many people who support false years of Christ's death are either badly informed of the facts surrounding these years or are just unaware of the inconsistencies with their beliefs. The more knowledgeable of these heretics, when faced with conflicting facts, tries to get to the real answer and in doing so creates fabrications to deal with these inconsistencies. For example, if you believed Christ died in a year that the day before the Passover was on a Wednesday (like 30 A.D.), then you need to have an explanation for this. Let's examine in detail this false doctrine by reconstructing the faulty logic they use.

First, they need to believe Christ was in the tomb seventy-two hours (three whole days and three whole nights) exactly. Why? Because they know the Bible clearly says Christ rose on the first day of the week (Sunday). By starting on Wednesday, the only way to reach Sunday (which on the Hebrew calendar begins with Saturday's sunset) is to begin exactly at sunset on Wednesday. If one began counting when Christ died around 3 pm, this would yield seventy-six hours... more than three days and three nights. Additionally, if you just counted exactly seventy-two hours from his actual death around 3 pm, you would end up with the resurrection on Saturday afternoon. So you see their dilemma of trying to get from Wednesday to Sunday.

By using their approach, one should realize that Christ would have risen on the fourth night if he were buried at sunset on Wednesday. Furthermore, Christ would have had to rise at "sunset" not "sunrise" which is what scripture says. At least other Christians who try to sell this false idea start on a Thursday (which we have discussed) because they recognize that "dawn" means sunrise instead of sunset! I believe these are the main points of their story.

I will limit my reciting of scripture verses since there is only one verse in all the Gospels that specifically says three days and three nights.[b] All the other verses and the actual detailed accounts of Christ's death speak to something else entirely. Therefore, on the Wednesday (Nisan 14, 3 pm) Jesus died, we will start our analysis from there. Let's count:

Wed.	3 pm to Thursday 3 pm	(1 day and 1 night)	Passover after sunset
Thur.	3 pm to Friday 3 pm	(1 day and 1 night)	Normal day after sunset
Fri.	3 pm to Saturday 3 pm	(1 day and 1 night)	Sabbath after sunset
Sat.	3 pm to 7:04pm / April 4, 33	(4 hours)	Sabbath until sunset

This is roughly seventy-six hours in total. Now they state "if" Jesus was buried exactly at sunset, then the extra four hours can be subtracted so that there are only seventy-two hours remaining. There you have it: three days and three nights. Let's look even closer at this type of accounting and rational.

Wednesday was the fourteenth day of Nisan and the day the Passover lambs were prepared (killed) for the Passover meal that was to be eaten after sunset on the fifteenth. Understand the fifteenth of Nisan is a

Sabbath, just like Saturday, but this Sabbath required the Jews to be in their homes after sunset to celebrate the Passover meal[c] and that they couldn't do work (burials) as well.

So Christ's friends (I don't know exactly everyone involved with the burial other than Joseph of Arimathea, Nicodemus and some women,[d] but they were all Jewish) had to finish placing Jesus' body in the tomb exactly at sunset to make their math work out and also needed to be in their homes at the same time. Interesting. I guess they ran really fast (in sandals) back in those days or their homes must have been very near to the tomb.

Next we see that from Thursday sunset (around 7 pm – the end of the first day of Passover) until Friday sunset there is a twenty-four hour break from the Sabbath requirements when Jesus' friends have a chance to go and prepare his body properly during daylight hours because there are no restrictions for doing any work on this day. But we find Christ is not important enough to take care of this task expeditiously, so they let his body decay another two days until sunset (darkness) of the first hour of Sunday. Clearly it is better to see and work in the dark. Not to mention running as fast as you can (Mary and Peter) is always a good thing to do at night. But there was a good source of moonlight around this time so I'm sure they wouldn't trip or fall. Interesting ideas...

No scriptures need to be quoted to prove how incorrect these ideas are. This is what they want Christians to believe? That Jesus was not important enough to finish preparing his body properly, as soon as possible (on Friday), in full daylight when this could be done with no Sabbath restrictions. It would be better to wait and do this in the dark after letting his body decay another two days. If you are a Christian who believes this version, you should reread all the scripture accounts of Christ's death with an open eye instead of believing those people who have developed a whole doctrine on one verse from Matthew.

I have traded messages on the Internet with some of these so called Christians and they truly believe they have figured out some hidden secret that the mainstream Christian churches have gotten wrong for hundreds and hundreds of years. One told me the reason why the women did not go to anoint the body of Christ on Friday was because it took the women all day to find the spices and prepare them. They claim their version of Christ's death is the only one that makes all the scripture accounts true. I guess they don't read much scripture[e] or understand that when Jesus himself said he would rise on the third day, he did not mean the fourth night.

Christ's Death

I could not understand why anyone would come up with such non-sense, let alone teach this doctrine to kids and new Christians who don't know any better, unless they believed Christ died in a year other than he really did—like 30 A.D. as many self-professing biblical experts believe these days. Most Christians who believe this lie are taught it by someone who thinks Christ was born in a year before 4 B.C. due to Whiston's footnotes.

Let's see what day of the week the fourteenth of Nisan fell on in the time period around Christ's death for all must believe Christ died in one of the years listed in table 3-1 below.

Table 3-1

Year	Day & Date	Comments
25 A.D.	Monday, April 2nd	
26 A.D.	Friday, March 22nd	This is very close to the vernal equinox and may have been on Monday April 22nd
27 A.D.	Wednesday, April 9th	
28 A.D.	Monday, March 29th	
29 A.D.	Saturday, April 16th	
30 A.D.	Wednesday, April 5th	Now this really has to be Thursday April 6th because the full moon was on April 6th and the Passover begins the 1st day after the full moon.
31 A.D.	Monday, March 26th	
32 A.D.	Monday, April 14th	
33 A.D.	**Friday, April 3rd**	
34 A.D.	Monday, March 22nd	Isaac Newton claimed this was the vernal equinox in 34 A.D. and the 14th of Nisan would have then been pushed to April 23rd (Friday)
35 A.D.	Monday, April 11th	

Realize that the fourteenth of Nisan (the day Christ was crucified) is the day of the first full moon after the vernal equinox (March 21st) while adjusting for "lo badu Pesah."

31

Obviously, computers were not used back then and the Jews made manual adjustments to their calendar to make sure the fourteenth of Nisan was always the first full moon after the spring equinox. However, computers are very accurate at knowing which days the full moon fell on all the way back to the beginning of Adam and so there can be no doubt about the days listed. The problems come in when the full moon hits on or near the spring equinox in any given year (like 26 A.D. & 34 A.D.). But this did not happen for any of the other years I have shown you.

So according to those who say Christ died on a Wednesday, there were two years this was possible: 27 A.D. and perhaps 30 A.D. (which really has a problem as I commented in the table). So which of these years did Christ die in? Neither! Just so you know, those who say Christ died on a Thursday also use 30 A.D. as the year of Christ's death so our next analysis will address their beliefs too.

Luke states that John the Baptist began his ministry in the fifteenth year of Tiberius Caesar. So when was that? Tiberius Caesar assumed the throne in August 14 A.D. when Augustus died. Check any website or history book. Any. Go ahead and investigate this fact for yourself if you are unfamiliar with history. Let's count:

Table 3-2

Years of Tiberius' Reign	Year
August 14 A.D. to August 15 A.D.	1st year
August 15 A.D. to August 16 A.D.	2nd year
August 16 A.D. to August 17 A.D.	3rd year
August 17 A.D. to August 18 A.D.	4th year
August 18 A.D. to August 19 A.D.	5th year
August 19 A.D. to August 20 A.D.	6th year
August 20 A.D. to August 21 A.D.	7th year
August 21 A.D. to August 22 A.D.	8th year
August 22 A.D. to August 23 A.D.	9th year
August 23 A.D. to August 24 A.D.	10th year
August 24 A.D. to August 25 A.D.	11th year
August 25 A.D. to August 26 A.D.	12th year
August 26 A.D. to August 27 A.D.	13th year
August 27 A.D. to August 28 A.D.	14th year
August 28 A.D. to August 29 A.D.	**15th year**

Thus we see the fifteenth year of Tiberius Caesar was August 28 A.D. to August 29 A.D. John the Baptist began preaching in the spring of the fifteenth year of Tiberius Caesar. This could only have been in the spring of 29 A.D. since it was the only spring in the fifteenth year of Tiberius. Jesus began teaching when he was "about" thirty the following autumn (six months later) and so this confirms the time would have been the fall of 29 A.D.

John the Baptist was a Levite and Levites were required to serve the Lord starting after their thirtieth birthday when their course was scheduled to work in the temple and they were to continue their service until they were fifty years old according to the Mosaic laws. Clearly John would have fulfilled this requirement just as Jesus would. As best I can determine, Jesus started about a month early of his thirtieth birthday or "about" thirty as it is written. Since no one knows Christ's exact birthday with certainty, this was based on calculations from my first book in which I concluded Jesus was born in October.

We now know for sure the year 27 A.D. is out as the year of Christ's death because it was before the fifteenth year of Tiberius. If you believe it was 30 A.D., since it is the only possible choice left, then you have to believe Christ died on the first Passover after he started preaching which scripture doesn't support. His ministry was 3½ years long. This realization that Christ's ministry lasted 3½ years is important for future work so please commit this fact to memory.

So all evidence points to the reality that Christ could not have died on a Wednesday or in 30 A.D.; otherwise Luke's account of the fifteenth year of Tiberius in the Bible was untrue. Furthermore, Jesus couldn't have died on a Thursday because the fourteenth of Nisan is never on a Thursday (which Table 3-1 supports). One can also see that Jesus could not have died in any year before 33 A.D. like 31 or 32 A.D. as some claim because the fourteenth of Nisan fell on a Monday in those years and this would have Christ lying in the tomb for six days. An error so outrageous that it's not worth even considering.

There are many who claim the fifteenth year of Tiberius was in some other year, like 26, 27, or 28 A.D. even though all history books say otherwise. Historians do not make these claims, but rather theologians or biblical students who are manipulating the date to reconcile what they believe the Bible says with what William Whiston wrote about Herod's death. They need to start in another year to make their math work out. They believe this so strongly they are willing to spread more untruths to support their ideas. I don't believe most do this intentionally, but rather that they are poor logical thinkers and mathematicians.

If we were to put ourselves in Dionysius Exiguus' shoes we could see more evidence of their follies as well. Since Dionysius was the one who ultimately reset history's timeline to Christ's birth we need only to think like him to see the truth. To do this work, Dionysius only needed to know the year Tiberius assumed the throne using the prior dating system and the verse from Luke that says Jesus was "about" thirty years old. With only this knowledge we can deduce what Dionysius actually thought about the fifteenth year of Tiberius.

Since 1 A.D. is the time he indirectly assigned as the first year of Christ's life we can logically conclude he thought that Christ was 29 years old in 29 A.D. (about thirty), thirty years old in 30 A.D. and so on and so forth. By back calculating we see that this line of reasoning is consistent with what history books claim was the date Tiberius took the throne. Since his work was done in 525 A.D., we also know that Dionysius concluded the fifteenth year of Tiberius was equivalent to 29 A.D. at that point in history.

Furthermore, if Dionysius believed anything else, like "about thirty" meant thirty-one years old, thirty-two years old, or some older age, then there would have been more time between when Tiberius took the throne and when Christ was born and not less which is what those who change the date are claiming. Basically, the math Dionysius used was simple. He took the date he believed Tiberius assumed the throne, added fifteen years and subtracted how old he believed Christ was during Tiberius' fifteenth year of ruling. This resulted in Jesus' first (birth) year.

Those who claim the fifteenth year of Tiberius was in 26, 27 or even 28 A.D. and claim Dionysius used the wrong starting year of Tiberius reign, are in error because the math does not allow for this conclusion. What they are really saying by claiming these false years is that Dionysius believed "about thirty" meant 26, 27 or 28 years old or that he could not do simple math!

So these false ideas are not supported with scripture, logical reasoning, or calendar mathematics. Clearly they are fabrications. Common sense is enough for most people to see these falsehoods for what they are. Christ died on Friday, April 3, 33 A.D. I find people try to make things harder than they really are. God hides the truth right in front of their faces and yet they cannot see it. Why is that? I do not know why people choose not to think for themselves and rely on others to think for them. If they really knew how God uses numbers, they wouldn't dispute the April 3, 33 A.D. date.

Do you realize that Christ died around 3 pm, three hours before the end of the Jewish day, after three hours of light and three hours of darkness while he was nailed to the cross? This was the third day after three months had passed in the year 33 A.D. (on the Julian calendar) and he was thirty-three years old? Additionally he rose on the third day and not the fourth. Do you really think all these three's are a coincidence? God's number is three.

Furthermore, if you understand that Christ died around 3 pm on Friday and rose around 7 am on Sunday (sunrise) as the scriptures and mainstream Christian denominations contend, you would realize Christ was dead for forty hours exactly! A most curious number that most Christians are familiar with. Forty days after his birth, Jesus was presented at the temple as required by Mosaic Law, Jesus fasted forty days in the desert prior to starting his ministry, waited forty days after his resurrection before ascending and you now know he was bodily dead for forty hours (1.666 days… 144,000 seconds). Now is seventy-two hours in the tomb or forty hours supported as the "right" amount of time after examining all this other evidence?

Christ's Second Coming

I had always been taught from the time I was a child to be ready at all times for Jesus' return because he might come at any moment. This is the doctrine of the major Christian denominations. However, I have come to realize that this idea is not "exactly" true as it is preached.

Christ will not come at any instant (relevant prophecies have not been fulfilled) and not only that, he *cannot* come at just any time as these preachers would have you believe because there is an appointed time for his return.

Nevertheless, it is very good advice to live by for another reason— death. When a person physically dies the time for choosing Christ as their personal savior is over. Since we have no idea when death will over take us, we must be ready at all times. Having been at death's doorstep on more than one occasion, I can attest the spirit of this doctrine is most certainly true. At our death we will all face God at the final judgment and so it could be said, "Christ can come at anytime," but these teachings if taken literally are just not factual.

> *Concerning the coming of our Lord Jesus Christ and our being gathered to him, we ask you, brothers, not to become easily unsettled or alarmed by some prophecy, report or letter supposed to have come from us, saying that the day of the Lord has already come. Don't let anyone deceive you in any way, for that day will not come until the rebellion occurs and the man of lawlessness is revealed, the man doomed to destruction. He will oppose and will exalt himself over everything that is called God or is worshiped, so that he sets himself up in God's temple, proclaiming himself to be God.*
>
> *2nd Thessalonians 2:1-4* (NIV)

Paul tells us that Christ's return will not just happen at anytime, but only after certain conditions (signs and prophecies) have been fulfilled. So how do those who believe Jesus can return without any further signs reconcile this problem? Well they need to say these "requirements" have been fulfilled in the past. This position is known as preterism, which is another faulty doctrine. Between the time of Paul's prophecy (about 51 or 52 A.D.) and the destruction of the temple (70 A.D.) these teachers of preterism say these prerequisites were all fulfilled when the Roman general Titus burned down Jerusalem and the temple.

However, this future leader was not Titus, who died of a fever in 81 A.D. Paul further clarifies:

And then the lawless one will be revealed, whom the Lord Jesus will overthrow with the breath of his mouth and destroy by the splendor of his coming.

2nd Thessalonians 2:8 (NIV)

So this future leader described will die at the second advent of Christ. Did this happen between 51 A.D. and 70 A.D. as they contend? Of course not! Why were the prophecies in Revelation given to St. John around 95 A.D. if all was fulfilled years earlier? What we can conclude is Christ cannot come back at this very moment because there are things that must take place before his return. There are signs that must come first. Signs that have not taken place yet and if we are to believe that the temple Paul writes about is a real place, then we can conclude that Christ cannot come back to rule the earth until the temple is rebuilt in Jerusalem. Remember what Jesus said after his resurrection and before his ascension,

This is what I told you while I was still with you: "Everything must be fulfilled that is written about me in the Law of Moses, the Prophets and the Psalms."

Luke 24:44 (TNIV)

This is what Jesus told two disciples of his (not apostles), which were traveling on the road to Emmaus, after he had gone through and explained how he had fulfilled many of the prophecies. Even after hearing his sermons, witnessing miracles, and hearing the testimony of their women of his resurrection, these two men still did not understand how all these events fit together in God's plan foretold by the prophets.

How much harder is it for Christians, two thousand years later, to understand and agree on what is written when we do not have the gift of first hand knowledge as they did? When will Christ return? When all that was foretold "in the Law of Moses, the Prophets and the Psalms" is fulfilled and not one day before… or one day later. You now know when that time will be and coming at any other time, such as this very moment, would render parts of scripture false.

Chapter 4

The Rapture Question

THE RAPTURE IS A WORD USED to describe an event in biblical prophecy that the apostle Paul writes about to the Thessalonians who were concerned about the return of Christ back in the first century. Modern day theologians and eschatology students understand this to be a description of these verses,

> For the Lord himself will come down from heaven, with a loud command, with the voice of the archangel and with the trumpet call of God, and the dead in Christ will rise first. After that, we who are still alive and are left will be caught up together with them in the clouds to meet the Lord in the air. And so we will be with the Lord forever.
>
> 1st Thessalonians 4:16-17 (TNIV)

So the issue for a Christian cannot be whether this event takes place because if it doesn't, then the Bible cannot be trusted. These thoughts must be rejected since scripture is the one source that can be trusted completely. Hence, those Christians who don't believe in the rapture are really saying they do not like the label, the word, being used to describe this event. Furthermore, these Christians claim this event is nothing more than a description of Christ's second coming: his return to earth. What logically flows from their position is the timing of this event. If we know when Christ will return, then we know when this event will occur! Most churches that teach there will be no rapture are not saying this event will not happen, they are only saying it will happen at the coming of Christ. A time they add is unknowable. They believe the time is unknowable because they teach Christ can come back at any time of the Father's choosing. But you should know by now this is not the case. There is a plan and within that plan there is a window of about a year that Christ can return that still allows all the prophecies of the Bible to be true.

Therefore, the real question is not whether the rapture will occur, but "when" will the rapture transpire? When will this event that Paul speaks of take place? When will we be "caught up" in the air? This issue is the heart of many differences between Christian denominations. So for purposes of discussion, I will continue to use this "word" to describe the event that Paul speaks of and for those who don't believe in the rapture I will label their position a post-tribulation rapture.

Of this position there is no need to debate. If they are right, we already know when the rapture will happen: In the autumn of the year 2028 A.D. From this doctrine, it logically falls out that if there is a time of tribulation as many Bible prophecies foretell, then believers in Christ's saving grace and unbelievers alike will share in that wrath. However, there are many evangelical Christians who argue this event will occur prior to Christ's return. Why do they teach a pre-tribulation rapture? Because there are many verses in the Bible that say Christians will be spared the coming wrath of God. Let's see what else Paul wrote on this very topic.

> ***For God did not appoint us to suffer wrath,*** *but to receive salvation through our Lord Jesus Christ. He died for us so that, whether we [Christians] are awake [alive] or asleep [dead], we may live together with him.*
>
> *1st Thessalonians 5:9-10 (NIV)*

Here is a clear example where scripture indicates Christians will not suffer the wrath of God if they accept Christ as their king and savior. If the old prophets are right and the wrath of God will be unleashed at the end of time, at the end of this age, then those who accept Christ are not destined to suffer this wrath. What does this really mean?

Given that the year is 2009 A.D. and the Lord will return in 2028 A.D., this means that Paul's prophecy must occur sometime between now and when the Lord's wrath will occur. This is the crux of the problem that scholars waste endless hours debating. This problem gets easier to understand when you know what God's calendar looks like. Just as the Christians who don't believe in the rapture now know Christ will return in 2028 A.D., those Christians who believe they will be saved from the coming wrath now have an idea when this event will happen.

So the question before them is, when will the Lord's wrath occur? We will look into this in great detail in the next chapter, but for now most scholars who debate these ideas claim it is either three and a half or seven years long. They label this the time of the "tribulation" and we will use their terminology for now. If these Christians are right and they will be exempt from the tribulation timeframe as Paul writes, they can expect the rapture to occur sometime between now and the year 2021 A.D. This is a period of roughly thirteen years. This viewpoint is known as the pre-tribulation rapture.

We have looked at the post-tribulation theology and the pre-tribulation doctrine. But the real issue is that both these positions highlight the same shortcomings of most Christians! That is to say, all Christians who believe there is no rapture, or the rapture will occur before anything bad happens to them have the same outlook on life. Their actions in these last days are identical!

These two positions make up the overwhelming majority of all Christians. There are very few who are left who believe the third position. That the rapture will occur sometime during the recognized timeframe of the tribulation period.

The fallout from their beliefs is that both groups claim Christ can come at anytime for them personally. One believes there is no tribulation period as prophesied and they need to be ready at all times and the other thinks there is a time period of judgment prior to Christ's return, but they will be spared from it by being taken to heaven before it occurs. Since the Bible does not say how much sooner before the tribulation unfolds, they must also be ready at all times. The conclusion of these two diverse ideas is both must be ready at all times and nothing bad will happen to them personally… only to the other guy… to the non-believers. Hmm…

Who is the other guy? It is our friends and family. We are the other guy! I find these ideas only promote one truth—no urgency for Christians to tell friends and family of the coming wrath. To build the kingdom of God by finding Christ's lost sheep. For if there is no belief in the coming wrath or that the wrath won't happen to us, then most find no compelling reason to testify that Christ is their personal savior. Basically, if you are one of these Christians are you concerned about things getting real bad? Not really. Because they don't involve you—right? In Revelation 2 and 3, we are told in the letters to the churches this is how it will be at the end. The church will be filled with a bunch of lukewarm believers. This is the letter to the church of Laodicea. Read it for it applies to the vast majority of Christians today.

Let's look at the pre-tribulation situation just a little closer. Logic would dictate that since Christians are the only ones fighting Satan and trying to save people from an eternity of suffering (whether we are good or bad at making disciples for Christ), that removing Christians many years before the tribulation time would be counter productive to saving as many as possible from the coming wrath. So we can deduce that those who believe in a pre-tribulation rapture can expect to be taken just prior to the devastations so that the harvest can be maximized for Christ. This logic would suggest a rapture around 2020 or 2021 A.D.—in other words; the later this happens the better.

Now I have spent many years studying what others have written on this subject, but very little self-study with just the Word of God. I find compelling arguments for both ideologies. However, after coming to the revelation that God has a plan He is working from, I have concluded both of these positions are in error. That during the "end" of the end-times God's plan will be clearly revealed and all those who watch for Jesus' return will know when things will take place so that they will be prepared to do their part. Let's see what else Paul wrote about this event.

> *Now, brothers, about times and dates we do not need to write to you, for you know very well that the day of the Lord will come like a thief in the night. While people are saying, "Peace and safety," destruction will come on them suddenly, as labor pains on a pregnant woman, and they will not escape. But you, brother, are not in the darkness so that this day should surprise you like a thief.*
>
> *1st Thessalonians 5:1-4* (TNIV)

We learn here that those who keep watch will not be surprised. They will not be complacent, but will be ready. They will not be kept in the dark, but will know what is going on around them. The Thessalonians knew very well that the Day of the Lord would surprise those who knew not the truth. Doesn't a pregnant woman know when her appointed time is and about when to expect labor pains? Are we not at the end of the third trimester of God's plan? Should we not expect trouble as well? We further learn, that destruction will come on "them" and by this we can infer again that those Christians who know the truth will somehow be spared God's wrath. But other prophecies imply Christians will not be spared from the tribulation.

If any one has an ear, let him hear. If any one is to be taken captive, to captivity he goes; if any one slays with the sword, with the sword must he be slain. Here is a call for the endurance and faith of the saints.

Revelation 13:9-10 (RSV)

Then the dragon was angry with the woman, and went off to make war on the rest of her offspring, on those who keep the commandments of God and bear testimony to Jesus. ...

Revelation 12:17 (RSV)

He will speak against the Most High and oppress his saints and try to change the set times and the laws. The saints will be handed over to him for a time, times and half a time.

Daniel 7:25 (NIV)

With flattery he will corrupt those who have violated the covenant, but the people who know their God will firmly resist him. Those who are wise will instruct many, though for a time they will fall by the sword or be burned or captured or plundered. When they fall, they will receive a little help, and many who are not sincere will join them. Some of the wise will stumble, so that they may be refined, purified and made spotless until the time of the end, for it will still come at the appointed time."

Daniel 11:32-35 (NIV)

These are just some of the many verses that indicate Christians will not be secure from trials and tribulations. So we have two groups of Christians who believe that God will save them from bad things and we have these verses that declare otherwise. Which understanding is true? The answer is the third position is the only way all the verses can be true and best fits God's plan. That position is the rapture will occur during the tribulation period!

If we examine all the verses carefully and not read extra meanings or words into them that are not there, we will realize that God doesn't promise us no problems… no tribulations. He only promises that those Christians who hold to His word, keep the faith, and testify to Christ's authority will be spared from the wrath that He has planned for the rebellious unbelievers! Looking closely at the verses I quoted that speak of Christians being persecuted, they all speak to a time under the rule of the antichrist.

This position is labeled the mid-tribulation or pre-wrath rapture. That is to say, God doesn't promise we will be spared from the antichrist's hatred, but only from His wrath to be poured out after many have proven, or decided, whether they are for the truth (Christ) or against it. Christians will not skate free as post and pre-tribulation Christians believe. Since the majority of all Christians fall into one of these two categories this means very few will be prepared for the trials that will confront them during the last days. This means those who are wise and keep watch will know the truth and be able to instruct those who are asleep at the wheel. So when will this happen. Well the best guess is the spring of 2025 A.D.

Why this time? Because the scripture from Daniel 11 referenced was taken right after the abomination of desolation is set up and this event occurs at the midpoint of the classical tribulation timeframe which is the last one of the seventy 'weeks' prophesied in Daniel chapter nine.

Chapter 5

The Tribulation Timeline

LET'S MOVE ON AND INVESTIGATE THE timing for the period theologians label the "tribulation" because there are two questions of importance to answer: when will the tribulation period commence and how long will it last? This block of time immediately precedes the end of the present age and the return of Christ. This timeframe is more complex than many of those who claim an understanding of these events realize.

We will begin our study in the book of Daniel with his prophecy of the "Seventy Weeks (KJV)" or "Seventy Sevens (NIV)." Those who study prophecies or have read other articles I have written on this subject will know what some of the specific events foretold by God are. We will concentrate on the timing aspects at this time. This prophecy is the one that gives us the basic framework in which to begin laying out the tribulation boundaries.

Gabriel told Daniel there would be seventy 'sevens' before man would live in "paradise" again—the paradise that Adam and Eve experienced in the Garden of Eden prior to eating fruit from the tree of the knowledge of good and evil. When you understand the deepest meaning of this prophecy, you realize the 'weeks' or 'sevens' are seven Sabbath years and since a Sabbath year occurs every seven years, then the term 'sevens' means forty-nine years in length.

This revelation, in actuality, translates into fifty-year cycles when you include the related Jubilee year that follows.[a] Therefore, the total time prophesied is thirty-five hundred years[17] and when you start counting from the midpoint of history (Ezra's decree in 458 B.C.) you arrive in the year 3043 A.D. We know from the earlier discussion in chapter one studying God's plan that 3043 A.D. is the time when heaven and earth will be remade and so Ezra's decree is the correct starting date foretold by Gabriel to apply to this prophecy.

These seventy 'sevens' are also interpreted for Daniel on another level of understanding that when they are finished Christ will begin ruling the nations of the earth. We know that after sixty-nine of the seventy 'sevens' Christ was "cut-off"—died. In chapter two I showed you this calculation and how it predicted Jesus' death to the exact day using "prophetic years," but this technique left one 'seven' (seven P-years) unaccounted for.

These remaining seven prophetic years are the seven years of the tribulation period that still await fulfillment in the future. What is of further interest in this prophecy is the use of the word "cut-off" in place of the word "died," which would have been much clearer. Since Christians know Christ is not dead and was bodily dead for only a few days, maybe this was why Gabriel chose his words carefully. I would like to think this was not God's only reason for using this term because it supports another belief.

Christ was "cut-off" (came up short) from completing the seventy 'sevens' and therefore, for this prophecy to be completely true (fulfilled), Christ will need to complete the seventy 'sevens' exactly as prophesied before he can begin ruling on earth. Since we are seven years short using this method of calculation we see this period of time must take place before Jesus can come again or else these verses in Daniel would be untrue.

There is a third way to view this prophecy and calculate the timing! This method requires just counting Sabbath years, skipping over the Jubilee years, from the time of Ezra's decree. As luck would have it, or as I would say the way God planned it, the year 459/458 B.C. when Ezra received Artaxerxes' decree was a Sabbath year. Starting from this Sabbath year and counting into the future, including 459/458 B.C. as the first year, we get to the end of the sixty-ninth Sabbath year in 27/28 A.D. The following year 28/29 A.D. was a year of Jubilee (which we skip counting using this method) and at the very end of this Jubilee Christ began his ministry. This was effectively the start of the seventieth 'week'.

Most scholars agree that Christ's ministry was about 3½ years long, but I believe it was 3½ years long exactly! So if you add the 3½ years to the year 29 A.D. you get to his death date in 33 A.D. The completion of the seventieth Sabbath 'week' would be the next Sabbath year in 35/36 A.D. You can see from these dates that Christ came up short 3½ years from reaching the seventieth Sabbath. He was "cut-off" exactly at the midpoint between the sixty-ninth and the seventieth Sabbath year. Now most theologians who teach the tribulation is 3½ years long, get this belief from misinterpreting passages from Revelation and are unfamiliar with the last method I used to calculate the timing of Daniel's prophecy. They get this idea because they notice the five time periods mentioned in Revelation (twelve hundred and sixty days twice, forty-two months twice, and time, times and half a time) and view them as all speaking about the same period of time and therefore Daniel's prophecies must be unclear or unfamiliar to them. If they knew of this last calculation using just Sabbath years, they would reference it as further proof of their beliefs.

We have one analysis from the book of Daniel that supports a remaining seven years and one that indicates there are only 3½ years left. Hmm… What a dilemma. By accepting the Bible as completely trustworthy, then both must be true at the same time on some level of understanding. Just as the "Seventy Sevens" prophecy is true on three levels at the same time, so must any interpretation be true for both the 3½ years and the seven years for the length of the tribulation to be the correct explanation.

Those who only study Revelation and believe in a 3½-year time-frame get mixed up because some of the timing is really talking about the last half of a seven-year tribulation period while other timing is speaking about the first half. When you divide seven in half you obviously get the same number and these people who only teach 3½ years do not examine the passages closely enough to determine which half is being discussed. Since they almost certainly have not studied Daniel or some of the other prophets, or are unable to see the borders of the overall picture of history from God's plan, they arrive at wrong conclusions and interpretations.

Here is a good rule to remember when studying Revelation and the end-times: Anyone who tells you the tribulation period is only 3½ years is in error and their sequence of events and understanding of those events must also be questioned. They may list a few events that in the final analysis will be correct, but since they really are missing the big picture the Holy Spirit hasn't revealed all the details of this period to them as of yet. You can bet many of the details they provide are false. In fact, this belief in a 3½-year tribulation period is consistent with what the antichrist will want you to believe when he comes and I predict he will deceive many who share this theology!

Now I want to retrace a few steps and cover some points I skipped over. If you remember I pointed out some of the heresies of Christ's life in a previous chapter. This was important because if you are one of those who believe any of them, then you'll have no logical explanation for what I just calculated. The pieces of God's plan for mankind need to fit together on all sides. I just showed you using one of Daniel's prophecies that the only year Christ could have died was the year 33 A.D. This was exactly sixty-nine and a half Sabbaths from Ezra and 3½ years from reaching the end of the seventieth Sabbath year.

This was also the only date within a ten-year range the Passover fell on a Saturday. Those who believe and teach Christ died in some year other than 33 A.D. cannot make their math work out on all levels! Ask them to show you Daniel's calculations.

A few have tried and say the correct decree to start counting from is not Ezra's decree, but rather the decree issued from Artaxerxes to Nehemiah in 445 B.C. (some claim the year was 444 B.C.). When you count sixty-nine 'sevens' into the future from either of these dates you get to 38/39 A.D. Clearly these years are outside the realm of possibilities for Christ's year of death. But they too recognize these are prophetic years and make an adjustment for this understanding (although an incorrect one) as follows:

$$483 \text{ P-yrs.} \times \frac{360 \text{ P-days}}{1 \text{ year}} \times \frac{1 \text{ year}}{365.25 \text{ days}} = 476.1 \text{ P}^2\text{-years}$$

Adding these prophetic adjusted years results in 32 (445 B.C.) or 33 A.D. (444 B.C.). So far so good? But look at the units. What are P²-years? I don't know, but they are not normal years. Let's continue their logic to its conclusion. If 444 B.C. is the correct starting year, we arrive at the correct year of Christ's death, but not the very day (just close) and if you believe the correct starting year was 445 B.C., you will arrive in 32 A.D. A year in which the Passover was on a Tuesday and so Christ had to die on a Monday. If 445 B.C. is the real year of this decree, then this method cannot be right and if 444 B.C. is the actual date, then this prophetic conversion method might be correct (realizing it does not give the exact day of Christ's death as the correct conversion method does and the calendar units "P²-years" are suspicious).

Nevertheless, using this conversion technique on other parts of God's plan only works for this one calculation if you ignore the problems I mentioned. Proof of this is if God's overall plan is seven thousand

P-yrs. long as I say it is, then making a conversion to regular years would make this time period shorter and each trimester of God's plan would not be 2,029.1666 years long, but 1,971.25 P^2-years[18] in length. If you add these "years" to the date of Christ's birth in 2 B.C. you would arrive in the year 1970 A.D. as the date of Christ's return. Did this occur? If you believe for some reason that this time should be added to the year Christ died in 33 A.D. then you would get to the year 2003 A.D. as his return date. Again, this did not transpire.

The point I am stressing is that using this incorrect conversion method does not allow anyone to find other pieces of God's plan! It provides a piece that fits with nothing else. The math does not work out on any other level (like the Sabbath and Jubilee levels). The scholars who do this work are unfamiliar with the overall plan God has in store for us all. But just because their math is poor does not make them bad Christians. In fact, those who search for and keep watch for Christ's return are doing what Jesus instructed them to do. Those servants who do not keep watch are the ones not following Jesus' instructions.

We have seen some Christians who agree Christ died on Friday, which we know is the correct day of his death, but then claim his year of death was a year that cannot support a Saturday Passover. They have no clue that the year they believe Christ died in is inconsistent with the day they say Jesus was crucified. Their calendar mathematics are weak.

Others believe different days of the week for his death to support a year that cannot be true. I have shown you that they are inconsistent with the fifteenth year of Tiberius' reign. Now you see from Daniel's prophecy, another independent method used to check the date, that 33 A.D. is the only year possible. Remember, God's plan has to agree with all scripture for it to be the correct interpretation... the correct understanding... the absolute truth.

Those who come up with the wrong years and days only need to look to the math provided to see the truth. They don't need to continue making up excuses, stories, rationalizations, or twist historical facts when they know the truth! Many times the simple truth is right in front of our eyes and yet we believe it must be "hidden" or it would have been found long ago. Truths that were never hidden in the first place, but when twisted allow Christians to deceive themselves and keep them from seeing God's majestic plan.

From our analysis of Daniel's prophecy of the "Seventy Weeks" so far, we realize there is a missing period of time of seven prophetic years in length and a missing three and a half P-years that Jesus needs to finish to fulfill the prophecy as well. Daniel is given some details of this period that will help with our tribulation timing exercise. Let's see what they are.

He will confirm a covenant with many for one 'seven'. In the middle of the 'seven' he will put an end to sacrifice and offering. And on a wing of the temple he will set up an abomination that causes desolation, until the end that is decreed is poured out on him.

Daniel 9:27 (NIV)

And he shall confirm the covenant with many for one week: and in the midst of the week he shall cause the sacrifice and the oblation to cease, and for the overspreading of abominations he shall make it desolate, even until the consummation, and the determined shall be poured upon the desolate.

Daniel 9:27 (KJV)

When we examine these verses, one from the NIV and the other from the KJV, we can learn a few things about the last seven years before Christ returns. The "he" mentioned is the "prince" who will come from the people (Romans) who will destroy the city (Jerusalem) and the sanctuary (temple). A future leader (not Titus who destroyed Jerusalem and the temple in 70 A.D.) who will come from the people who made up the Roman Empire in the first century and will broker a treaty between the Jews and other nations. This agreement will be the "key" sign that will signal the countdown of the final seven prophetic years before Christ returns!

Since the Jews were scattered and removed from Israel in 70 A.D., we can be sure this agreement was not signed between that time and when Israel became a nation again in 1948 A.D. We can further deduce that it was not signed between 1948 A.D. and seven years ago or Christ would have returned. So we need only to concern ourselves with searching for a treaty within the past seven years (2002 to 2009 A.D.) to determine if we are in the tribulation period as some false prophets are already claiming. A careful examination of this time interval reveals no European leader has signed an agreement with the Jews and their enemies (many) within the past seven years. So we are left to conclude the fulfillment of this seven-year period is still in our future.

This future leader we have been discussing is labeled by most who study these things as the "antichrist" and so we will also use this term when referring to him to be consistent. Now something is going to happen in the middle… around the middle… in the midst of this seven-year period. There is a little controversy with what is to take place. That controversy has to do with the exact nature of the events. Those events are not important at this time and will be discussed later. What is important is that there is an event or a few events that will occur at the midpoint that will be the trigger of much destruction.

The end of the verse is fascinating in that the word "pour" is used to signify the end of this period. The end that is planned (decreed, predetermined) is poured out on the antichrist (NIV) or the desolate (KJV) and this act will signify the completion of the seven years. This clue is important because when we study Revelation you will see the seven "bowl judgments" are poured out at the end and when they are completed Christ will return. Both Daniel and Revelation are consistent with terminology and confirm that the bowl judgments are in the latter half of the seven-year period. God is not haphazard with His choice of words.

If there are a remaining seven years, and there are, to complete Daniel's "Seventy Sevens" prophecy and there are also 3½ years that Christ needs to finish before he can rule on earth, which there are, then Christ's missing 3½ years has to be the last half of the tribulation period! This is the only way possible for both conditions to be true. This arrangement would allow both calculations and the prophecy to be completely true.

Furthermore, Christ cannot finish his missing 3½ years on earth because then he would have to come somewhere in the middle or the beginning of Daniel's missing 'week' (seven P-years) and that would render the prophecy false. Consequently, Christ must complete his missing 3½ years without touching the earth, or in other words, from heaven! We now comprehend that during the seven-year tribulation period, that Christ will be finishing his "required" 3½ years necessary to make Daniel's prophecy completely true from heaven. Another way to state this is, he will begin actively reigning from heaven at the midpoint of the tribulation period.

One last thing worth mentioning about these verses is the antichrist is always trying to imitate God's plan. Why? I am not exactly sure, but my best guess is the closer he mimics the plan, the more people he will be able to deceive because the "forgery" plan will be very close to the real plan. Only God's people, those whose names are written in the book of life, will be able to distinguish the fake from the real plan!

This is the same difficulty most of us would have if we saw a real work of art and a good forgery side by side. Only an art expert who studies those things would be able to distinguish the difference. This will also be the case with Satan, the antichrist, and the false prophet's (the fake triune deity) plans and the only way for Christians to know the difference is to be educated in The Word and what the end-time prophecies hold. Since Jesus has to fulfill 3½ years, then the antichrist is going to try and fulfill a forged 3½ years as well. The only difference is he will try to complete his 3½ year period earlier than when Christ's 3½ year period is planned!

Clearly he cannot wait until after Christ comes because Jesus will already be ruling and he would have no power to deceive those individuals Christ plans to allow entrance into the Millennial Kingdom. This mock 3½ years will actually take place during the first half of the seven-year tribulation period. At the end of that 3½ years (tribulation midpoint) the antichrist will declare he is God, just as Christ will when he finishes his 3½ years and requires all the inhabitants of the earth to begin worshiping him!

This is the abomination (the final straw) that causes desolation (God's wrath) to come upon the earth and its remaining inhabitants. This is also the same point in time when the antichrist is stripped of authority and this earthly authority is given to Christ in heaven so he can actively manage God's wrath i.e. when God's judgments are administered to those who rebelled from the truth of His authority. From a purely human perception there will be no change in the antichrist's leadership, but only from God's perspective. Let's review what we have discussed so far on a chart so that we can commit these realizations to permanent memory.

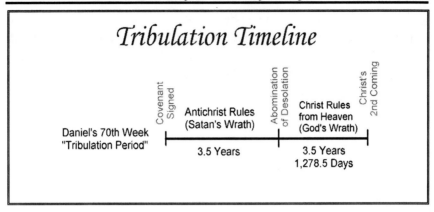

There is just one piece of unfinished business for the forgery plan to be complete. We know that Jesus' 3½ years are the last half of the tribulation period and for the antichrist to copy this requirement (since his 3½ years are in the first half of the real tribulation period) he will need to manufacture a fake 3½ years to put in front of the real seven-year tribulation period! He won't need to do this for Christians who believe the tribulation is only 3½ years long because they have already bought into the deception, nor will he need to for those Christians and unbelievers who believe in no tribulation period and are clueless. The antichrist will need to arrive on the scene 3½ years before the covenant is confirmed only to mislead those Christians who know that there is a seven P-year period remaining.

Listen, Satan has no need to deceive unbelievers (rebels) because they are already his by default. Then what is his goal? To deceive Christians! For the few Christians who keep watch for the signs, most will be unable to recognize him, but it is unrealistic to believe anyone can assume powerful positions in a government (that is not a monarchy) in the blink of an eye. It takes a little experience and track record to move up the ladder in politics. With Satan's help, his son, the "lawless one" will start with a higher social position than most (son of a high ranking politician) and will climb faster (everything he does will go the right way for him and he will do it quicker than anyone would have believed possible) in politics because of it. My best guess is this will take about three years if you read my previous book. This will be enough time to fabricate the fake early years of the tribulation. This will also be a time of great turmoil, which we will discuss later.

If we review what has been put forth so far, we have about a ten to ten and a half year future period of time that we need to be aware of. Since Christ will return in 2028 A.D. this means the signs of the antichrist will be visible to a few Christians who keep watch as early as 2018 A.D. Less than ten years from now!

Two Thousand Three Hundred Days

In Daniel 8 there is a prophecy that causes many problems for scholars. I have studied this prophecy extensively and have come up with many possibilities for the correct interpretation that fits within God's plan. Some say these days really are years using the prophecy rule of "a year for a day" found in Ezekiel.[b] If you use the starting time from Ezra, which is the midpoint in man's history, you arrive at the year 1843 A.D.[19] This date is of no significance, although Seventh-day Adventists would disagree, and so this interpretation must be rejected.

If this prophecy is to be applied substituting "a year for a day," then the best interpretation is as follows: first the years are prophetic years and so twenty-three hundred days are actually 2,333.5 years[13] and the beginning time was 306/305 B.C. when the four horns in Daniel 8: Cassander (Macedonia), Lysimachus (Thrace), Seleucus (Mesopotamia/Persia), and Ptolemy I Soter (Egypt), officially sprang up from the broken great horn's (Alexander the Great) empire with the murder of his sons. Using this understanding and this starting point gets us to the year 2028 A.D.[20] Another possibility is the twenty-three hundred P-days are in error and they really should be twenty-four hundred P-days as discussed in chapter two. That interpretation led us to the date 2028 A.D.[10,11,12] as you well know.

We have seen from Daniel's "Seventy Weeks" prophecy that interpretations can be true on multiple levels and this prophecy may be no different. When we look closely at Daniel 8 we find out that the twenty-three hundred P-days encompass the following events.

1. The "daily sacrifice" was given over to it.
2. The rebellion that causes desolation occurs.
3. The surrender of the sanctuary and of the host that will be trampled under foot.

These three things would take place during the twenty-three hundred P-days and then at the end of these twenty-three hundred P-days "the sanctuary would be reconsecrated." Did these events occur over this time in history? Not that I can tell, so they are best left interpreted as prophetic days just as they are written. Consequently we have a key event prophesied that signals the end of the twenty-three hundred P-days and some events that take place during this time period, but with no starting time foretold.

When I arrived at the year 2028 A.D. substituting a year for a day and making a few assumptions about the starting times I believed this was justification that the term "the sanctuary would be reconsecrated" meant Christ's return. But this term could mean a number of things besides this. Some scholars look at this verse and claim the sanctuary is the temple and is evidence the Jewish temple will be rebuilt. They might be right, but I am unconvinced. Look at some of the possibilities I have come up with when brainstorming this text. You may think of even a few more yourself.

Table 5-1

Phrase	Interpretation	Timing
Sanctuary will be reconsecrated	Christ's second coming.	End of the tribulation period.
Sanctuary will be reconsecrated	Temple rebuilt and reconsecrated.	Middle of the tribulation period.
Sanctuary will be reconsecrated	The antichrist will decimate the sanctuary of Christ (Christians) and the reconsecration will be the rapture of the Church.	Middle of the tribulation period (mid-tribulation rapture).
Sanctuary will be reconsecrated	The temple will be rebuilt and dedicated to God and then shortly afterwards the antichrist will enter it and reconsecrate it (rededicate it) to himself.	Middle of the tribulation period.
Sanctuary will be reconsecrated	The antichrist has been given authority to rule Christ's sanctuary (earth) and this authority will be taken away and given back to Christ.	Middle of the tribulation period.

We can see with all these interpretations that the end of the twenty-three hundred P-days is most likely the middle of the seven-year tribulation period with the exception of the first interpretation. If you believe the correct interpretation is the first one in the table, you must realize that twenty-three hundred or even twenty-four hundred P-days falls short of the required seven years prophesied for this period and so the first under-standing cannot possibly be correct if these are to be viewed as actual days.

In depth inspection of this prophecy reveals this is a prediction of things that will occur during the antichrist's reign and further confirms these days should not be considered prophetic years as we have done here, but rather interpreted as prophetic days since they were revealed by a prophet of God. Hence just as prophetic years need to be adjusted, so do prophetic days. For example, twenty-three hundred prophetic days are twenty-three hundred P-days divided by 360 P-days or 6.388 years… real years. Translating these back into normal days yields 2,333.5 days.[21] This is the same absolute number you get when converting twenty-three hundred prophetic years to real years.

Next, we need to decide whether twenty-three hundred or twenty-four hundred P-days is the true number. Both have interesting arguments for why they are the true answer. Remember, twenty-four hundred can be reduced down using numerical factors God constantly uses whereas twenty-three hundred cannot. This makes a strong case for thinking twenty-four hundred may be the true number. However, this prophecy is really a prophecy about a period of time under the antichrist's influence and so this number doesn't need to match the way God uses numbers, but the way Satan mimics him.

Do you remember when I said the antichrist would need to have a period of time that mimics the seven-year tribulation period so he could more easily deceive Christians who support this version of the tribulation timeline? Well this is it! That's right. The twenty-three hundred days are the fake seven-year tribulation. They are essentially the false last week of Daniel still unfulfilled! Satan has no need to deceive non-believers because his work is already done. They are already his. So the only remaining thing he can do during his last days is to try and win over (deceive) Christians and Jews so he can take as many as he can to their doom. This will be the main goal of the antichrist. Just as Christ's work is trying to save non-believers through the testimony of Christians before his return (because the true believers are already his), the antichrist will be doing just the opposite working on those Christians and Jews who are lukewarm fence sitters. In my first book I showed how twenty-three hundred days were equivalent to sixty-nine of Daniel's seventy 'weeks'. Let's see how that math works again for those readers who have not read that book.

If the twenty-three hundred "days" are interpreted as years from God's perspective, then the twenty-three hundred "years" are prophetic years (P-yrs.). We also know from chapter one God's plan has twelve thousand P-years in one Season (S) of Mankind… in His whole plan. There are 360 "days" in that one season on God's calendar. These are called seasonal days or S-days. Therefore, one S-day on God's calendar is actually 33.333 P-years[22] long. Let's look at a transformation to see how twenty-three hundred P-years match up with Daniel's sixty-nine 'weeks'. Watch:

$$2{,}000 \text{ P-yrs.} + 300 \text{ P-yrs.} = 2{,}300 \text{ P-yrs. Therefore:}$$

$$2{,}000 \text{ P-yrs.} \times \frac{1 \text{ S-day}}{33.333 \text{ P-yrs.}} = 60 \text{ S-days} \,\&$$

$$300 \text{ P-yrs.} \times \frac{1 \text{ S-day}}{33.333 \text{ P-yrs.}} = 9 \text{ S-days and thus}$$

$$2{,}300 \text{ P-yrs.} \times \frac{1 \text{ S-day}}{33.333 \text{ P-yrs.}} = 69 \text{ S-days!}$$

Converting this result into normal years gives 2,333.5 years. This translated time is actually equivalent to seventy S-days.[23] By making the conversion from prophetic years to actual years, we gained one full S-day just like when you convert sixty-nine 'weeks' from Daniel to man's perspective gains one extra 'week'!

Now you can see the parallelism of these two numbers and Daniel's two prophecies. What this reveals is that if you convert this number into real years or real days from prophetic times, it can be viewed as the equivalent of completing seventy 'weeks'. So from the antichrist's perspective it can be said at the end of twenty-three hundred P-days he will have completed the required seventy 'sevens' prophesied just as Christ must!

If you also remember, there were other ways to view Daniel's prophecy of "Seventy Sevens." So too with the antichrist's forgeries of these calculations. Daniel is told there are seven 'sevens' (weeks) and sixty-two 'sevens,' but Daniel is never told why the seven 'sevens' are separated out in the prophecy.

Just a thought, if these separated out 'sevens' are Jubilee cycles, then in actuality they mean three hundred and fifty total years. What do you suppose happens when you multiply three hundred and fifty years by an intermediate number of man (6.666), which Revelation says is associated with the man of perdition (six hundred and sixty-six)? You get 2,333.33 years... extremely close to the 2,333.5 number previously calculated and the real time mentioned in Daniel 8 pertaining to the antichrist.

Similar comparisons apply using twenty-four hundred P-days as well. When these prophetic days are converted you get 2,435 days exactly or 6.666 years.[11,24] Fascinating. After looking at all these twists to the numbers I am resigned to the fact the twenty-three hundred P-days may be the best answer for now given all the information, but I wouldn't rule out twenty-four hundred P-days as having a high probability of being right either. What this ultimately means is that it may be feasible to recognize the antichrist sooner than what I am going to first show.

Where does this timing fit into God's plans? The most probable answer is the 2,333.5 actual days end at the middle of the "real" tribulation timeframe. If you add 3½ years worth of days (1,279 days) to this total, you get a new total of 3,612.5 days. This happens to be forty days short of ten years exactly! Interesting. Calculations that yield this number always grab my attention. As you know, God uses this number a lot and there are reasons for it.

So how did I actually arrive at this understanding? Well I had studied these numbers for weeks and played with all kinds of combinations and ideas much like a puzzle. Some people put puzzles together by studying the shape and colors of the missing pieces and then go searching for the pieces that match the closest. They spend more time thinking and visualizing what the missing pieces look like and less time trying to fit pieces together.

Others use the opposite approach: the "brute force" method. They spend less time visualizing the missing pieces and more time grabbing all the pieces that might fit and then trying them until they find one that fits. These approaches both have an advantage over what I was trying to accomplish in that they know all the pieces belong to the puzzle, what the final puzzle looks like, and where to look for the pieces. I had none of these advantages. The method I used was to gather all possible pieces, some of which may not even belong, and basically use the "brute force" method and try connecting them in every combination until I could see mathematically if they fit. Then I tested what I had connected to see if scripture supported what was connected.

By understanding how God uses numbers, it is safe to assume that these forty days must be included in the overall timeframe and the total time accounted for by Daniel's prophecies is now ten years exactly. This idea will be proven later.

Recognizing that after Christ rose from the dead, he waited an additional forty days before his ascension to heaven, we can infer that after he returns there will be an additional forty days before the "official" start of the Millennial Kingdom. So we see that the Bible gives us information not only for seven years before Christ returns, but also for ten years. Let's see what our updated chart looks like now.

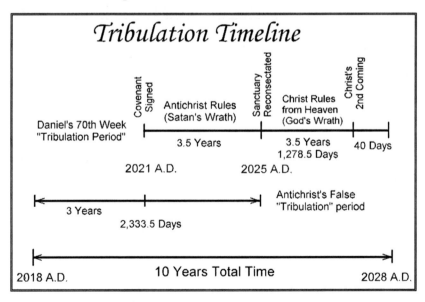

So how does this timeline compare with all the times forecasted in Daniel and Revelation? Let's thoroughly examine these times prophesied by Daniel and the apostle St. John.

Detailed Timing of the Tribulation

We begin this detailed mathematical examination by comparing the prophecies that pertain to the timing of the tribulation period with what we have learned so far. As we scrutinize the timing for this future period of history, we know there are two basic viewpoints held by theologians. Some eschatology experts claim the length of time for the tribulation is seven years while others say it will only take three and a half years. I believe that using mathematics and solving problems are the skills needed to come to the right understanding about these last days (along with the Holy Spirit's leadership), and that these are not the talents God has given everyone. I will do my best to keep to the goal of explaining the timing of the tribulation while minimizing the boring analytical details as much as possible.

The first issue is to define what the tribulation period means. Many who argue about the timing are really arguing about semantics. For our purposes, my use of the term "tribulation" is not limited to just the time of God's wrath, but covers the time period from when the antichrist begins his work at deceiving the nations to the start of the Millennial Kingdom. This is a broader interpretation than the "classical" understanding of this period of time.

Why do I expand this definition of the tribulation from our earlier work? Because the Lord has provided the timing in the Bible that allows us to do so! As a Christian, don't you want to know everything you can about the last days? To not only know about the wrath of God's judgments poured out on the unbelieving and unrepentant sinners of the world who reject Christ and accept the antichrist, but to also know about the things that will happen during the antichrist's leadership prior to the coming wrath? Of course you do.

There are several verses that speak to the timing of events and the length of times for this period. Many of these verses I believe have a duality to them that allows both a literal interpretation and a deeper interpretation. In this section we will concentrate on the literal interpretations and piece together the most logical understanding by meticulously examining all the times. We have looked at the outline of this period so I will skip over many of the textual proofs so you can grasp the detailed timing of this period.

For those who are knowledgeable on this subject and have a different viewpoint, this approach will be insufficient. But for those who are unknowledgeable, this approach will work best because I have painfully found that giving too much information overwhelms the majority of readers and the message gets lost doing this.

We will begin by listing the pieces from scripture that speak to this timeframe and see if we can assemble them afterwards so they fit together in a unified understanding of this period.

> *I was given a reed like a measuring rod and was told, "Go and measure the temple of God and the altar, and count the worshipers there. But exclude the outer court; do not measure it, because it has been given to the Gentiles. They will trample on the holy city for **42 months**. And I will give power to my two witnesses, and they will prophesy for **1,260 days**, clothed in sackcloth."*
>
> *Revelation 11:1-3 (NIV)*

> *The beast was given a mouth to utter proud words and blasphemies and to exercise its authority for **forty-two months**. He opened his mouth to blaspheme God, and to slander his name and his dwelling place and those who live in heaven. He was given power to make war against the saints and to conquer them. And he was given authority over every tribe, people, language and nation. All inhabitants of the earth will worship the beast—all whose names have not been written in the book of life belonging to the Lamb that was slain from the creation of the world.*
>
> *Revelation 13:5-8 (NIV)*

> *The woman fled into the desert to a place prepared for her by God, where she might be taken care of for **1,260 days**.*
>
> *Revelation 12:6 (NIV)*

> *But after the **three and a half days** a breath of life from God entered them, and they stood on their feet, and terror struck those who saw them.*
>
> *Revelation 11:11 (NIV)*

*The woman was given the two wings of a great eagle, so that she might fly to the place prepared for her in the desert, where she would be taken care of for **a time, times and half a time**, out of the serpent's reach.*

Revelation 12:14 (NIV)

*... The people of the ruler who will come will destroy the city and the sanctuary. The end will come like a flood: War will continue until the end, and desolations have been decreed. He will confirm a covenant with many for **one 'seven.'** In the **middle of the 'seven'** he will put an end to sacrifice and offering. And on a wing of the temple he will set up an abomination that causes desolation, until the end that is decreed is poured out on him.*

Daniel 9:26-27 (NIV)

*"From the time that the daily sacrifice is abolished and the abomination that causes desolation is set up, there will be **1,290 days**. Blessed is the one who waits for and reaches the end of the **1,335 days.**"*

Daniel 12:11-12 (NIV)

*Then I heard a holy one speaking, and another holy one said to him, "How long will it take for the vision to be fulfilled—the vision concerning the daily sacrifice, the rebellion that causes desolation, and the surrender of the sanctuary and of the host that will be trampled underfoot?" He said to me, "It will take **2,300 evenings and mornings**, then the sanctuary will be reconsecrated."*

Daniel 8:13-14 (NIV)

The vision of the evenings and mornings that has been given you is true, but seal up the vision, for it concerns the distant future.

Daniel 8:26 (NIV)

These are all the verses of scripture that appear to speak about the timing of the end-time tribulation period. I believe all these verses use prophetic accounting just as the timing for other prophetic events I have discussed in God's plan. There are no exceptions! So how do we easily make the conversion from God's time to our time? By using the same method we have been employing throughout the book. Multiplying all the relevant timing that deals with the tribulation by the number 1.01458333,[25] which is the conversion factor you get when changing from 360-day years to 365.25-day years.

Most Bible scholars incorrectly think that a "week" mentioned in Daniel 9 means seven years. However, with this understanding they are unable to mathematically calculate from that prophecy the correct date Christ was "cut off." We know some eschatology scholars get close and still others use their incorrect calculations as justification for the heresies they spread about Christ's death. But I have already proven by converting Daniel's "weeks" from prophetic years to normal years, that Daniel's prophecy predicts Christ's death to the very day! Some people who see this calculation call me all sorts of things—even claiming what I have calculated is not possible and yet the math is right there in front of them to be challenged. I ask them to show me where I have erred, but almost all decline and choose to continue with the lies they were taught or developed from their own belief system. How disheartening it is and how naive I am to believe that if you show someone "one plus one equals two," that they would continue to believe it could be some other number. Many self-professing Christians really do have a blinding faith and I hope it helps them when times get tough.

When converting prophetic days to normal days, they must first be converted to regular years and from there, back to normal days. So the seven years from Daniel's 'week' are really prophetic years (P-yrs.), which translates into 7.1021 years[26] when measured on earth. We also need to realize that fractions of a day should be rounded when counting days.

For example, 1,259.0417 days really means the first hour of the twelve hundred and sixtieth day or twelve hundred and sixty days when counting in days with no regard to the hour of the day. So it is possible that the times given by God could all be up to one day short as in the example I provided. Now I am not saying they are, I am only saying that mathematics allows for it. If you want to believe these are exact numbers, then you need to believe you can calculate exact times of the day down to the very

second, which I do not believe is possible. This may be why Jesus said, "no one knows the day or the hour" because that information is not given in most cases. So let's convert all the tribulation times from scripture in the table 5-2 below.

"Prophetic Conversion" Table 5-2

Timing Given	*Conversion (365.25/360)*	*Rounding*
1,260 days	1,278.38 days	1,279 days
42 months	1,278.38 days	1,279 days
3.5 days	3.55 days	4 days
Time, Times & ½ a Time	N/A	N/A
1 'week' or 'seven'	7.102 yrs or 2,594.04 days	2,595 days
1,290 days	1,308.81 days	1,309 days
1,335 days	1,354.47 days	1,354 days
2,300 days	2,333.54 days	2,334 days
2,400 days *	2,435 days	2,435 days
10 years	3,652.5 days	3,653 days

*Twenty-four hundred days has been shown as a possible replacement of the two thousand three hundred days and is the reason why it is listed in the table.

Now that we've had a look at the timing of end-time events, let's do some preliminary timeline assembly. Using the scripture verses from: Revelation 11:1-3 (42 months and 1,260 days), Revelation 12:6 (1,260 days), Revelation 13:5-8 (42 months), and Revelation 11:11 (3½ days) with their corresponding real times from table 5-2, I will add this information to a timeline that will allow us to see a visual picture of what the timing of these texts mean. You may want to go back and reread these texts on pages 61 and 62 before continuing.

Reading these verses and the verses that accompany them in the Bible, which I have not listed for space reasons, one realizes that the twelve hundred and sixty days spoken of in Revelation 11 through 12 are the same period of time and the forty-two months listed in Revelation 11:1-3 are a separate but equal amount of time. Here is what this information looks like on a timeline.

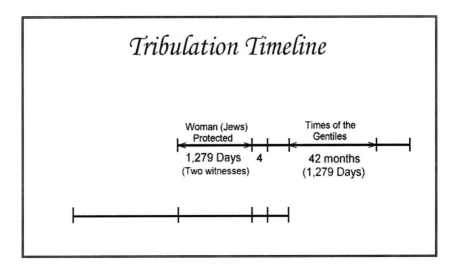

We can see from this incomplete timeline the beginnings of how all these times fit together. Since the timeline is not finished it may be hard to visualize the completed timeline, just as looking at an incomplete puzzle would be impossible to visualize if you had no idea what the final picture looked like. Bear with me because as we move along, the time-line will become clearer and clearer and come into focus.

Also, the forty-two months the Christians will be persecuted will fall during the same time the two witnesses are testifying and the Jews are protected and not during the Times of the Gentiles. This was not noted on the timeline. Let's examine more pieces for the tribulation timeline.

The woman was given the two wings of a great eagle, so that she might fly to the place prepared for her in the desert, where she would be taken care of for **a time, times and half a time**, *out of the serpent's reach.*

Revelation 12:14 (NIV)

... The people of the ruler who will come will destroy the city and the sanctuary. The end will come like a flood: War will continue until the end, and desolations have been decreed. He will confirm a covenant with many for **one 'seven.'** *In the* **middle of the**

'seven' *he will put an end to sacrifice and offering. And on a wing of the temple he will set up an abomination that causes desolation, until the end that is decreed is poured out on him.*

Daniel 9:26-27 (NIV)

The "woman" in this prophecy represents Israel and she will be protected for a "time, times, and half a time." This is confirmed in Daniel 9:27 where it says a covenant will be confirmed with many for the first half of the tribulation period. As best as can be determined, this covenant will be an agreement that will protect (deceive) the Jews during this time. This piece is added to the timeline as follows:

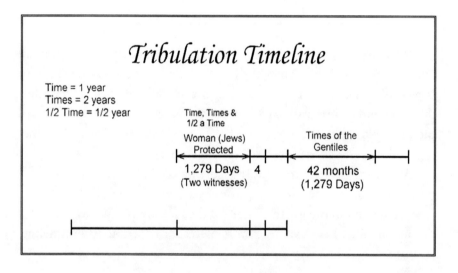

We have finished adding all the times mentioned in the book of Revelation that deal with the tribulation. Now we will start adding the remaining times from the book of Daniel.

*"From the time that the daily sacrifice is abolished and the abomination that causes desolation is set up, there will be **1,290 days**. Blessed is the one who waits for and reaches the end of the **1,335 days**."*

Daniel 12:11-12 (NIV)

The first amount of time that will be added to the timeline is the 1,290 P-days, which we find on the conversion table 5-2, is really 1,309 total days. These 1,309 days are represented as 30 days plus 1,279 days and are added as shown to the timeline.

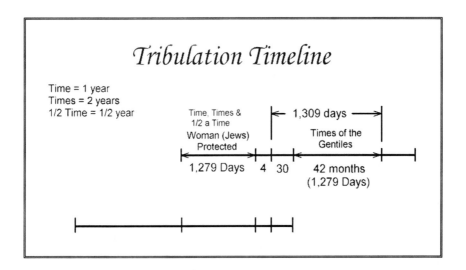

Let's now add the 1,335 P-days. But before doing this we need to examine this data more closely. When one looks at Daniel chapter twelve in detail it is easy to get the impression there are forty-five days between when Christ returns after the 1,290 P-days and when the 1,335 P-days are completed. Why? Because obviously 1,335 P-days minus 1,290 P-days is forty-five P-days.

But this number is not how God works. How many times does the number forty-five crop up in the Bible? Well at least once. If there are more, there are not very many more and are unimportant. If you study the Bible, the number you should be familiar with that God uses is "forty." This is the real duration and this revelation was first noticed from our previous work. Secondly, God knows that fractions of a day really mean hours, minutes, and seconds. So the days listed as whole numbers can be up to almost one full day less.

Daniel's last 'week' is actually 2,594.04 days (review table 5-2 pg. 64) long when converted from years into days and they are also represented as the 2,595th day... effectively 2,595 days. Remember that 1,260 P-days plus 1,335 P-days (2,595) is the same number as one of Daniel's 'weeks'!

But what if 2,594.04 days were the exact amount of time down to the minute? Assuming this is the case, we could use this number instead and not the full 2,595 days. Applying this understanding allows us to substitute 1,334 P-days for 1,335 P-days. This idea in effect makes our actual error smaller. Then these 1,334 P-days translate into 1,353 actual days instead of the 1,354 days shown in the conversion table 5-2. So let us use this understanding of this timeframe and incorporate it into our chart to see what it looks like.

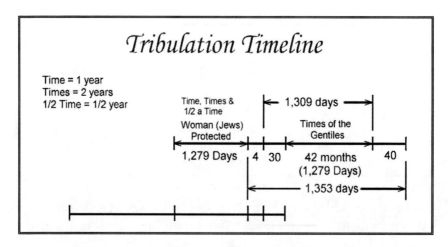

By doing this we now realize that the "abomination that causes desolation" spoken of by Daniel is not a one day event, but in all probability four days. Daniel writes,

> *His armed forces will rise up to desecrate the temple fortress [fortresses are cities and therefore this means Jerusalem or if interpreted spiritually, it might mean Christian strongholds like the United States] and will abolish the daily sacrifice. Then they [his army] will set up the abomination that causes desolation.*
> *Daniel 11:31* (NIV)

Since an army is involved we can infer that whatever the "abomination" is, implementing this will not be something that will be done easily and will require some military intervention. My best guess is when the antichrist tries to set himself up in the temple in Jerusalem at this time, many Jews will recognize him for who he really is and will put up some resistance.[c] That revolt will take a few days to deal with.

Moving on, we still have more pieces to discuss even though this is the point where prophecy experts would stop, but we are just beginning. We need to add the two thousand three hundred days as was done in the prior work.

*Then I heard a holy one speaking, and another holy one said to him, "How long will it take for the vision to be fulfilled—the vision concerning the daily sacrifice, the rebellion that causes desolation, and the surrender of the sanctuary and of the host that will be trampled underfoot?" He said to me, "It will take **2,300 evenings and mornings**, then the sanctuary will be reconsecrated."*

Daniel 8:13-14 (NIV)

The vision of the evenings and mornings that has been given you is true, but seal up the vision, for it concerns the distant future.

Daniel 8:26 (NIV)

Now where does this prophecy fit in the tribulation period? We know this is one of the most hard to decipher prophecies of Daniel to comprehend and interpret. But it need not be, for it speaks about the time under the antichrist's leadership as we have already learned. What is the total real time? From our conversion table 5-2 we know it is twenty-three hundred and thirty-four days. What a strange number. Why? Well, by converting it, we got thirty-four extra days. Looking at our timeline, do these thirty-four additional days look familiar? Yes they do. Let's break apart this time under the antichrist's influence as follows: 2,300 days + 4 days + 30 days.

Next, let's break apart the remaining twenty-three hundred days even further by separating three and a half years out so that it can now be represented as one thousand and twenty-one days plus twelve hundred and seventy-nine days. Now we have enough information to place this data on our timeline so that we can see what it looks like.

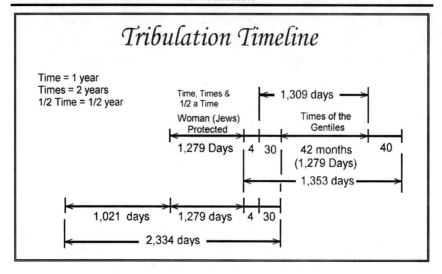

We have all the times recorded in the prophecies of Daniel and Revelation now shown on the timeline, as they will really occur. However, I would like to make just a few more minor adjustments. Changes to improve the accuracy and comprehension of the tribulation period. Now what is currently shown is really the amount of time there will be, but by making these last adjustments this will allow tribulation students a closer look so that they might grasp the unlocked mysteries of these important times.

For example, the four days will actually be four "whole" days when counting by days, but with God telling us there will be three and a half prophetic days He has informed us of the hour of the fourth day... the time of that day!

I don't know the beginning point from where to count the three and a half P-days, so I am currently unable to calculate the exact hour of the fourth day other than to speculate. Let's do that before continuing because it is interesting. Using the Gregorian calendar (new days begin at midnight) and counting from when a new day starts (midnight), three and a half P-days would end at noon. But the real time is 3.5486 days.[27]

Taking away the full days and concentrating on only the decimal portion, yields 13.1666 hours[28] into the fourth day. This is effectively military time and is thirteen hours (is the thirteenth hour Satan's hour?) and .1666 hours (ten minutes exactly or one sixth of an hour) from midnight or 1:10 pm. If you understood the way God uses numbers, this time of the day might ring bells.

Now when the twenty-three hundred P-days were converted to real days, we got 2,333.54 days. The last part of this unrounded number (2,300 days + 30 days + **3.541666** days) is very close to the number **3.5486111** we were just examining. Using that number would get us to 1 pm exactly! There is a minor difference of exactly ten minutes between the two. Effectively, ten minutes amounts to no error at all. So let's go back and replace the rounded four days with the unrounded numbers and see what that looks like.

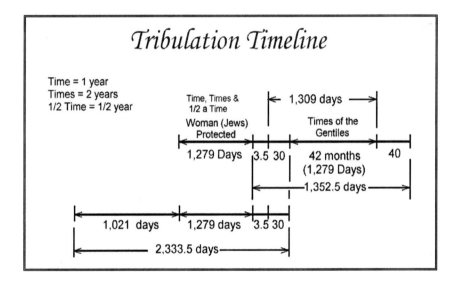

I wish to make one final observation about the timeline. The number ten is the "number of completion." This concept is taken from countless studies on how God uses numbers. This is the number God uses to finalize things.[d]

You have seen the logic, methods, and how this timeline was systematically assembled. You have witnessed that by converting all the prophetic times to real times, allowed for things to be put in their logical places. Can all these mathematical manipulations be mere chance that the pieces fit? What is the probability that converting twenty-three hundred P-days would give the exact number of extra days revealed in the books of Daniel and Revelation?

How about converting the twenty-three hundred P-days and the three and a half P-days into real time and coming within ten minutes of each other? What about the actual time after Christ's return really being forty days instead of forty-five P-days? Isn't this how the Lord really works and gives proof that this understanding is the true understanding? Or is it just some masterful mathematics on my part? I think not. God is the master mathematician. I am nothing in comparison.

Now if you look very closely at the timeline there is something else going on that all, but a few are unable to see. But you already know what it is. The hidden truth of God's power… the power of how God works. Did I not tell you the number of completion is ten? Have you found the hidden truth or remembered what I said previously? Let's see what this secret is.

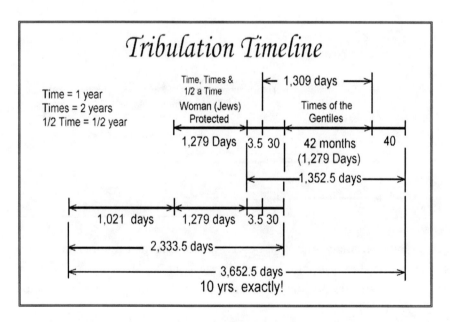

After all the mathematics and manipulation, the total time came to exactly ten years… not a day more and not a day less! The same as before. Wow! Our God is an awesome God. I could not have come up with this plan in a million years and here we have it buried within the prophecies of Daniel and Revelation. What is even more important are the events associated with these times.

It will be too hard to cram it all on this diagram so that will be covered in later chapters. But the important thing this chart reveals is the time when the antichrist comes into the picture will be ten years exactly from when we enter the Millennial Kingdom.

Based on other things I wrote about him in my first book, he will begin his rise in politics at the age of thirty-nine (just a few months short of forty years old) and for the first one thousand and twenty-one days (just under three years) he will move up the political ladder gaining power. At the end of these one thousand and twenty-one days he will confirm the covenant spoken of in Daniel.

After another 1,279 days (3½ years) he will break that covenant when he sets up the abomination during 3½ P-days in the middle of Daniel's last 'week'. After a bloody period of thirty days enforcing the new laws to worship only him as the world's dictator, the rapture will occur and the last 3½ years of Christ's rule from heaven and God's wrath will commence.

Lastly, Christ will return at the end of the remaining 3½-year period (completing Daniel's seventy 'sevens' by reigning from heaven) to wage war on the antichrist, the false prophet, and the unbelievers of the world. This war and the removal of those things that cause sin and the people destined (weeds) for the Lake of Fire will take forty more days and nights to complete this cleansing. Then the time of Christ's earthly reign will begin.

This is pretty impressive, but is this the final truth of this matter? I think it is very close and may be the final answer. It definitely is close enough for Christians to use as a guide without further refinements during the last days. But I am bothered by a couple of small errors that are not easily rectified. Errors no one would find, but I know they are there.

It is like using a piece in a puzzle that isn't quite right and forcing it to fit. It is off ever so slightly, but it is a counterfeit piece nevertheless. Only the expert eye can see it, but the problem is this piece came with the puzzle (came from the Bible). In other words, it is suppose to fit!

Let's see what I am talking about. Daniel's missing 'week' is really 2,595 days and if you remember I "jiggled" this piece mathematically

(2,594) to make it fit. Why? Because I ultimately knew I was going to add a future piece that didn't quite fit and I wanted to make that piece fit perfectly. Before some of the conversions were made, I realized that 1,260 days plus 1,335 days equaled 2,595 days exactly. I had only converted Daniel's missing 'week' to real time and since this was the correct conversion method, then this meant the 1,260 and 1,335 days had to be real times and could not be converted at all.

This was because Daniel's 'week' was now real time and equaled those two timeframes. This idea was evidence that the times written in the prophecies were actual times as they were written.

But I said at the very beginning of our studies that they were not real times, but prophetic times. By converting the times (1,260 P-days to 1,279 days and 1,335 P-days to 1,354 days) I had now increased the total time to 2,633 days and so now it didn't match the converted time of Daniel's last 'week'. Theoretically, we now had thirty-eight days too many.

However, this situation is really a case of the duality of God's numbers being right on multiple levels. Daniel's last 'week' covers the period of time from the signing of the covenant to the return of Christ and not all the way to the beginning of the Millennial Kingdom. So the last forty days shown on the timeline really can't be included.

What does this mean? Forty days needed to be removed from the calculation and when this was done, there was now only 2,593 days. So instead of being thirty-eight days too long, Daniel's last 'week' was now two days short.

However, our timeline is really short three days from what Daniel predicts it should be (1,279 days + 4 days + 30 days + 1,279 days = 2,592 days). Most people would ignore these errors, but they trouble me because God is precise. I could possibly live with a one-day error because that could be rationalized away with "not knowing the hour or the day."

I have wrestled with the problem for months and this idea just came to me. It is Satan's work! Let's see what I mean. First I have been making conversions using the "easy" method. That is to say I multiply by 365.25/360 to get the final time and always round up to the next full day.

However, there is another method that may be more precise but is more complicated. Let's look at this method. First, Daniel's seven year period is converted to real time using a more precise number for days in a year (365.2421199 instead of 365.25) and so we get 7.1019301 real years as opposed to 7.10208. This is a small difference of an hour and twenty minutes.

Next, we need to convert years back to days. This is where the method deviates. We convert based on what actually happens on a calendar taking into account leap year rules. So every four years is four times 365.25 days or one thousand, four hundred and sixty-one days exactly. This leaves us with 3.10193 years left of which no leap years are involved and so these are only 365-day years.

Multiplying the remaining years by three hundred and sixty-five days results in an additional 1,132.2045 days. Adding these times together yields 2,593.2 days. This is effectively two less days than what we got using the easy method. This leaves us only one day off. Now I said I would be happy if I could get the error down to one day, but I am not because there is a long standing problem that has troubled me for months which I wrote about earlier and in laborious detail in my first book.

The problem is that there is evidence of the twenty-three hundred P-days in Daniel being incorrect and that it could be twenty-four hundred P-days. What is the converted time for twenty-four hundred P-days? Looking back to the conversion table 5-2 we find that the actual time is two thousand four hundred and thirty-five days exactly! How many years is two thousand four hundred and thirty-five days?

It is 6.6666 years![24] What is the number of the antichrist? Six hundred and sixty-six. Do these numbers look familiar? They are variations on the number of the beast.[e] Let's break apart this time so that it can replace the twenty-three hundred P-days on the timeline: 2,400 days + 5 days + 30 days. We need to make another adjustment by undoing the one-day reduction from the thirteen hundred and thirty-five P-days we made earlier. Here is what those timeline changes look like after substituting the twenty-three hundred P-days with two thousand and four hundred prophetic days.

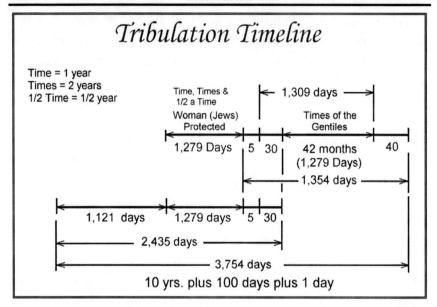

Tribulation Timeline

Time = 1 year
Times = 2 years
1/2 Time = 1/2 year

Time, Times &
1/2 a Time

Woman (Jews)
Protected

1,279 Days | 5 30

←— 1,309 days —→

Times of the
Gentiles

42 months
(1,279 Days)

40

←— 1,354 days —→

1,121 days | 1,279 days | 5 30

←— 2,435 days —→

←—————— 3,754 days ——————→

10 yrs. plus 100 days plus 1 day

Now this version of the timeline corrects the one-day error in Daniel's final week (2,593 days) and makes both numbers (1,290 P-days or 1,309 days & 1,335 P-days or 1,354 days) in Daniel 12 also match without mathematical manipulations. But this adjusted timeline has new problems. The three and a half days in Revelation has picked up a one-day error in the opposite direction. In other words, we just pushed the error to another place on the timeline.

If we move the error to the thirty days and create a thirty-one day month this will only push the error into the 1,309 (1,290) days time-frame. Whereas the first timeline with two thousand three hundred P-days was one day short, this version appears to be one day too many! In addition, the timeline grew another one hundred and one days. Interesting. If the extra day could somehow be eliminated from the overall timeline, then there would be one hundred days more than ten years exactly.

Since we are using twenty-four hundred P-days (6.666 years) in the timeline spot where the antichrist will rise to power it does seem like this time would be more fitting. But this raises another question, "if two thousand four hundred P-days is the real number, then why did it ever get changed?" Obviously it had to be the work of Satan if it did. Since this time is effectively one hundred days more, we would get more of an advanced warning of the antichrist's movements and plans. Maybe he didn't want this information to be known, but God does.

So is this one-day error fixable? Because if it were, this would give more credence that this is the correct version of the timeline. Presently, I cannot see a clear path to this answer. I have made multiple calculations trying to eliminate this error and all this work does is move the new error around on the timeline.

The best answer may lie in the beginning time and the ending time. For instance if we subtracted twelve hours each from both these times we could get our overall time down to ten years and one hundred days exactly which is almost certainly the true number if this is the correct timeline. Doing this would not affect the timing of Daniel's final week (2,593 days) or Daniel's 1,290 prophetic days. It would change the 1,335 prophetic days back to 1,353.5 days, which we have seen is possible and make the forty days really thirty-nine and a half days.

Now this highlights another dilemma. For instance, did Jesus wait forty whole days before ascending into heaven so that he actually ascended on the forty-first day or did he ascend on the fortieth day, which is thirty-nine whole days and some hours? If the truth is the latter, then it could be only thirty-nine and a half days, but if it is the former, then it cannot. This is the first problem.

An equal problem occurs at the beginning of the timeline when twelve hours are removed. We get 2,434.5 days instead of 2,435 days. This block of time is the only piece, when converted that came to a "whole" number of days. It was exactly 2,435 days and needed no rounding. Furthermore, it represented 6.666 years exactly. By manipulating this timeframe so there is twelve hours less, this technique throws off the revelation the math provides.

Since we cannot remove the one-day anywhere in Daniel's last week we are left with this option of removing half the error at the front end and the backend of the timeline. One day really doesn't matter in the grand scheme of things and so assuming this is the correct time I will arbitrarily split the one-day error at the beginning and ending of the timeline and leave it at that. Recognizing the fact that the true timeline may be the one that is ten years exactly.

Let's remove the extra time from our overall timeline, but keep the detailed times the same due to rounding. What does our final iteration of the tribulation timeline look like with these changes and the added information from our beginning analysis?

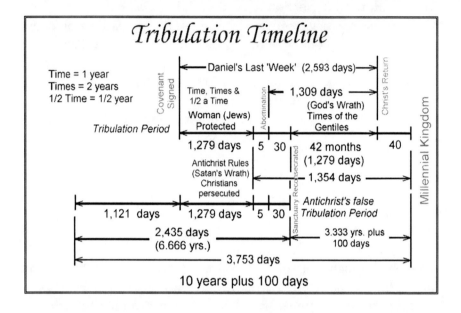

So the reader is left to make his or her own choice. Are there ten years or ten years plus an additional one hundred days that are prophesied in the Bible at the end of the age? Future history will ultimately testify to which one of these timelines is the correct version.

Actual Dates

Now some hints and mathematical techniques may allow this timeline to be aligned to real dates on a calendar so that potential dates could be provided. For instance, let's pick the last half of the tribulation. The forty-two months the gentiles (unbelievers) will be under Christ's rule from heaven and God's wrath. If the time is really 1,279 days long as I portrayed it, then there must be a leap year within this block of time.

If there weren't, then the time would only be 1,277.5 (3½ x 365) days instead of 1,278.5 days. This same logic holds true for the first half under the antichrist's yoke. So even though it is possible to have three leap years within a ten-year period (by either starting with a leap year in the first or second year or by ending with it) it may also be true that only two leap years occur within our overall timeframe. This means there are four leap year patterns possible to examine. Let's see what these leap year (LY) patterns look like in Table 5-3.

Table 5-3

Yr. 1	Yr. 2	Yr. 3	Yr. 4	Yr. 5	Yr. 6	Yr. 7	Yr. 8	Yr. 9	Yr. 10
LY				LY				LY	
	LY				LY				LY
		LY				LY			
			LY				LY		

From our table of patterns we see the first two rows show it is possible to have three leap years in this ten year span, while the last two rows show it is possible to have only two leap years. By superimposing our timeline on this leap year table, and sliding it back and forth, it might be possible to align it so that actual dates might be determinable.

Since 2028 A.D. is a leap year, we can see that the pattern in row two should be the only pattern of interest to examine further. But this exercise, if done for other years, might reveal possible errors in the projected end date of Christ's return that may require our 2028 date to be reexamined. Why? Because the tribulation time must fit within the overall plan and if it doesn't, then we must find the reason why.

However, to perform this analysis we must speculate on some possible dates and brainstorm why the antichrist and God would want to do certain things on certain dates. I said that the most logical time Christ would return would be the Day of Atonement in my first book. If this were true, then what other days would be predicted on our timeline and would they make scriptural sense? If they did make sense, then we might conclude we had found the true dates. If they didn't, then we would need to repeat the analysis using another starting point.

This analysis is not complete and so actual dates are not provided here. It may come to pass that this analytical exercise may be unsuccessful, but again without completing these analyses we cannot know with certainty. Only God knows right now. There are still other ways we can investigate the tribulation timeline to glean more knowledge of God's plans for the end of this age. Another of these has to do with examining the Jewish calendar.

Calendar Mathematics

So why are there three and a half P-days plus thirty days in the middle of the tribulation timeline? Is there more to it than what was prophesied in Daniel and Revelation? The simple answer may be that's the way God planned it. But there may be another answer that is more basic than even that response. It's because that's the way the calendar mathematics work out! This is the answer that came to me around 3 am on the morning of June 16, 2008. As it turned out, by the time I got up and turned the computer on, it was 3:33 am when I began typing.

I was on vacation and had noticed unusual cloud formations and colors around sunset the evening before. These clouds were so unusual that I got my camera and went out and took pictures. Something I had never done before in my life. Even the local nightly news on television reported this phenomenon.

As it would happen, this was also an evening approaching a full moon. I only make this observation because when I was writing my first book I had received similar impulses, urges, or revelations on days there was a full moon. One such episode occurred on an August night in 2007 A.D. when there was an eclipse of a full moon and so now I am more aware of my surroundings and the dates and times I get these inspirations, which I believe come from the Holy Spirit.

Now I wrote in my first book the reason why Daniel used twelve hundred and ninety days instead of the twelve hundred and sixty days that were prophesied in Revelation was because he was using the Jewish calendar and the additional thirty days was signaling that the midpoint of the traditional seven-year understanding of the tribulation would occur in a year that had a thirteenth month in it. To grasp what I am saying requires a working knowledge of the Hebrew calendar.

The Jews use a calendar based on the phases of the moon. Why? Because it's very hard to distinguish one day from the next when using the sun as a marker of time. It looks the same all the time. But at night the shape of the moon (phase) is always changing. So when they were learning how to account for time thousands of years ago, it was easiest to watch the moon.

Using this method, they came up with twelve months in a year just like the current solar calendar we use today. However, it turned out the average number of days in a lunar cycle is twenty-nine and a half. This gave them three hundred and fifty-four days for one year.

We know the real length of a year is very close to 365.25 days and so by using their calendar after a few years the seasons would begin to drift. It wouldn't take very long before winter would come in June for instance. Therefore, to correct for this problem with their calendar, the Jews would add an extra month... a thirteenth month, every so often, around the end of winter and the beginning of spring. This practice in modern days is accomplished by adding the additional month seven times for every nineteen years. This is about every three years on average. What does this lesson have to do with the tribulation timeline? Well it was needed so we can compare the two calendars.

Let us presume that the year Daniel's last 'week' begins in is a year just after an extra Jewish month is added. If we were to count forward in time three years we would reach the next time when another month must be added. By the way, this additional month on the Jewish calendar is called Adar II and follows the month of Adar in our month of March/April. So how much time will pass from the last calendar correction until the next scheduled correction? Since there is normally an 11.25-day difference between the two calendar systems we need only to multiply by the three years that have passed to get our answer of 33.75 days or thirty-four days!

Does this number look familiar? It should. It matches the times in the middle of the tribulation timeline exactly. What are the ramifications of this revelation and what conclusions can be drawn from this knowledge? The most important conclusion is that if this is the reason why there are thirty-four extra days in the middle of Daniel's last 'week,' then this would mean the midpoint will fall in a year the Jews add an extra month. This would be an additional clue, along with our incomplete leap year analysis, that would pin down the timeline and would rule out the timeline with thirty-five days that was derived using twenty-four hundred P-days.

Another point worth noting which is not consistent with this concept is; even though there are thirty-four days, these days do not line up perfectly with the timeline. This is because the month of Adar II is always twenty-nine days long and not thirty. So this breakdown would be five days plus twenty-nine days instead of the four and thirty days the analysis of the tribulation times predicts it should be. This flaw may be enough to disprove this theory. Let's continue to see where this trail leads.

I have said repeatedly that Christ would return in the fall. This realization forces the tribulation midpoint to occur in the spring around March/April. This belief is consistence with where the Jewish calendar predicts these days should fall and is further evidence that the midpoint may coincide in a year with a Jewish thirteenth month. Isn't this how Satan really works? Declaring he is God at the thirteenth hour of the thirteenth month (Adar II).

If we were to expand this idea even further, realizing that Satan is always copying God's plan, then we might deduce that since Christ first came to the Jews as a king on Palm Sunday, the antichrist will do the same. He might enter Jerusalem on the same day (March 29) Palm Sunday occurred in the year Christ died.

However, this logic would not be supported by the Jewish calendar idea. Why? Because the tribulation timeline indicates the antichrist will claim he is God on the fourth or fifth day in the middle of Daniel's last 'week' and before the extra thirty days. This means, using the thirteenth month understanding that this historic event must occur on the last day of the month of Adar or the first day of the month of Adar II and these days never fall on March 29th.

If we were to apply these ideas and examine the calendar, we would find many of these data points converging in the spring of the year 2024 A.D. This is also a leap year on the Gregorian calendar. This new information causes problems for the year 2028 A.D. as the true ending date because three and a half years from March 2024 A.D. gets us to September 2027 A.D., which is one year short. What a dilemma! This information also leads us to realizing there is no extra leap day to account for in the latter half of the tribulation timeline as currently predicted. This is the kind of logical reasoning exercise that needs to be completed for leap years I mentioned earlier.

So which theories are true? The data using the timeline and the Jewish calendar that predicts a year for Christ's return in 2027 A.D. or all the other methods used in chapter two that indicate Christ will return in 2028 A.D.? All my earlier work indicates it will be 2028 A.D., but this new evidence is hard to ignore.

Does a one-year discrepancy make that much difference in the grand scheme of things? Not really, but I believe God is more precise than this and there must be an explanation on why I was awoken after a few hours of sleep and shown this information.

At this time I am at a loss for an answer and so you will have to make your own conclusions realizing the time may be even shorter than I have projected in the past.

144,000

Some final observations from the very last timeline that was developed on page 78. They have little to do with the timing of the tribulation, but are interesting nevertheless because they have to do with the way God works and uses numbers.

From my first book it was discovered that God's number in its completed form was 333.333 and that man's number was 666.666. In Revelation Christians are told that the "beast" will have the same number as man.[f] We see from the timeline chart that the time of the antichrist's rule will be an intermediate form of this number: 6.666 years. We can also see that the time of Christ's rule from heaven will be an intermediate form of God's number: 3.333 years plus an additional one hundred P-days.

When the twenty-three hundred P-days was replaced at the front of the timeline with twenty-four hundred P-days, the perception of the extra one hundred P-days somehow mysteriously shifted to the latter part of the timeline! I say "perception" only because these one hundred days were always there in the latter part during God's wrath. It's just that they were more easily seen when the twenty-three hundred P-days was substituted with twenty-four hundred P-days.

Another important thing to note is one more number in prophecy is hidden within this extra one hundred days. These are the things students of eschatology find fascinating, but I have no explanation for. We know from chapter three that Christ was physically dead forty hours and those forty hours, when converted to seconds, are 144,000 seconds exactly. This is a number mentioned in a Revelation prophecy that I have no clue as to why God uses this particular number. When you make a similar conversion from one hundred days into minutes you get the identical number: one hundred days is really 144,000 minutes! Realize that the time of God's righteous judgment occurs during the last third of the tribulation period and is 3.333 years plus 144,000 P-min.

Also, we know that God uses the conversion factor "a year for a day" in His plans and, by multiplying this factor with the number of completion (ten) raised to the third power, we get the final form of this prophetic conversion factor: "one thousand years for a day." Obviously if someone told you a year was equal to a day, you would tell them they didn't know what they were talking about. This kind of math doesn't work out the way we do math on earth. It is symbolic math and needs further conversions if we wish to get to real numbers that can be understood.

Yet, if we were to follow the logic that God uses, we can learn more about the way He uses numbers. Since a year can be substituted for a day, then a year can also be substituted for twenty-four hours. So six years is equivalent to 144 hours.[29] If we continue this logic to its final conclusion by multiplying by the final form of the number of completion (one thousand or ten to the third power) we see that six thousand years equates to 144,000 hours.[30]

We know from God's plan that Christ will return after six thousand P-years of history and so it can be said as well, that Christ will return after 144,000 P-hours! Now I really do not comprehend what P-hours are, but in Revelation it says there was silence for half an hour in heaven.[g] If we were to apply our normal conversion factors to get from prophetic time to normal earthly time we might be able to determine what P-hours meant.

I am unsure if the half-hour in heaven mentioned in Revelation is this type of hour or is half an S-hour,[31] but it's surely not a normal hour on earth. Realizing this truth and assuming the half hour in heaven is half a P-hour, then this time is equivalent to either seven days fourteen hours and 37.5 minutes (365.25 days), seven days fourteen hours and thirty minutes (365 days), or seven days and fifteen hours exactly (366 days) depending on whether this half a P-hour occurs in a leap year, a regular year, or if we just use the average.[32] Since Revelation says "about" half an hour, we might deduce the time is really seven days fourteen hours exactly. I show you these things so you can see how God works and how we can relate His methods to real time.

We have examined the timing of the last ten years before Christ's second advent. However, the Holy Spirit provides other insights into the Lord's plan that need to be taken into account to get to the whole truth. We will explore more of God's methods using other techniques and in so doing, try to get to the complete truth of the timing for the last days of this age.

Chapter 6

God's Ball of Yarn

Observations from God's Plan

WHY DOES GOD SAY, "A DAY is like a thousand years" or "a day is Like a year?" Why does the Creator of all things use the number 144,000 in the book of Revelation, but nowhere else in scripture… or does He? Why did our Master instruct the Israelites to observe Jubilee years? Why did the prophets Daniel and St. John specify the numbers they did in the books of Daniel and Revelation? We have seen how many of the numbers from the end-time prophecies interlock in chapter five, but not why God chose the numbers He did. Why, Why, Why…

The answers to these questions may shock you when you see them and begin to open your eyes to the truth of God's existence and His plan. Many Christians, Jews, and even non-believers have pondered these very questions throughout the ages along with other questions relating to God's existence. The difference between these questions and other unanswerable questions from ages past is that these will be answered! God is pouring out His Spirit on the world and the mysteries of His plan for mankind are being unlocked so that no Christian will be kept in the dark who chooses to search for the truth.

I have shown you God's plan. Shown you multiple ways from scripture that point to Christ's return in 2028 A.D. I have revealed the timing for the tribulation period and how the numbers fit together in these last years before the coming wrath. For those of you who read the first book, I have shown many examples of how God uses numbers like forty days is really four S-minutes and forty-four S-seconds on God's plan.[a] Or that man's number, six hundred and sixty-six, can be used for time as well as identifying the beast of Revelation.[b] But I really have never answered the question "why" because I didn't have the answers before.

Well the answer to all those questions is because they were designed that way. Why does the sun come up in the morning? Because it must. The laws of physics govern its motion and it cannot be changed unless

God changes the rules. The same can be said for why things must take place as prophesied in the Bible. There is a plan God is working from and many of these things are "built into" the plan. They are the result of God's design and won't be changed until God remakes the heavens and earth. So how do I know this? Because I have seen the overwhelming evidence within the plan that the Holy Spirit has unlocked for me.

A Day for a Year

Why does God use the method "a day for a year" when punishing His people or prophesying about future events? Here is what scripture has to say on this topic.

*Your children will be shepherds here for forty years, suffering for your unfaithfulness, until the last of your bodies lies in the wilderness. For forty years—**one year for each of the forty days you explored the land**—you will suffer for your sins and know what it is like to have me against you.' I, the LORD, have spoken, and I will surely do these things to this whole wicked community, which has banded together against me. They will meet their end in this wilderness; here they will die.*

Numbers 14:33-35 (TNIV)

*"Then lie on your left side and put the sin of the house of Israel upon yourself. You are to bear their sin for the number of days you lie on your side. **I have assigned you the same number of days as the years of their sin.** So for 390 days you will bear the sin of the house of Israel." "After you have finished this, lie down again, this time on your right side, and bear the sin of the house of Judah. **I have assigned you 40 days, a day for each year.** Turn your face toward the siege of Jerusalem and with bared arm prophesy against her. I will tie you up with ropes so that you cannot turn from one side to the other until you have finished the days of your siege."*

Ezekiel 4:4-8 (TNIV)

We know God has a plan that He is working from. He uses certain rules and methods consistently and repeatedly to implement His purposes for mankind. We see from the book of Numbers that God uses this technique to punish Israel's rebellious ways. Again, in Ezekiel, the Lord is applying this method when prophesying about the punishment that will befall the Jews with the coming destruction of Jerusalem in 587 B.C. So why does God use this specific technique when dealing with mankind?

If we look back to God's overall plan for mankind, we will find the answer we seek and be able to unlock many more mysteries as well. However, finding the reason why will require a more formal mathematical analysis; something I have been trying to keep to a minimum without much success. In fact, much of this chapter will require heavy use of mathematics to reveal how God uses numbers. We know from our previous studies that there are twelve thousand prophetic years in God's overall plan (one Season... one S-year) and these translate into 12,175 real years.[33] Therefore:

$$\boxed{\text{1 S-year = 12,175 years}}$$

If we make further observations from the plan shown in chapter one, we realize there are three hundred and sixty seasonal days (S-days) that are also equivalent to both twelve thousand P-years and 12,175 years. So how many years are there in just one S-day? There are 33.333 P-years per every seasonal day.[22] How many days are there in 33.333 P-years? We just need to multiply this number by three hundred and sixty prophetic days in one prophetic year to get our answer—twelve thousand P-days.[34] If we convert these twelve thousand prophetic days into real days, by using our conversion factor of 1.01458333, we find there are 12,175 real days in one S-day on God's calendar[35] and so,

$$\boxed{\text{1 S-day = 12,175 days}}$$

Interesting. After our analyses we calculated the same absolute number for both. As a result, we now know that:

$$1 \text{ Seasonal year (S-year)} = 12{,}175 \text{ years \&}$$
$$1 \text{ Seasonal day (S-day)} = 12{,}175 \text{ days}$$

Using a mathematical technique, let's set these two equations equal to each other, since God said they are equal (a day for a year), and see if they really are. By going ahead and doing this, it can be proven mathematically if they are equivalent or not.

$$\frac{1 \text{ S-day}}{12{,}175 \text{ days}} = \frac{1 \text{ S-year}}{12{,}175 \text{ years}}$$

Cross-multiplying we get:

$$(12{,}175 \text{ years}) (1 \text{ S-day}) = (12{,}175 \text{ days}) (1 \text{ S-year})$$

We notice these equations are very similar. The absolute numbers are identical, but the units appear to be slightly different. So let's divide each side by 12,175 to reduce this equation down to something more manageable.... Now we have:

$$1 \text{ year S-day} = 1 \text{ day S-year}$$

or restated,

$$(1 \text{ year}) (1 \text{ S-day}) = (1 \text{ day}) (1 \text{ S-year})$$

To complete this mathematical proof and determine whether God's plan supports or refutes this idea, we need to convert all units to the same type on both sides of the equal sign to verify whether they balance out. Any of the four units can be used for this exercise so I will arbitrarily pick the unit "day" to finish this analysis since people can easily grasp what "days" are as opposed to "S-days." So the "day" unit needs no conversion, but the other three do. Let's go ahead and do this. Everyone knows there are 365.25 days in one year so this fact can be used to convert the "years" unit into "days."

Next we need to look back at God's plan for the other unit conversions. How many real days are in 1 S-day? We have already calculated this number earlier. It is 12,175 days. What does our equation look like so far when this data is substituted in our last equation?

$$(365.25 \text{ days}) (12{,}175 \text{ days}) = (1 \text{ day}) (1 \text{ S-year})$$

or

$$4{,}446{,}918.75 \text{ days}^2 = (1 \text{ day}) (1 \text{ S-year})$$

Now we need to find out how many real days are in 1 S-year (God's whole plan). We know there are 12,175 years in God's overall plan for mankind and since a year has 365.25 days in it, we only need to substitute this last piece of information:

$$4{,}446{,}918.75 \text{ days}^2 = (1 \text{ day}) (12{,}175 \text{ years} \times \frac{365.25 \text{ days}}{1 \text{ year}})$$

or

$$4{,}446{,}918.75 \text{ days}^2 = 4{,}446{,}918.75 \text{ days}^2$$

After all the mathematical manipulations, both sides are in fact identical! The original premise therefore is true.

$$\frac{1 \text{ S-day}}{12{,}175 \text{ days}} = \frac{1 \text{ S-year}}{12{,}175 \text{ years}}$$

It is also mathematically true that if each numerator is equal, then each denominator is equal too and so…

$$1 \text{ S-day} = 1 \text{ S-year}$$
$$\text{and}$$
$$1 \text{ Day} = 1 \text{ Year}$$

Having determined that "a day is like a year," we need to understand that the real equation is "x" days are like "x" years where "x" can be any number of God's choosing to achieve His purposes. This rule also applies to "an S-day is like an S-year."

What other things have we learned besides the fact "x days are like x years" (which we should already have known if we just listen to God)? That the prophetic descriptors used in God's plan to distinguish real times versus prophetic times can be moved between units without affecting the mathematical truths.

This means that "1 year S-day" and "1 day S-year" are not similar, but identical when "using" prophetic math! One can just simply move the "S-" descriptor (or "P-", or "H-") from one unit to another in the equation as mathematically needed. Another way to say this is, we can transfer the "S-" prescript from in front of the "day" units and place it in front of the "year" units (or vise-versa) without affecting the end result. The prophetic descriptor acts like a multiplier, divisor, or scaling factor for real measurable units much like "milli, micro, centi, and kilo" descriptors are used for scaling metric numbers. Test… who made the metric system? Answer… God! God made everything. It is only for man to uncover God's truths.

If we had known this rule existed, this could have simplified the analysis and we would not have had to go through the rigorous unit substitutions to verify the original premise. This is a helpful rule to remember and saves a lot of time when using and working with prophetic math and God's plans. Let's label this newly discovered rule the "transitive property" of prophetic mathematics (the math of the heavens). As we continue studying the Overall Plan for Mankind, it will be found that these methods are being used in other levels of the plan even though they are never specifically mentioned in scripture.

In addition, the overall plan of God's shown in chapter one must be a very good rendition of the real plan in heaven. If this statement were untrue, then it wouldn't have been possible to find the truth of "a day for a year," taken from scripture, hidden within the plan. Actually, this truth was always in God's plan from the beginning of creation and was only revealed via the prophets in bygone years, as God deemed necessary to warn the unfaithful Israelites of their coming destruction for continuously prostituting themselves by worshiping idols and other Gods. The same things we are guilty of in today's world.

These calculations are more evidence that God is using the plan for His purposes and has now chosen to reveal it to mankind in these last days before Christ returns. When the plan is compared to God's Word, it highlights how numerous pieces of the Bible fit together in ways that many theologians would never have dreamed possible. And by uncovering the confirming truths of the Bible revealed through studying the plan, one should also understand the times and date predicted must also be true! Whenever I discover one of these secrets, I have **never** found one instance that contradicted what has been written in scripture.

God's Number

If we look back to the analysis of "a day for a year," we see a number we calculated that was quite large. That number was 4,446,918.75 days[2] and was calculated to prove that God's Word can be trusted. This number doesn't appear on the surface to be anything of further value in revealing how God operates, but it is. The problem is in its recognition. It is disguised as many of God's numbers are.

If you remember, I arbitrarily chose the "day" units to perform the mathematical proof and could just as easily have chosen another unit. Because I did this, I could not straightforwardly see more proof of the Creator's hand at work. I have had this problem many times and sometimes it takes a lot of study before I am able to see things clearly that are very simple to understand and recognize when they are finally revealed.

This number kept bothering me as I worked on other end-time things. I spent some additional time trying to understand what it meant and finally put it aside. It was an itch that went unscratched. However, after coming back to it and deciding to see what would happen if I had chosen one of the other units (instead of days) for my analysis, I realized that the number just had to be converted into one of the other units to unlock more of God's handiwork.

So what were all the answers for the equation (1 year) (1 S-day) or (1 day) (1 S-year) which are identical after applying the "transitive property" of prophetic mathematics?

Table 6-1

Unit	Final Answer
Days	4,446,918.75 days^2
Years	33.333 years^2
S-Days	0.03 S-days^2
S-Years	2.2487481 x 10^{-7} S-Years^2

We see after investigating all the possibilities that a familiar number is in fact revealed... God's number (33.333) in one of its intermediate forms. What other knowledge can be gleaned from these additional calculations? Well since we know a day is like a year we should understand that 4,446,918.75 days^2 is equivalent to 33.333 years^2. Let's see...

$$\sqrt{4,446,918.75 \text{ days}^2} = 2,108.7718555 \text{ days} \ \&$$

$$\sqrt{33.333 \text{ years}^2} = 5.7735027 \text{ years} \quad \text{so:}$$

$$(2,108.7719 \text{ days}) (2,108.7719 \text{ days}) = (5.7735 \text{ yrs.}) (5.7735 \text{ yrs.})$$

If we convert each 2,108.7719 days into years by dividing by 365.25 days in a year we verify again that "a day is like a year" and that this large figure was derived from God's number of 33.333.

$$(5.7735 \text{ years}) (5.7735 \text{ years}) = (5.7735 \text{ years}) (5.7735 \text{ years})$$

Summarizing what additional information has been uncovered from God's Plan:

1) "A day is like a year" and "a year is like a day."
2) "An S-day is like an S-year" and an "S-year is like an S-day."
3) The transitive property states, "the operation of a prophetic descriptor works just like metric scaling units."
4) God's number is used in its various forms to determine key prophetic truths within all levels of His plan.
5) Many of God's methods can be traced back to God's overall plan that is based on the "math of the heavens."

The Year of Jubilee

> 'Count off seven sabbath years—seven times seven years—so that the seven sabbath years amount to a period of forty-nine years. Then have the trumpet sounded everywhere on the tenth day of the seventh month; on the Day of Atonement sound the trumpet throughout your land. Consecrate the fiftieth year and proclaim liberty throughout the land to all its inhabitants. It shall be a jubilee for you; each of you is to return to your family property and to your own clan. The fiftieth year shall be a jubilee for you; do not sow and do not reap what grows of itself or harvest the untended vines. For it is a jubilee and is to be holy for you; eat only what is taken directly from the fields. " 'In this Year of Jubilee everyone is to return to their own property.'
>
> Leviticus 25:8-13 (TNIV)

What are Jubilee years and why does God tell us to observe them? In the book of Leviticus, Moses was told the Israelites were not only to observe the Sabbath every seven days, but also a Sabbath year every seven years (a year for a day). Basically moving from one level (scale) of God's plan to another. In this case, moving from the lowest level to the next highest level. If we were to repeat this process by moving up another level in God's Plan, we would then get to the Jubilee year level described above. That is to say, just as the Sabbath day is the seventh day of the week and the Sabbath year is every seventh year, a Jubilee year falls after seven Sabbath years are completed.

Continuing this line of reasoning would suggest that every forty-ninth year was a Jubilee year and some scholars believe and teach this. Why? Because they see the beauty in the symmetry of this kind of logic... but this is not what the Word of the Lord says.

The Bible says after seven Sabbath years passed, the Jews were to proclaim the next year, the fiftieth year, as a year of Jubilee. This statement alone should be enough to convince anyone that the time span between Jubilee years is fifty years and not forty-nine years, as the numerical sequence would predict. It is a special case that breaks the pattern.

However, many theologians and Rabbis still wish to believe this is a forty-nine year cycle and so they get around this problem by claiming the first year after the forty-ninth Sabbath year (which is the beginning year for the next seven Sabbath year cycle) is the Jubilee year to observe. Hence it is the fiftieth year.

Great so far. But this idea only works for the first fifty years... the first Sabbath year cycle. When this pattern is repeated, what happens? It turns out the next Jubilee year is not the fiftieth year from the last, but the forty-ninth! In other words, it doesn't match God's Word. The only way it can match is by double accounting the previous Jubilee year (which had been already counted in the preceding cycle) with the current Jubilee year thereby treating it as if it were just a regular year in the current cycle count. Are we to double account Jubilee years to make them fit our preconceived forty-nine year cycle or are we to treat each Jubilee year as something special and unique as God intended? I think most already know this answer if they let the Holy Spirit speak to my words.

Now if common sense isn't enough to convince you (because even the scholars who teach this idea either know this fact and ignore it or they are unaware of what the pattern looks like after the first fifty years, which I think is the most likely reason for their misunderstanding), then there is another way to prove Jubilee years are fifty years long. Using God's plan.

Clearly Sabbath days, Sabbath years, and Jubilee years are important to God and in my first book it was shown how many events on God's calendar are planned around these important days and years including Christ's return on a year of Jubilee. So if they are this important and the Lord uses them for planning purposes, then they must be somewhere within His plan, and they are!

Having said this, they are not very obvious when studying the overall plan. Why? Because they are at a lower level in the overall plan... a deeper level. How do we find lower levels? By scaling the plan just as God scales His punishments and numbers using the number of completion. Just as God's number in its most elementary form is .333 and when scaled up, using the number of completion, is 333.333 in its highest form (with 3.333 and 33.333 the intermediate forms), so too with other numbers God uses and levels of God's Plan. Let's look at a few examples in table 6-2 to better illustrate the point I am making.

Table 6-2

Level	# Of Completion	God's #	Man's #	# Of Perfection	God's Plan
3	1,000 (10³)	333.333	666.666	700	12,175
2	100 (10²)	33.333	66.666	70	1,217.5
1	10 (10¹)	3.333	6.666	7	121.75
		3	6	7	12
0	1 (10⁰)	.333	.666	.7	12.175

Note: Levels are shown in descending order. Highest level (God's Overall Plan) first row to lowest level last row. Notice that the number of completion is God's number plus Man's number.

So to find lower levels within God's plan all we need to do is to keep reducing the plan by the number of completion (a factor of ten). Since God's overall plan is 12,175 years long, the next level down is ten times smaller or 1,217.5 years. Now these could also mean "days" as well as years since we know "a day is like a year" in God's plan. The next level lower is 121.75. We can continue down even further, but for our purposes of finding evidence of Jubilee years within the plan, we will stop at this level. As far as I have been able to uncover, this is the Jubilee plan.

So there are 121.75 "somethings" at this level of the plan. Are these days or years? Neither. They are Jubilee years! I said in my first book that God has been using Jubilees to mark important events in man's history since the creation of Adam even though He didn't instruct mankind to start observing them until the Jews entered the promise land in 1415 B.C. How do I know? Well, the flood came in a year (2403/2402 B.C.) that coincided with God's number 33.3. It was the 3rd Sabbath year after the 33rd Jubilee from when God started counting. This is a very strange coincidence if this idea were untrue.

In view of the evidence from "2028," it appears God has been using Sabbath years and Jubilee years from the creation of Adam (and maybe even before Adam as the plan suggests). Let's assume this is true and start this level of the plan from Adam instead of day one of creation where the overall plan begins. This means this portion of God's plan is only six H-days long as opposed to twelve H-days long and covers the three trimesters of man's history, the Times of Adam, the Times of the Jews, and the Times of the Gentiles.

I said that the 121.75 in this plan represent Jubilee years. How did I figure this out? After rigorous calculations and study, the answer came to me just as others have. You see, if you take six thousand P-years and divide them by fifty years between each Jubilee you get one hundred and twenty Jubilees.

In the book of Genesis, after the flood, God said He would limit man's time to one hundred and twenty years.[c] In today's world I know of none who have ever attained this age, but there may be a select few in the 6.6 billion people alive today that might or have. I personally have known at least one person who lived to be over one hundred years old and a few others in their mid-nineties still actively enjoying life. All believe or believed in Christ as their savior. However, after the flood, the Bible records many who lived long past one hundred and twenty years of age. I believe they still had some of the genes of long life and this was needed to repopulate the earth. But by the time the first trimester ended in 2031 B.C., this limit of man's lifespan was in full effect.

However, clearly God did not limit man's age immediately after the flood as many scriptural texts in Genesis indicate. So we see that this limit might represent more than just the literal meaning just as other things God says are true on numerous levels. So by knowing the Lord's methods, we can deduce on another level of understanding God was also limiting man's self-rule on the earth to prophetically one hundred and twenty Jubilees! Since this was a prophecy given by Moses (a prediction of man's future lifespan), these one hundred and twenty years (Jubilees) should have in fact been one hundred and twenty P-years or in this case P-Jubilees. It seems from this line of deductive reasoning that not only can a "day" substitute for a "year," but in God's eyes, a "Jubilee" can substitute for a "year" as well!

Having written all this, we should realize that from this intermediate level within the Lord's overall plan, that there are 121.75 Jubilees and not one hundred and twenty Jubilees. This is because there are not really six thousand years in this part of the plan, but six thousand P-years in the plan and, as we know, when six thousand P-years are converted to real years they become 6,087.5 years. Since Jubilees are also real years from man's perspective, we can now make the correct calculation as opposed to the "symbolic" one made initially when six thousand years were used hastily. When this is done, we do in fact get 121.75 Jubilees[36] as God's plan predicted we would.

Now if the scholars who say the time is forty-nine years knew God's other methods and His plans, they would probably not continue to support their viewpoint. Is this just another chance of fate that the number of Jubilees between Adam and Christ's return is identical to the number of years in God's overall plan... just one hundred times smaller? Can you get these kinds of coincidences using a forty-nine year cycle for the time between Jubilee years? I have tried and they don't work out.

So how many Jubilee years are there for each day in heaven (H-day) on the Jubilee level of the plan? We know there are one thousand fourteen years and seven months for every H-day. If we divide this time by fifty years we get 20.29166 Jubilees for every H-day. This resulting number highlights a problem I was unaware of for many months when doing this work, but stumbled onto it only after the Holy Spirit had removed enough prophetic splinters from my eyes.

Most people cannot see it just as I could not see it. I have tested a few people for this problem and they were blind to it as well. It has to do with recognizing numbers. Do you know what the number 3,652.50 represents? How about 36,525? How about 365.25? Most know what this last number represents immediately when they see it. It is the number of days in one year. Knowing this, look more closely at the first two numbers I asked you about. They are really the same number just scaled by the number of completion. So the first number listed represents the number of days in ten years and the second one the number of days in a hundred years exactly.

This was the problem I was having. Seeing familiar numbers and not recognizing they were one of God's numbers within His plan scaled by a factor of ten, one hundred, or one thousand. This is important when working with numbers within His plan. Realizing this phenomenon is going on when moving between levels we can look closer at the 20.29166 Jubilees and see that this number is one hundred times smaller than the familiar number 2,029.1666 (the length of time for one trimester of God's plan).

This is the very first number God revealed to me that started me on my journey doing the Lord's work in these last days. Just as the total number of Jubilees (121.75) is one hundred times smaller than the 12,175 years in the overall plan, the number of Jubilees (20.29166) in 1/6th (one H-day) of this level of the plan is one hundred times smaller than 2,029.166 years which equates to 1/6th of God's overall plan for mankind! That was a mouthful of knowledge to spit out at once. Let's see what this level of the plan looks like on a diagram with all the information covered included.

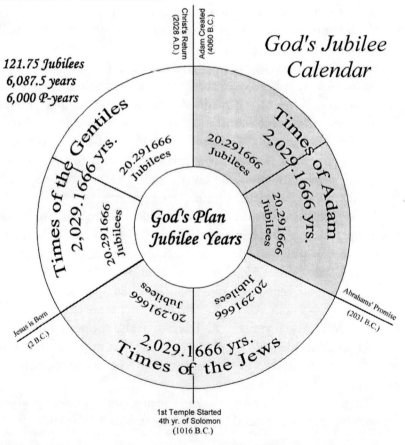

We can see from this diagram that when God's overall plan from chapter one was scaled by two levels (10² or one hundred), the total number of years in the plan went from 12,175 years to 121.75 Jubilees. Furthermore, the number of Jubilee years in each of the six H-days of man's history shown on God's Jubilee Calendar is exactly one hundred times smaller than the number of years in each trimester of God's plan. Not to mention that if 121.75 were days, then it would be exactly one third (.333) of a year and if you multiply 121.75 by God's number 33.333 you get 4,058.333 years which is one third of God's overall plan! Can these mathematical coincidences continue to be mere chance? Not likely.

3,043.75 (Three Days)

We have only begun to discuss and show how God uses number's repeatedly within His plans. Some are truly amazing. How many coincidences does a person need to experience before he or she can start to believe they are not coincidences at all? Do you realize that if Dionysus had accurately reset history's timeline for Jesus' birth (to 2 B.C.), that Christ's second advent would occur in 2029 A.D. and this date would have matched the number 2029 shown on this detailed level of God's plan! My first book would have been titled "2029" instead of "2028" which, by the way was the working title until the Holy Spirit taught me more things about this number. This was the original number that turned on the light of God's mysteries that I used to first calculate Noah's flood and ultimately Christ's return.

But this was not God's will for these things to align, but part of His plan. Why? For two reasons if you understand how God works. The first reason is a numerology reason. The date 2029 when reduced down to its basic number (by adding the digits together until they are ten or less) equates to four (2 + 0 + 2 + 9 = 13 & 1 + 3 = 4) and the date 2028 reduces to the number three using the same reduction method. Three is God's number while four is not.

If you remember from chapter one, there were some strange coincidences with the number 3,043. This was the number of years between Adam and the fourth year of Solomon when the first temple to the Lord was started. It was also the number of years between the first temple and Christ's second coming as well as the number of years between Christ's birth and the remaking of heaven and earth in the year... you guessed it 3043 A.D. This is also the exact number for the average number of days in a month (30.4375)!⁽³⁷⁾

Now if Dionysus had gotten the calculation right this would have moved the end date to 3044 A.D. for the remaking of heaven and earth. When this date is reduced down we get the number two, which is not a number God uses, instead of the number ten for the year 3043... the number of completion which really is the appropriate number for the remaking of heaven and earth when man's history will be finished... completed.

So if Dionysus had gotten it right, then we would have ended up with the dates 2029 A.D. and 3044 A.D. which do not appear to match the way God works as opposed to 2028 A.D. and 3043 A.D. which are supported by God's methods. Additionally, it appears the symmetry for the number 3,043.75 is more important than the symmetry of the number 2,029.1666. Why? Because the number 3,043.75 is really three thousand years—three H-days which symbolize three days. Whereas the number 2,029.1666 represents symbolically only two days and all Christians should be able to see these three H-days characteristically parallel the three days mentioned in the Bible of Christ's death and resurrection.

Expanding on the earlier mysteries uncovered, we can include these additional discoveries:

6) In addition to a "day is like a year," God uses other "time" analogies to achieve His purposes. We see that a "Jubilee is like a year or a day" as well.

7) There are different levels within God's overall plan and they are used for different purposes with respect to God's interaction with mankind. All these levels are scaled by the number of completion from some other level of the plan.

8) God uses the same ideas and numbers over and over in different levels of the plan depending on whether they involve an individual, nation, or all of mankind. The only difference being the scale of the numbers being used.

144,000 Revisited

As I continue studying God's plan and Word, and work toward finishing my writing pertaining to the last twenty years before our savior's return, I occasionally get sidetracked. These forays into other areas never come back empty handed as you have already seen with the things I have brought back ("a day is like a year" and Jubilee years are important to God). I cannot always predict where these trips will lead me, what they will uncover, or when the distractions will bear fruit, but they always seem to lead back to God's plan providing more details, extra information, and additional evidence that confirms there is a plan the Lord is using.

It has been over a year since I uncovered the plan and I have had many of these sidetracking episodes. The current one is leading me in a direction of learning why God uses the number 144,000, as well as other numbers, and the math and rules our Creator uses when administering His will for mankind.

At times I believe I am just on the verge of reaching the finish line, but a strong gust of wind comes up and pushes me back from obtaining the necessary knowledge. It seems my journey is always two steps forward and one step back. My mind has so many numbers and ideas twisting through it that many times I cannot focus on just one. And when I try to, like the number 144,000, it leads me through a labyrinth to other important numbers. They are all so intertwined it is like trying to unravel a gigantic ball of yarn. But here is what I have untangled so far with respect to the mysterious number 144,000.

We learned from the first book that God uses important numbers to set times on His plan. We have seen this in the things we have unlocked already in this chapter. In addition to God's number being 333.333 in its highest form, this number is used in calculating events on His calendar. Not only that, but scaled versions of God's number are used as well like .333, 3.333, 33.333; all based on the number of completion "ten." This is how God works when applying judgments, determining when Christ came the first time, when he will return, etc.

Many believe we can never know when Christ will return or that God has a plan. But they never study the Bible enough to find out. Many believed years ago that the world was flat and so they never sought to find the truth. Some believed man could not travel faster than the speed of sound, or ever walk on the moon. Why? Because they never studied to find out if these things were true or possible. One should realize two things from these false ideas of man's. First, if the time on the Creator's plan for this information to be revealed is not the right time, then the truth will not be uncovered and second, if you never search for the truth you will never find it. You will always believe what other badly informed people tell you who have not searched for the truth either.

We know that all numbers in prophecy need to be adjusted to get to the true number. For example, the number twelve hundred and sixty days in Revelation is not twelve hundred and sixty real days, but three and a half real years. Why? Because God's use of numbers is precise and He is using prophetic accounting methods. That is to say, He is using 360 P-days for a year as opposed to 365.25 days. When we convert these days to real days by multiplying by the prophetic conversion constant 1.01458333 we get 1,279 days.[38] Now I have been trying to figure out why God uses the number 144,000 in Revelation for an incredibly long period of time with very little success. I have uncovered clues at various times, but still could not see the big picture. One of the problems I guess could be it was not the "right" time. I had not learned enough of the other things necessary to get to the answer or the Lord had deemed it was not the right time. I comprehend now my training from the Holy Spirit was not far enough along.

This difficulty can be best understood like this: a baby needs to learn to crawl before it can walk, it needs to learn to walk before it can jog, and it needs to learn to jog before it can run. Math is like this as well. One needs to learn arithmetic (add, subtract, multiply, and divide), before they can learn fractions, and fractions before learning algebra, and algebra before geometry, and geometry before trigonometry, and trigonometry before calculus, and…. you get the picture.

I believe I have been having this very problem. I needed to learn first that God had a plan, and from that plan other events could be determined. But having done that, I thought I knew all there was to learn and had uncovered all there was to uncover, and so I stopped searching and studying the plan. Although, I was still using the overall plan as a reference tool as my work progressed on uncovering the details of the last days of this age. The hitch was I had stopped searching and studying when there was more to be learned just like those people I described earlier who never study at all.

Another difficulty has been, that although I know some of the rules God uses, I haven't always been applying them correctly. For example, I stated that all numbers given in biblical prophecies had to be converted, but when I applied this rule, I only applied it to numbers dealing with time. I thought that since 144,000 wasn't dealing with time, but the sealing of Jews, the calendar conversion factor didn't need to be used... right? Wrong. I finally realized there were two rules working here simultaneously. God uses prophetic numbers not only for describing things (like the number of the antichrist, Satan, and man's number), but also for time. I should have realized that 144,000 was not only being used in Revelation for sealing the Jews, but it was being used for timing purposes as well—just like six hundred and sixty-six was. I knew this was a technique God uses, but I was not applying it properly with respect to this number.

We know from the study of the tribulation timing that the forty hours Christ was physically dead was 144,000 seconds long. And those one hundred days used within the tribulation timeline corresponded to 144,000 minutes. But by knowing this information, I felt the final understanding for why God uses this number was still incomplete. These discovered facts were just clues to a "bigger" truth.

I next was led down some end-time calculations that unwound more of the tangled 144,000 yarn and learned that 6,000 P-years "was like" 144,000 P-hours. And since 6,000 P-years is the time from the creation of Adam to the return of Christ, we see this number symbolically used again for time on God's calendar. But by using this number in God's designs these ways, what is God in reality trying to tell us? What do all these applications have in common? This was the question that still eluded me.

Now I have seen more of these mathematical wonders and continue to slowly unravel God's ball of twisted yarn. Realizing that there are scales, levels within the plan, and at each level key numbers are being reused—much like Daniel's prophecies being true on various levels. This is how calculus works as well. For instance, if you take the meas-urement of length and apply the first derivative, with respect to time, you end up with the unit of speed (velocity) as the answer... the next higher level. If we apply the next derivative (the next level up) we end up with units of acceleration. So too with God's plan and His use of numbers revealing different aspects of God's blueprints by providing redundancy and reinforcing evidence of His existence.

We know the highest level so far uncovered in God's Plan for Mankind is 12,000 P-years (12,175 real years) long. One of the lowest levels is the tribulation period of ten years. But there appears to be other levels in between (like the Jubilee level) to uncover and determine what those levels mean in real terms. These are the things to be investigated more fully, but I have already seen glimpses of these and how the numbers relate, just not their full ramifications for mankind. Always more things to do.

This incomplete effort is where I have seen the number 144,000 begin to appear with surprising regularity along with others of God's numbers. Where other numbers standout more easily, this number is always concealed much like 144,000 seconds was concealed as forty hours.

Let's look at some of the other places God uses this number on His plan. Finding this number hidden within the plan only adds to the insurmountable evidence for the genuineness of the plan. Some still argue there is no plan, even when they see the plan. However, can anyone devise a plan that is based on the use of biblical numbers, confirmed by God's Word written by forty authors separated in time over many centuries, and still have hidden things within the plan, that when searched, surrender more corroborating evidence of the plan almost two thousand years after the last writer made his prophecies?

Beginning with the big picture, in God's overall plan we know there are 12,000 P-years or 12,175 years. This number, 12,175 is another very important number. We have witnessed its use in various hidden ways, as are many of the numbers God uses. We have already seen one of its uses in the Jubilee plan and will see even more. We will use this number to prove that "12,175 years is like 144,000" from God's vantage point.

So 12,175 years equals 144,000. Hmm… The only question now is 144,000 what? Just as forty hours equals 144,000 seconds; there is a conversion that I was unaware of. But I have uncovered it as my endless searching for God's truths push me forward. Now you do realize that 144,000 seconds is forty hours. They are the same exact amount of time, just represented using different units of measurement. We just don't usually think of large amounts of time using such small units of measure. We all know that one dollar equals: four quarters or ten dimes or twenty nickels or even one hundred pennies because we use these concepts regularly; we just don't think of time quite this way.

I am sure everyone can grasp what 12,175 years means… a very long time from mankind's perspective. It is also represented other ways, just as the dollar was in our example, and we are already aware of one of the other representations… 12,000 P-years. Other representations from the plan are 360 S-days and 12 H-days, which all cover the same exact amount of time. All are equal and seen rather easily when examining the plan.

Let's look at the deeper things of God that the Spirit searches and reveals. How many months are in one year—twelve? Everyone knows this. How many months are in 12,175 years? There are 146,100 months.[39] We are almost there. We need to do some reverse engineering on this number. Remember the rule "all numbers in prophecy must be converted" if they deal with time? Well if 144,000 is used as time, then it must be converted using our conversion factor 1.014583 as well. What number do we get when we convert 144,000? We get 146,100.[40] This is the same number we got from our month calculation above.

What this is really saying is when 144,000 is used for prophetic (symbolic) timing, the real number is 146,100. Therefore, 146,100 months are equal to 144,000 P-months! There you have it. God's overall plan is not only 146,100 months long (12,175 years), but can be understood to be 144,000 P-months (12,000 P-years) long. Basically twelve times 12,000 P-years… so simple, but yet unseen. The number was right in front of our eyes, but most could not see it. It took me a very long time following a convoluted journey to get to this extremely simple equation and understanding. Funny how God hides simple things right in front of us.

The Holy Spirit continues to remove my blindness in stages. I still am unable to see everything with clarity, but I am beginning to make steady progress using the light of Christ. How are you doing? Are you studying God's Word regularly, yearning for the advent of our Lord, and spreading the good news of Jesus Christ? I pray my work is adding to your thirst for the truth.

So on the highest level uncovered, God's plan equates to 144,000 P-months. What next? Well God's plans on other levels equate to this number as well or a scaled form of this number. What is the next lower level? Well we just looked at the overall plan that included creation. What about just that portion that has to do with mankind? The three trimesters of mankind: the Times of Adam, the Jews, and the Gentiles. Six thousand P-years or half the plan. Surely this cannot equal 144,000 if the whole plan equals 144,000. This kind of logic would dictate that if 144,000 equals the whole plan, then half the plan should be 72,000... 72,000 P-months and it is. But this part of the plan is viewed from God's perspective as 144,000 as well. I discussed this concept at the end of the previous chapter. This answer also took a twisting road through a maze of numbers to uncover something so simple.

I will review this winding road for you so that you can reach the end faster. We know that "a day is like a year" and even like "a thousand years?" This rule is strong evidence of God using different scales within His plan; plans inside of plans inside of still other plans.

So a day is like a year and since a day is twenty-four hours long, then twenty-four hours are also like a year. If we multiply by six we see that six years are akin to 144 hours. Now let's multiply by a thousand (the number of completion in its highest form) because a day is like a thousand years as well. We can now make out that six thousand years is like 144,000 hours. Realizing that when 144,000 is used as time it is really prophetic, then these have to be P-hours and if they are prophetic hours, then the six thousand years has to be prophetic in nature as well. Subsequently, we end up with "six thousand P-years is like 144,000 P-hours!"

Let's look at our last example on how God hides the number 144,000. Remember, God broke up the six thousand P-years of man's history into three parts (symbolically three days), which are the three trimesters of His plan. If we were to use this same approach for His entire plan, breaking it up into three trimesters as well, we would find more amazing evidence of this number. There are a number of ways to do this, but I will start from a number you are familiar with... 4,446,918.75 days^2. We already know this number represents 33.333 years^2, but it also represents 144,000! We know that:

$$\sqrt{4{,}446{,}918.75 \text{ days}^2} = 2{,}108.7718555 \text{ days} \ \& \text{ so:}$$

$$(2{,}108.7719 \text{ days}) \times (2{,}108.7719 \text{ days}) = 4{,}446{,}918.75 \text{ days}^2$$

Let's convert 4,446,918.75 days^2 back into a prophetic number since 144,000 is really a prophetic number and not a real number. We do this by dividing each 2,108.7719 days by our conversion factor. This gives us 2,078.46097 P-days.[41] Re-multiplying out these numbers:

$$(2{,}078.46097 \text{ P-days}) \times (2{,}078.46097 \text{ P-days}) =$$

$$4{,}320{,}000 \text{ (P-days)}^2$$

Now 4,446,918.75 days are the total number of days in God's plan[42] and thus, 4,320,000 P-days are the total number of prophetic days in God's Overall Plan.[43] If we divide this number by three (three trimesters) we get 1,440,000 P-days[44] —or ten times 144,000. In other words, there are three equal blocks of time of the scaled number 144,000 P-days in God's overall plan for mankind and they translate into our number of 4,446,918.75 days^2! I bet you never expected it was possible to calculate this important number from the book of Revelation using this number developed for another purpose. What does all this new information we have been investigating look like together on one diagram?

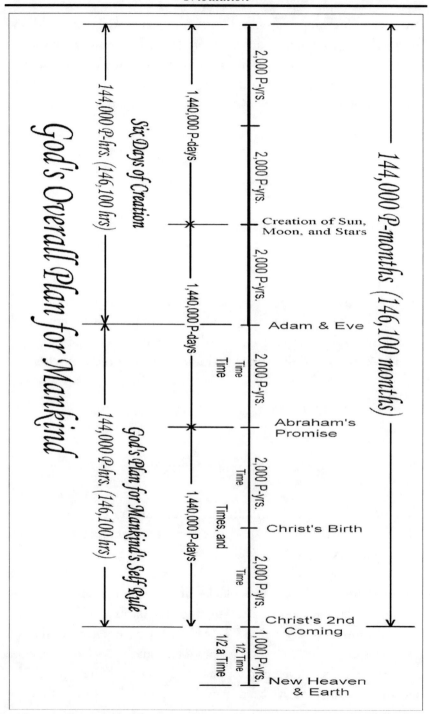

Notice the overall plan is 144,000 P-months long and the portion pertaining to man's self-rule (half the plan) is viewed by God as having 144,000 P-hours in duration. Further verification being uncovered of God's methods, His plan, and what the number 144,000 might ultimately mean. If we were to apply the conversion factor to the number of Jews sealed in Revelation, we would find there are twelve tribes of 12,175 Jews prophesied instead of twelve tribes of twelve thousand Jews. Presently, I do not believe converting tribes with a conversion number developed for time is suitable, but seeing how God works on many levels I would not rule it out. I have not come to the full understanding if these tribes are literal or are symbolic and mean something else. And until I do, I will leave this analysis for that time shall it ever come.

We now have five instances where God uses the number 144,000 for time spans within His plan all dealing with different units of time (P-months, P-days, P-hours, P-minutes, and P-seconds). The first three instances are at the "top" of the plan and the last two deal with day-to-day planning at the lowest levels of the Lord's plan. What do they have in common? I am still unsure. But I now realize that some of my earlier work may have been in error. Since 144,000 is the "prophetic" time from God's perspective it probably never represents real time. When this time is used within God's plan, the right number is 146,100.

Numerology would support this conclusion as 144,000 reduces down to the number nine while 146,100's reduced number is three... God's number. So what can be deduced from this effort is that Christ was not dead forty hours as I have claimed on many occasions, but that he had to be dead forty P-hours because this is the prophetic time that is equivalent to 144,000. For this event to be real time and symbolize 144,000, it had to be 40.58333 hours[45] or 146,100 real seconds. This translates into forty hours and thirty-five extra minutes. Furthermore, we now see another familiar number emerge— 4,058.333 (one third of the Overall Plan for Mankind): just a hundred times smaller.

As for the extra one hundred days (144,000 minutes) I calculated in chapter five as a possibility for the tribulation timeline, the same line of reasoning applies. It has to be 146,100 real minutes or 101.45833 days if it is to symbolize 144,000. We see again that 101.45833 is one tenth of the number 1,014.5833, which is equivalent to one H-day.

In light of this new information and understanding, I believe the data still supports that Christ was dead the equivalent of 144,000 P-seconds, but the possibility of the extra one hundred P-days needing to be included in the tribulation timeline now seems remote since this would add another half a day to the one-day error already identified. This new information is evidence that the twenty-three hundred P-days seems to be the correct choice as opposed to the twenty-four hundred P-days.

However, remember when twenty-four hundred P-days was converted, it resulted in an extra thirty-five days exactly? Now we just learned that converting forty P-hours results in an additional thirty-five minutes. Interesting. Is a "day like a minute?" Also, this number of twenty-four hundred and thirty-five is hidden within God's plan. How many Jubilees did we calculate for the six thousand P-years of mankind's self-rule? There were 121.75 Jubilees and if this idea were extrapolated to include the six H-days of creation, this would double the number to 243.5 Jubilees for the whole plan. Guarding against number recognition blindness, we see this number of Jubilees is exactly ten times smaller than 2,435! What does this mean? Typically it would be additional evidence that this is the true number and not twenty-three hundred, but with the additional half-day error coming from the converted difference of one hundred days I still have no idea how to reconcile these contradictions at this time.

Since I know there are other levels between the overall plan and the detailed portions of the plan (lowest levels), I am confident the middle levels of the plan that need to be refined will have this number imbedded within them as well. They only need to be discovered. As for what all these applications of the number 144,000 have in common and why God uses them, I make the following educated guess. The number 144,000 appears to be the number God uses to "seal up" parts of His plan that cannot be altered.

Seven Minutes of Conflict

We have examined many interesting things and I am convinced there are many more that still need to be uncovered. As I was writing this book and thought I had finished the tribulation timing detailed in chapter five, I began to focus on the end-time events associated with these important dates and times. But a few weeks ago I got sidetracked again from that work and began another journey that had very little to do with identifying specific future events and everything to do with God's plan and timing. Things I thought I had a good grasp of I found I had more to learn. These are the issues I have been discussing in this chapter. A chapter that was never originally planned for the book, but has forced its way in.

These ideas have uncovered more of the way God operates and built upon the truths uncovered in the first book allowing me to see minor errors from the first book... like Christ was not dead forty hours exactly, but forty P-hours exactly. Bringing greater understanding of the numbers, the mathematical techniques, and the rules that God uses to govern His creation. Over the past year and a half I have taken my first baby steps and now walk with the knowledge of Christ's return with greater confidence. Looking forward to the time when I can run with greater speed and wisdom that will be needed in the years ahead.

When I began my career as an engineer, my supervisor assigned me small projects to work on. As the years passed and my experience grew; larger, more costly, and complex projects were assigned to me. After the early years, I was promoted to different areas in the company with greater responsibilities. I was required to manage and interact with various groups of people having diverse abilities. All the while gaining experience and knowledge in every facet of the company's operations: improving my skills while gaining wisdom that was to my benefit and to theirs.

After many years, I came full circle back to engineering where I started and was assigned to a large project with others who had similar and complementary skills to my own. We were to basically rebuild the plant I worked at from the ground up. To validate everything I had learned during the twenty years I worked there by applying the entire knowledge, experience, and wisdom I had gained all at once!

This is where my last path has been leading me. What I am compelled to try and do now. To relate how everything I have been showing you has been nothing more than training exercises for me to accomplish one of God's greater purposes. Understanding this, I have dreaded and procrastinated writing this section from the fear of not presenting the information clearly enough so that you would not get lost in the maze of ideas I have been traveling. I pray that you will find the end and attain the blessed assurance of knowing, without a doubt that Christ's return is near. This maze is really a circle—a sphere of God's design details and within this sphere are other spheres of knowledge and details that contain yet deeper spheres within.

So how does one transfer this revelation of why God's plans are the way they are so that you and others may benefit from the knowledge? This is my fear. I know not where to start—at the top, at the bottom, or some place in between. But today is the day and there will be no more delay and so I have started just by writing about my fears and objectives. And as has happened on many occasions like these, I write as the Spirit moves me. I find myself later surprised at some of the things I wrote, or how they were presented, when I go back to review and edit them.

My goal is to apply all the rules I have learned (all at once) so that we can get to the bottom line of "why" God chose the numbers He did for the end-time prophecies. I have already written about some of these things in this chapter, but some still remain hidden and can't be seen until this question is answered.

These ideas are extremely complicated and I hope to systematically use a step-by-step approach to guide you to the answers. The problem is God is employing many rules simultaneously and not sequentially as my approach suggests. Nevertheless, I have no choice but to start some-where and move step-by-step toward the final destination of "why."

So let's begin at a place you are familiar with—at the highest level of God's plan for mankind… 12,175 years. Using our understanding of moving within levels of God's plan, let's move down one level in the plan by reducing this level by the number of completion. This gives us 1,217.5 years. But a year is like a day and so these can be days and for our purposes they are. We now have 1,217.5 days. We will hold here and review something I wrote about in the first book about God's number. Remember God's plans are based on His number and Satan's plans are based on man's number and forgeries of God's number.

"God's number is three and it represents the trinity. God is also one… whole… unity… When you take the wholeness of God (1) and divide it by the trinity of God (3) you get God's number .33333 in its most basic form. Much like an atom is the building block for you. Taking God's most basic number and multiplying it by the number of completion (10) raised to the 3rd power (10 x 10 x 10… one 10 for each of the trinity) results in 333.3333333. This is the number of God in its completeness. Notice the repeating three in this number. This repeating three represents the eternity of God: His endlessness, His going on forever and ever."

2028 (God's Plan for Mankind) pgs. 35-36

This understanding was developed when I was trying to calculate the number of man, which is also the number of the beast of Revelation. What was learned was God's number and forms of that number are used for timing within His plan. From the quoted material we can learn something else we haven't discussed before… that God is one and God is three. In other words, three can equal one!

Now if you are a Christian you can sort of grasp the concept of the trinity, but from a purely mathematical understanding three can never equal one. It is from this spiritual (symbolic) understanding that we take our next step forward. That there is not 1,217.5 days on this specific level of God's plan we are investigating, but three times this amount. Three trimesters or three "days." Let's see how this is represented graphically.

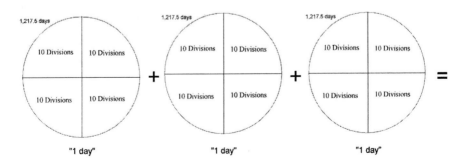

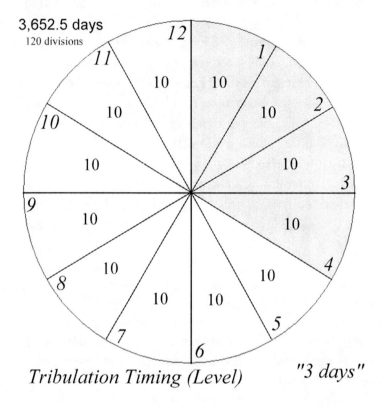

Tribulation Timing (Level)   *"3 days"*

Looking at our first concept of this level we notice that there are 3,652.5 days exactly (1,217.5 days x 3)! This is ten years and since it comes to ten years exactly we can know with confidence that this level pertains to the tribulation timing.

Furthermore, I divided this level into twelve divisions and each of these divisions into ten smaller units for a total of one hundred and twenty divisions. Why? Because I remembered something from analyzing the plan in the first book that I had forgotten. That the plan was not only used as a calendar for planning purposes, but also a clock! This is how I arrived at the larger divisions. As for the smaller divisions, I divided each "day" into four blocks of ten divisions to signify forty: a number God uses time after time.

So what do these smaller units represent? If the larger units are hours of a clock, then the smaller units must be minutes. Since there are sixty minutes in an hour, then each smaller unit represents six minutes of time for a total of seven hundred and twenty minutes designed for this understanding of the plan.

How many prophetic days does ten years represent? Let's find out since all the times in prophecy are prophetic and this plan is currently in real days. By making this reverse conversion we can compare the times in Revelation and Daniel as written, which is just the opposite of what was done in chapter five. I think this will simplify a person's ability to understand things if simplification is even possible. Thus:

$$3{,}652.5 \text{ days} \times \frac{P\text{-}}{1.01458333} = \textit{3,600 P-days}$$

or ten prophetic years. Next we need to briefly put this clock and ten year understanding of the plan aside (we will come back to them eventually) so that we can develop another understanding of this same level. This version is based on God's number of 33.333. How many P-days are there if we divide the total time of 3,600 P-days for this level by God's number? There are one hundred and eight.[46]

This next version of the plan has one hundred and eight segments of 33.333 P-days in it. Why? Because I realized very early on when I was working with the number twenty-three hundred that this number was symbolically equal to sixty-nine of Daniel's seventy 'weeks.' And when it was converted to real time we got the equivalent of Daniel's seventy 'weeks.' What I failed to understand until now was that the number, one thousand three hundred and thirty five P-days in Daniel 12, also represents an important number using the same methodology. What number is it? Let's divide these P-days by God's number just as we did earlier with the twenty-three hundred P-days that pertained to the antichrist. What answer do we get? Take a guess... we get 40.05 P-days! Interesting.

We get slightly over forty P-days. The difficulty is you should discern by now that God is precise and we will look closer at these extra .05 P-days in just a bit as we move closer toward the truth. So this version of the tribulation level calls for one hundred and eight segments of God's number. Let's start laying out what this understanding of the plan looks like.

108 (33.333 P-days)

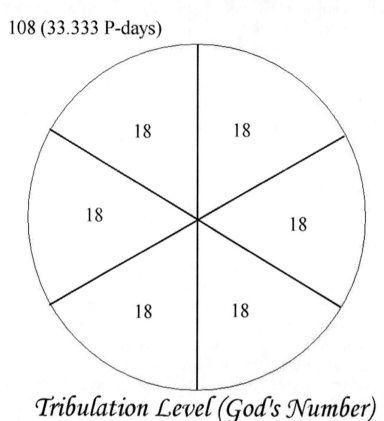

Tribulation Level (God's Number)

I arbitrarily divided this version of the plan into six divisions of eighteen 33.333 P-days because otherwise it would be too busy to divide it up into one hundred and eight individual segments that there really are. The number eighteen has no significance at all and was chosen because it was a multiple of one hundred and eight and could easily represent the one hundred and eight portions of 33.333 P-days.

Since this is the level of the tribulation period, let's go ahead and start superimposing the information recorded in the prophecies of Daniel and Revelation on the plan just as we did in chapter five—beginning with the twenty-three hundred P-days prophesied in Daniel eight. This is a prophecy of the antichrist and at the end of the twenty-three hundred days it says, "the sanctuary will be reconsecrated." We also know that these days are the equivalent of sixty-nine of God's number (33.333).

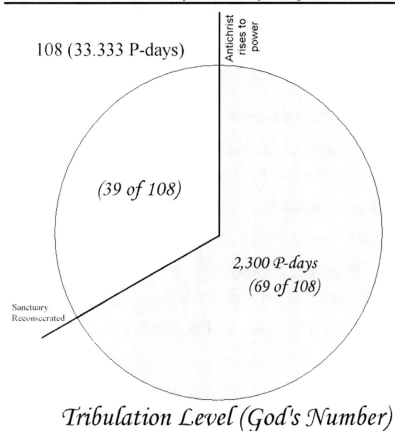

108 (33.333 P-days)

Antichrist rises to power

(39 of 108)

2,300 *P-days*
(69 of 108)

Sanctuary
Reconsecrated

Tribulation Level (God's Number)

We see from this ten-year plan the days are now divided not in six equal segments of eighteen as before, but into two unequal pieces of sixty-nine and thirty-nine. The larger piece represents the time period of the antichrist as prophesied by Daniel. We see symbolically that after sixty-nine "segments" he will be cut-off just as Christ was cut-off after sixty-nine 'weeks'. Remember, earlier I said many prophecies of Christ are also prophecies of the antichrist, but never the other way around? Here we see an example of this. In the book of Revelation it implies the beast (antichrist) will be stripped of the authority given to him by Satan when Christ begins ruling.[d]

By looking at this plan we can mark this spot for later use in determining where events fall as predicted in Revelation. Focusing on the thirty-nine portion of the plan, let's figure out how many prophetic days this represents. There are a couple of ways to do this: the simplest this represents. There are a couple of ways to do this: the simplest being to subtract twenty-three hundred P-days from the overall number of days, which are thirty-six hundred P-days. This results in thirteen hundred P-days. What do these thirteen hundred P-days mean since nowhere in prophecy does this number appear?

We must look back to our earlier work and the book of Revelation to find the answer. Revelation says there are some events that will take place that are twelve hundred and sixty P-days long. We also know that there will be forty days from the time Christ returns until those who survived the wrath years and are deemed worthy may enter the Millennium Kingdom. So lets add this information to our tribulation plan.

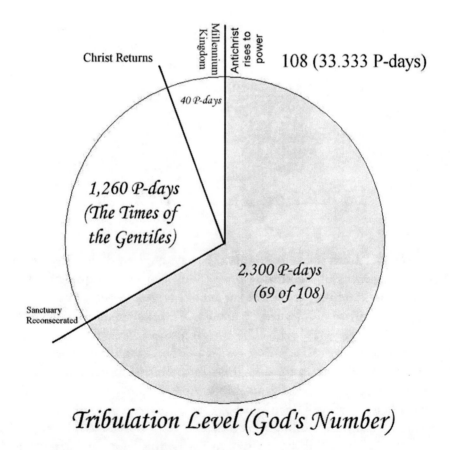

Tribulation Level (God's Number)

Things are beginning to get much more complex because of the symbolism of the numbers and the conflict between Satan trying to wrestle control of the world from God. But before we continue I want to quickly review. We have seen two versions of the tribulation level of God's plan. One based on God's number of 33.333 that is three thousand six hundred P-days long divided into one hundred and eight segments. The other is based on seven hundred and twenty segments and operates just like a clock on the wall with smaller units of six minutes as its basis.

Now know this, the antichrist's goal is to symbolically reach seventy "weeks" just as Christ's is. Christ is at sixty-nine and a half 'weeks' (Sabbath years) and holding and needs to complete three and a half years to begin reigning on earth. However, the antichrist has not completed any yet. Satan's objective is to finish the entire seventy all at once. To do this, the antichrist needs to keep control for the full twenty-three hundred P-days because if he can reach the end in control then he will win since twenty-three hundred P-days are really 2,333.54166 real days, which are effectively 70.00625 of Daniel's 'weeks' or seventy "days" based on God's number plus another symbolic 75 P-days.[47] This is just another case of Satan's forgeries. Do you realize there are 75 P-days between the twelve hundred and sixty P-days of Revelation and the thirteen hundred and thirty-five P-days of Daniel?

But Christians know that Christ has won the battle. The outcome is not in doubt. Nevertheless, there will still be a battle both literally and symbolically, on earth and in heaven. In everyway imaginable. So if Satan and his chosen, the antichrist, are to lose (to be kept from reaching the end of the twenty-three hundred P-days) then Christ must begin taking control before this time. How this is done is given in Daniel with the twelve hundred and ninety and thirteen hundred and thirty-five P-days mentioned. This information will be added to our current diagram in just a moment.

We know there is also another twelve hundred and sixty P-days mentioned in the book of Revelation. As it turns out, this time is going on concurrently with part of the twenty-three hundred P-days. This knowledge suggests there are another one thousand and forty P-days that need to be accounted for. This is a very interesting number indeed since one thousand is the highest form of the number of completion and forty days are repeatedly used by God for various reasons of which I have been unable to see what all the uses have in common. So these one thousand and forty days need to be accounted for on the diagram.

Back in chapter five they were placed at the beginning of the timeline, which seemed the most logical choice given the way the analysis flowed, and were not separated out. However, I am unsure exactly how they should be placed using prophetic accounting methods and what their significance is, but I will speculate on this matter after putting them on the diagram.

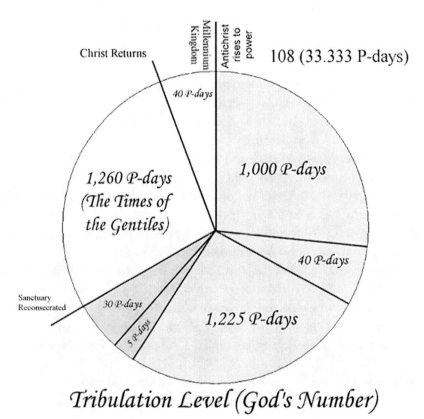

Tribulation Level (God's Number)

Let's first concentrate on the breakdown of the antichrist's twenty-three hundred days and why these numbers were placed on the diagram in the way they were. His twenty-three hundred days are comprised of the 1,000 plus 40 plus 1,225 plus 5 plus 30 P-days. This is not consistent with how they were portrayed on the tribulation timelines developed in chapter five. Those timelines were developed using real days whereas this level of God's plan has been developed using prophetic accounting. This difference causes some small discrepancies. Let's investigate the first one.

We know that there is a missing 'week' from Daniel's prophecy of the "Seventy Weeks." On one level of understanding this translates into seven P-years and seven P-years are 2,520 P-days.[48] If we divide this result by two, we get the two periods of twelve hundred and sixty P-days mentioned in Revelation. This math suggests the two periods must be back to back with no time in between to account for, so I indirectly showed it this way. It is the time from when Christ returns to the beginning of the 1,225 P-days. Thus the first twelve hundred and sixty days are presently represented on the plan as 1,225 plus 5 plus 30 P-days. The second half tribulation period of twelve hundred and sixty P-days is shown undivided on this tribulation diagram.

By doing this we accomplish two things. We kept the understanding of Daniel's prophecy unaltered and second we can account for the extra thousand and forty days separately within the twenty-three hundred days because of the symbolism associated with these numbers. If we accounted for them as we did in chapter five, which we cannot rule out as the correct understanding, then this would have messed up the math associated with Daniel's prophecy as well as the extra prophetic days would have been broken down differently so that no special significance could be assigned to them. If you want to see how they looked, then go back and study the timelines in chapter five.

Now I accounted for the one thousand days first followed by the forty days, but it could just as easily be the other way around. I have a fifty-fifty chance of getting this part right. Presently I can see why both possibilities may be true. If the forty days were at the start, this may be just another attempt at counterfeiting Christ. Since Jesus began his ministry with a forty day fast and was tempted by Satan we can infer something similar in this scenario. So what did Satan tempt Christ with? He offered him authority over the whole world if Jesus would just bow down and worship him. Maybe you can visualize the same thing happening with the antichrist. In Jesus' case, he rejected Satan's offer; whereas the antichrist accepts Satan's offer and does just the opposite of Christ and gives in to the temptation.

Revelation says the antichrist will be given authority from Satan,[e] but that authority will be stripped away by God.[f] In other words, God allows it for a period of time to achieve His purposes. And what are those purposes? To divide... to test... to separate believers from unbelievers.⁽ᵍ⁾ To cleave a dividing stake through mankind so that there will be no doubt as to which side of the fence a person is on.

Afterwards, the one thousand days would follow and encompass the "training" of the antichrist, his ministry so to speak. These days would mark the time when the antichrist rises to power and at the end of those days he will confirm the convent spoken of by Daniel,⁽ʰ⁾ which would signal the beginning of the classical seven-year tribulation period. Why? Because the Jews will spiritually prostitute themselves with this false messiah!

This was the first scenario and you can see how persuasive these ideas are. But I chose to represent the timing the other way around. That the one thousand days would precede the forty. During this time the same thing would take place that was outlined in the first scenario. The antichrist would rise to power. After the one thousand days would follow the forty days. I would like to think realistically these represent the period of time where the antichrist negotiates a treaty (covenant) with Israel and its enemies, and takes full control of the European Union with the aid of ten leaders from member countries. This is why I placed these P-days in the order I did.

If we keep true to the order of things when Christ first came and try to copy it as exactly as possible, we would find that Christ's forty days fasting in the desert came three and a half years before he was cut-off. Applying this same idea would place the antichrist's counterfeit forty P-days three and a half years before the end of the twenty-three hundred days just as I have shown them. There is nothing that would prevent this future dictator of the European Union from making a deal with the devil during this time rather than at the beginning of the one thousand P-days.

Now let's discuss the thirty and five P-days shown with darker shading at the end of the twenty-three hundred P-days. Why are these here? Daniel is given times for when Christ will return (after twelve hundred and ninety days from the abomination of desolation) and when Christ will be crowned at the start of the Millennium Kingdom (roughly one thousand three hundred and thirty-five days from the abomination). This is how these extra thirty-five P-days came to be on the diagram. By adding the twelve hundred and sixty P-days with the thirty P-days, we get the twelve hundred and ninety P-days spoken of by Daniel. Including the additional five P-days at the beginning of this period with the forty P-days at the end gives us the thirteen hundred and thirty-five P-days.

So there are thirty-five P-days that overlap the thirteen hundred thirty-five P-days and the twenty-three hundred P-days. Now these thirty-five P-days are the most important and controversial times on the entire plan. They can be verified many ways mathematically and we will spend extensive time examining how thorough God is with His planning. The real issue is that these days are not so mundane as to think of them as "just overlapping" because this overlap is the ultimate struggle between good and evil. The period of time where the antichrist will be trying to reach the end of seventy "God's number days" and knowing that Christ needs to add these same thirty-five P-days so that he can complete forty of God's number days. In other words, both need these days to win! Let's look at this additional information that has been discussed on our diagram.

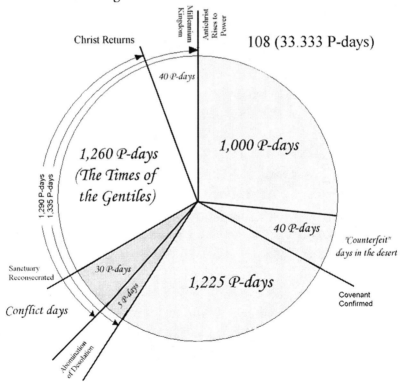

Tribulation Level (God's Number)

We can now see graphically how the times in Daniel and Revelation are laid out on God's plan. Let's not lose sight of the fact that these times are all prophetic times and would need to be adjusted if we want to understand the real times. How can we be sure this is how the times really will be? We can study the appropriate scriptures that pertain to these numbers and see if they make sense.

Conversely, by studying this plan we can get a clearer understanding of what scripture is telling us. Case in point, the term "reconsecrating the sanctuary" cannot mean Christ's second advent, as I once believed. Furthermore, it cannot mean that a physical temple is being reconsecrated as well. Why? Because to reconsecrate something it must first be desecrated. Our plan shows in fact that the "abomination of desolation" is the desecration and some think this desecration will be to a rebuilt Jewish temple in the last days. For the sake of argument, let's agree with them. Next, they maintain this desecration will result from the antichrist claiming he is God after entering the temple—just as other kings of the past did.

But if this is the case, what event would occur thirty days later that in God's eyes would constitute as "resanctifying" the temple? The answer is nothing. If there is a physical temple rebuilt, once it is desecrated by the antichrist nothing will be able to reconsecrate it to the Lord until Christ returns because of all the evil that will occur over the remaining three and a half years in Israel. So this term means something else in God's eyes, perhaps the rapture of the church after thirty days of bloodshed for Christians who refuse to take the mark of the beast or possibly the reestablishment of God's Word after the rapture (144,000 preaching the true gospel of Christ to those fence sitters who missed the rapture).

Let's dig deeper into the numbers of the Bible and see how God uses them. This is unavoidable if you want confirmation that this is the truth of the times mentioned in Daniel and Revelation. We know that thirteen hundred and thirty-five P-days represents the 40.05 "days" [49] derived from God's number and Jesus needs to fulfill these "forty days." But if they were exactly forty then there should have been only 1,333.3333 days prophesied by Daniel.

Taking into account rounding, Daniel should have prophesied thirteen hundred and thirty-three or thirteen hundred and thirty-four P-days as opposed to thirteen hundred and thirty-five P-days, but we know he did not. Why? Because thirteen hundred and thirty-five P-days minus 1,333.333 P-days leaves 1.6666 P-days remaining and what does 1.6666 P-days equal? They equal forty P-hours and forty P-hours equals 144,000 P-seconds! Don't lose sight of the fact we are still dealing with prophetic numbers that would have to be adjusted if we were discussing real time. Therefore:

$$1,335 \text{ P-days} = 1,333.333 \text{ P-days} + 1.666 \text{ P-days}$$

This equation can further be stated as:

$$1,335 \text{ P-days} = 1,333.333 \text{ P-days} + 40 \text{ P-hrs. And also as,}$$
$$1,335 \text{ P-days} = 1,333.333 \text{ P-days} + 144,000 \text{ P-seconds.}$$

If we divide the P-day equations above by God's number (Gn), we get the following truth:

$$40.05 \text{ Gn-days} = 40 \text{ Gn-days} + 40 \text{ P-hrs (or } 144,000 \text{ P-seconds)}$$

As a result, 0.05 Gn-days is equivalent to exactly forty prophetic hours![50] I really don't have a good term for describing symbolic days that are derived from God's number and have finally decided on the units "Gn (God's numbered)-days" to represent this idea. The whole point of this work is to illustrate that the time God picked to tell Daniel was precise and had no errors due to rounding. Furthermore, this time is consistent with the way God operates and uses numbers. The thirteen hundred and thirty-five prophetic days represent forty Gn-days plus an additional forty P-hours. This is interesting because the additional time is identical to the time Christ was physically dead.

Let's move on and see how these prophesied times look if they were superimposed on a clock—God's "Tribulation Clock."

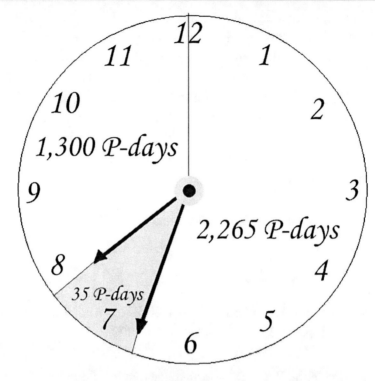

God's Tribulation Clock

I have shown the two major prophetic times from Daniel, twenty-three hundred P-days (2,265 + 35) and thirteen hundred and thirty-five P-days (1,300 + 35), and how they might be viewed as clock hands on a clock. The short hand representing the full twenty-three hundred P-days starting from twelve o'clock (clockwise) and the long hand representing the full thirteen hundred thirty-five P-days from 12 o'clock as well (counter clockwise).

These times are displayed with the disputed thirty-five days broken out separately as you can clearly see. How many hours and what symbolic time of the day do these prophetic days represent on God's tribulation clock? This answer can be found by determining the percentage of each block of time as compared to the whole amount of time, which is thirty-six hundred P-days and then multiplying by the total hours on the clock which are twelve. Let's go ahead and make these calculations.

2,265 P-days: $\dfrac{\text{2,265 P-days}}{\text{3,600 P-days}}$ x 12 Hours = **7.55 hrs.**

35 P-days: $\dfrac{\text{35 P-days}}{\text{3,600 P-days}}$ x 12 Hours = **.11666 hrs.**

1,300 P-days: $\dfrac{\text{1,300 P-days}}{\text{3,600 P-days}}$ x 12 Hours = **4.333 hrs.**

----

12 Hours

Therefore:

2,300 P-days: 7.5500 hrs. + .11666 hrs. = **7.666 hrs.**

1,335 P-days: .11666 hrs. + 4.3333 hrs. = **4.45 hrs.**

Next we need to represent these numbers in other ways so they can reveal God's awesome use of numbers and the redundancy of His planning on multiple levels. Let's convert them further to hours and minutes by taking the decimal portion of hours and changing it to minutes.

Table 6-3

Prophetic Time	Hours	Hours & Minutes
2,265 P-day	7.555 hrs.	7 hrs. & 33 minutes
35 P-days	0.11666 hrs.	7 minutes
1,300 P-days	4.333 hrs.	4 hrs. & 20 minutes
2,300 P-days	7.666 hrs	7 hrs. & 40 minutes
1,335 P-days	4.45 hrs	4 hrs. & 27 minutes

If we examine these numbers closely an interesting phenomenon is taking place. All these P-days when transformed came to exact times. No rounding was done when they were converted from strictly hours, to hours and minutes. An additional observation is that they are all numbers used by God, although at first glance a few do not seem so (four hours and twenty minutes e.g.).

So what time is it that the clock hands represent on our tribulation prophetic clock? If we begin at twelve o'clock and move the length of time that 2,265 P-days represents, we would get to 7:33 just as our table shows. Do these numbers look familiar? Of course they do. Seven is the number of perfection and thirty-three is God's number. Both are repeatedly used by God.

I conducted a quick experiment with a few friends and family. I took some dice and started shaking them in my hand and pretended to roll them out just as one might play the game of craps at a casino. Each person was tested independently of the others. As I began to roll the dice out I said, "come on baby, come on baby, lucky number…" and at this point I stopped, looked at them, and asked them to finish what I was going to say. There were seven people and six of them responded without hesitation with the answer "seven."

We see from this experiment that people have been programmed or trained that the number seven is lucky. That there is something special about this number and I am convinced this idea has been passed down from generation to generation. How did this idea ever come about? I'm not sure, but God has allowed the knowledge of how special some numbers are to be uncovered for His purposes.

So we are at 7:33 on the clock. This is the "symbolic" time the hidden battle begins. The transfer of power from the antichrist on earth to Christ in heaven. How long will it last… thirty-five P-days, which is seven minutes! Hmm. These thirty-five P-days will end with the reconsecration of the sanctuary. This will be 7:40 on God's tribulation clock. Again we see two numbers that God uses regularly. Can these discoveries be coincidences that the numbers in Daniel when viewed as time on a clock come to exactly these times? I think you know my answer, but we are not close to being through, for God is even more thorough than this. Most do not need more proof. For if they understood the mathematics, then they grasp the probabilities of these coincidences being merely chance are astronomical. They can only be more verification of God's detailed planning.

Before moving on, I want to sidetrack here and show you something else I discovered that relates to this as well as how God uses His methods over and over. If you were to look back at God's plan on page one and think of it in terms of a clock just as we are now, what would you notice? That Adam was created at six o'clock (six is man's number),

Christ would return at twelve o'clock (the midnight hour),[1] and Christ first came at the hour of completion ten o'clock. But look closer at man's first trial… first judgment… Noah's flood. What time was it on God's "master clock" for mankind? It is somewhere between seven and eight o'clock; and we can calculate the exact time.

It turns out that if you go to a real clock that is adjusted properly (where the minute hand is exactly on twelve and the hour hand is exactly on one of the "hour" numbers {I have found some clocks where the hands are off slightly probably due to them being mishandled during time setting}) and set the hands on this clock to the time where both line up exactly between seven and eight o'clock you will get the answer! That's right, the line on God's overall plan for mankind where the flood occurs is just that, a line. The hands overlap and they look like a line… a marker on God's master clock of human history.

I know not why or how the Holy Spirit shows these oddities to me, but he does. Go ahead and do that if you have a clock before you read on. Calculating this time is easier for anyone who read the first book, but for those readers who have not you need to understand a few things. This was the very first calculation I did and was based on the number 2,029.166 and how God administers punishments. This was the number the Holy Spirit had used to start opening my eyes. I thought this was the whole answer, the key to everything, but it was just the beginning.

It was revealed the flood came on the forty-ninth seasonal day of the first trimester in God's plan. Since there are sixty S-days in each trimester of God's plan we can calculate where this event fell on God's master clock. The equation is complicated for most people, but I will show it nevertheless because it is strictly based on proportions of the circle the clock is laid out on.

$$\frac{49 \text{ S-days}}{60 \text{ S-days}} \times \frac{2{,}029.166 \text{ years}}{1 \text{ trimester}} = 1{,}657.1528 \text{ years}$$

This is the time between when Adam was created and the flood came. Next we need to add the time from the beginning of creation until Adam's creation, which was six H-days or 6,087.5 years[51] to this number. The result is 7,744.6528 years. If we divide this time by the total time of God's plan, 12,175 years, we will get the percentage of the time it took for this event to occur. The last step just requires us to multiply by twelve hours and then convert to actual time in hours, minutes, and seconds. Let's do this:

$$\frac{7,744.6528 \text{ yrs.}}{12,175 \text{ years}} \times 12 \text{ hours} = 7.63333 \text{ hours}$$

Converting the decimal hours to minutes by multiplying by the conversion constant, sixty minutes in an hour, give us thirty-eight minutes exactly (no rounding). Therefore, the flood occurred at exactly 7:38 on God's "master clock." Is it strange that these calculations come out to exact times again and in this case, exactly where the hands line up on a clock? This was the time you found when you lined the hands up between seven and eight was it not? Not only that, this time fell between 7:33 and 7:40. The times we calculated for the thirty-five P-days of conflict between Christ and the antichrist…between Satan and God.

What are the odds of this happening? How many times does one need to be shown the truth before they can believe? Just once for those who have the mathematical skills and the Holy Spirit's guidance. For others who refuse to believe in Christ as their personal savior, they will never know for God blinds them from seeing the truth. Many probably just think I am some clever mathematician. I am good with numbers, but insignificant in comparison to God's awesome math skills. I am just the messenger of this news… nothing more. For those Christians who rarely study Bible prophecies or struggle with mathematics, these revelations may require being shown many times (along with personal study) before the knowledge of Christ's return is revealed to you.

Let's return from our sidetracking journey and look again to the tribulation numbers from Daniel and Revelation. If we backtrack a few diagrams earlier to where these times were first laid out on the tribulation level of the plan, we can learn more things. We know that the antichrist mimics God's methods. Like the forty P-days before the start of Daniel's last week—before the start of the first twelve hundred and sixty P-days in Revelation. So when is the starting time on God's tribulation clock for this false forty-day period? It is just after the one thousand P-days remember?

$$\frac{1,000 \text{ P-days}}{3,600 \text{ P-days}} \times 12 = \textbf{3.333 hours} \text{ or } 3:20 \text{ o'clock.}$$

If these days are laid out correctly on the clock and the forty P-days come after the one thousand P-days as I have showed them (remember I said it could just as easily be the other way around), then we are witnessing another instance of Satan mimicking God.

Let's look at another representation of the thirteen hundred and thirty-five P-days, which is 4.45 hours. This can also be represented as 14,400 seconds (four hours) plus .3333 hours plus seven minutes. All numbers God uses repeatedly. We know this time can also be viewed as forty Gn-days plus forty P-hours.

What time does forty days and forty nights from the great flood translate to on God's overall plan for mankind? It is four S-minutes and forty-four S-seconds.[a] Numbers eerily close to the 4.45 hrs. that thirteen hundred and thirty-five P-days symbolically represent. If we were to remove the extra forty P-hours and just focus on the forty Gn-days this would translate into 4.4444 hours. By further examining the twists of the numbers we can see how God's plan for the last days of this age, the tribulation period, duplicates His wrath poured out on mankind with Noah's flood.

Do you remember the total number of days in God's overall plan? There are 4,446,918.75 days or 4.44691875 when this number is factored down by the number of completion. If we round this number, the result is 4.45. The same number we got when viewing thirteen hundred and thirty-five P-days as time on a clock!

Many believe the scripture; "just as it was in the days of Noah so will it be in the last days" is talking about man's depravity, rejection of God and the truth of God's Word, and the continuous focus on the materialistic things of this world instead of the things of God. And these are all true. But on a deeper level of God's planning we now find timing parallels between the flood and the tribulation period are also evident.

I know how hard it is to keep all these numbers and the various understandings of tribulation plans in your head all at once since they inundate my mind constantly. But there are a few more I want to share, before concluding this discussion, as further indications of God's planning.

Final Observations from God's Ball of Yarn

If we look back to God's "master clock", the clock based on the overall plan of 12,175 years, we can see how close we are to Christ's return from a different perspective than years on a calendar. Since there are 43,200 seconds (three times 14,400... hmm)[52] on this clock, then 3.54825 [53] seconds elapse for every year that passes by.

This number is very close to three and a half when it is converted from prophetic time to real time—3.551042. In fact, if these numbers were rounded, they would both come to 3.55! Conversely, the 3.55 seconds viewed "prophetically" are three and a half P-seconds per year.[54] In view of the truth that these revelations of Christ's return are twenty years in advance of the actual event (it is now September 2008 A.D. as I write), we can now see that God is revealing these hidden mysteries with only seventy P-seconds[55] left on His master clock until the start of the Millennial Kingdom!

On God's tribulation clock we calculated that the thirty-five P-days of conflict corresponded to exactly seven minutes when viewed as time and I explained that seven was a special number the truth of which has been passed down from generation to generation. Now you should know by now that the number of completion, ten, is the sum of God's number (3.333) and man's number (6.666). But what does this concept look like if it were expanded to the clock analogies we have been investigating? It looks like the number seven... seven minutes! Watch:

7 minutes = 6 minutes and 40 seconds *plus* 20 seconds and,

6 minutes and 40 seconds = 6.666 minutes (man's number)
 plus   20 seconds = <u>.333 minutes (God's number)</u>
                  7.000 minutes (number of perfection)

And so the number of perfection when viewed as time on a clock is equivalent to the number of completion! When we look around we can see evidence of God's design everywhere we look. Christians only have to search with an open mind and an open heart to see the Creator's hands at work. God's plans working together on every level imaginable to achieve His purposes. What an awesome God we have!

Lastly, we know when ten is raised to the third power it becomes the final form of the number of completion. What number do we get when we raise God's number (3.333) to the third power? We get a most fascinating number—37.037037037! Why is this number interesting? Because if it is viewed as years, it would be equal to thirty-seven yrs. and fourteen days (13.5277 days).[56]

Remember Jesus was about thirty when he began his ministry in September of 29 A.D. and his earthly teaching lasted three and a half years. This would have made him about thirty-three and a half years old when he was crucified. Adding the additional three and a half years he must reign from heaven to fulfill Daniel's prophecy would get us to about thirty-seven years. If Jesus started roughly fourteen days after his thirtieth birthday, then it is possible to believe his time could be exactly equivalent to God's number in this higher form! By cantilevering this idea, it might be possible to uncover more evidence of why God chose the numbers He did in Revelation and Daniel.

For instance, 37.037037037 years is 13,527.777 days.[57] This new number is exactly ten times larger than 1,333.333 P-days (forty Gn-days) when it is converted to real time.[58] Now this idea gets even stranger. Why? Because if you subtract the 1,333.333 P-days (Daniel's 1,335 days minus forty prophetic hours) from the total years we should get the time Christ spent on the earth (his age) at the point he ascended into heaven. Although, presently I am unclear as to whether the forty days he spent on earth after his resurrection should actually be considered. Let's make this calculation now:

13,527.77 days (37.037 yrs.) − 1,352.77 days (1,333.33 P-days) =

12,175 days!

Wow! Is this the result you expected to see? What does this outcome really mean? First, it is forty prophetic hours too long from the total time because we used 1,333.333 as opposed to 1,335 P-days. However, we know that Christ was physically dead forty prophetic hours and was not on this earth. The apostle's creed says, "he descended into hell and on the third day he arose." Was he really in hell? I know his spirit was not on the earth and so I believe this to be a true statement of faith that all Christians should believe. So we can use this fact to offset the extra forty P-hours. Was this the reason why God added an additional forty P-hours on the tribulation timeline to get to thirteen hundred and thirty-five P-days? You be the judge.

I assumed in my book "2028" very early on that Christ lived 33.333 years and started with this idea until it appeared the evidence supported it was more likely thirty-three and a half years. But if you were to continue down this road to its final conclusion, you might appreciate that 12,175 days is 33.333 years exactly[59] and that Christ existed one day on this earth for every year in God's overall plan for mankind if this idea were true.

Speaking of hell, visualize these thoughts. One third (God's number) of forty P-hrs. is 13.52777 hrs. (one hundredth of Daniel's number)! This leaves 27.0555 hrs. remaining of the forty P-hours. Now I'd like to think Jesus' time spent in hell went like this: he died around 3 pm (2:58 pm according to this idea) on Friday and for the first 3.0555 hours (three hrs. and 3.333 minutes) he reasoned with the devil. At which time Satan would have none of it and Jesus was thrown in a cell around 6 pm, which was the start of the Sabbath day. The Lord rested for twenty-four hrs. in the cell until the start of the first day of the week... now 27.0555 hrs. (.666% of the total time) into his time spent in hell. At which time Jesus broke out (all hell broke loose) and the fight was on for another 13.52777 hrs. until he arose at 7:33 am Sunday morning which was ten minutes after the official sunrise on April 5 in 33 A.D.!

We have examined the timing for the tribulation period at great length focusing on the times foretold in the books of Revelation and Daniel. As for real events prophesied, very little has been discussed. In the remaining chapters we will investigate many aspects of the last days of the end-times to see what God has planned for mankind's second rebellion.

Chapter 7

The Next Ten Years

WE NOW KNOW WHEN TO EXPECT the return of Christ and how God works. We also know a lot about the timing of the tribulation period prior to Christ's arrival. The time when the antichrist will rule and God's desolations are planned for the rebellious world. These events will begin in the latter months of the year 2018 A.D. But what about the next ten years? What kinds of problems can we expect during the time leading up to the last days of this age?

We can expect tomorrow to be just like today—deterioration accelerating ever so slightly with each passing day. Imperceptivity overtaking many in today's fast paced world. Those not paying attention will never be the wiser. These changes growing acceptable as "just the way things are." We need only to look at the daily newspaper or the late night news to see what tomorrow brings on the winds of change for the next ten years—just more of the same.

These trends are quite clear and ominous and cannot be stopped or slowed. Why? Because of the rebellion of mankind to the truth of God's authority and existence. God knew that in the last days, knowledge and wealth would increase and as a result, man would begin to raise himself up above Him. Turning away from the Lord to the point of widespread disbelief in Him.

So what are the specific things to expect during this time? In the United States the standard of living will continue its decline. The poor will continue to get poorer and the middle classes and the number of wealthy families will continue to shrink. The chasm between the ultra-wealthy and the rest will expand as the concentration of wealth gravitates to the few elitists at the top of the food chain. Not only will this transfer of wealth continue from the poor and middle class to the ultra-wealthy, but also the destruction of wealth from greed/mismanagement and the shifting of wealth from the United States to other countries will increase unabated because of huge trade imbalances and poor fiscal restraint.

Population growth will continue its rise despite efforts to keep it under control. This trend will put pressure on natural resources like crude oil and natural gas on a scale the world has never seen at a time when supplies are beginning to shrink. Large countries like China and India will continue their escalation of using the earth's finite resources as they move to improve the standard of living for their large populations of over one billion people. These inescapable forces will cause the prices of everything to skyrocket.

Natural disasters will rise in frequency and magnitude. Floods, hurricanes, storms, earthquakes, volcanic eruptions, tornados, wildfires and many others will be commonplace in the news. We are already getting accustomed to hearing about them without much fanfare. In my lifetime there has been a drought that dried up parts of the Mississippi River, stranded barge and boat traffic, and two years where heavy rainfall has caused disastrous flooding along the river—one of those was 2007 A.D. Hurricane Katrina hit New Orleans with major damage and destruction in 2005 A.D. It was the second category five hurricane of that year (Wilma being the other and the strongest ever recorded in the Atlantic basin with winds reaching 185 miles per hour) and the costliest in U.S. history.

Tidal waves, flooding, and earthquakes have devastated other parts of the globe. No areas in the world appear to be safe from nature's wrath. Hotter and hotter days on average can be expected until the very end and these shifts in weather patterns will produce less rain in the crop growing regions of the world. Consequently, food shortages and famines can be expected in many places that do not have the monetary means to compete for food supplies.

In addition to the natural disasters and weather changes, man made disasters will increase in frequency due to armed conflicts and the decay of mankind's moral fiber. Attacks by terrorist groups with political and religious agendas as well as riots due to injustice, decaying economic conditions, and resource shortages are becoming commonplace.

Knowledge will continue to increase and with this increase, so does man's pride. God has been removed in schools and universities and with His removal this generation's "I know it all and need no one" attitude is just another sign of the rebellious mindset toward Christ and Christians at the time of the end.

These anti-Christian thoughts being cultivated in secular universities and schools are being witnessed in many of today's leaders. An example of this way of thinking is the United States began minting a presidential dollar denominated coin in 2007 without the words "In God we Trust." The only coin minted without these words and evidence of a greater trend that will continue. The European Union's currency (Euros) has no reference to God's authority on any of its legal tender as well.

During this time there will be an increase in false teachers and prophets. Finding the truth of God will become harder and harder. Many will be led astray and many will fall away from the faith because of the constant bombardment of lies that weigh on them. I am witnessing this trend first hand with those who contact me on the Internet to discuss end-time issues.

Freedoms in the United States are being taken away at an alarming rate all in the name of "safety." The government is doing this as part of its program to eliminate terror and unsuspecting Americans are giving away freedoms that were fought for by their ancestors without any resistance. We are being brainwashed into accepting personal searches at airports and sporting events when we have done nothing wrong. We permit the government to tap our phones, intercept and scan our electronic mail, and examine our financial records without just cause or due process.

Politicians are passing laws on everything from what kind of light bulbs we can buy (which are only made in China), shower heads, toilet designs, etc. all in the name of energy conservation to wasting time figuring out if Barry Bonds took steroids. The consequences of passing a law for everything are that freedoms are being taken away. More serious laws are being passed that restrict one's constitutional rights again in the name of "security." As this movement continues, you can expect it to spill over into freedom of religion and then there will be a backlash from Christians who are one of the few groups left fighting these restrictions. However, the United States government is planning for this resistance to complete control by building concentration camps to house those who refuse to go along with their plans.[a]

Are these the things we want politicians spending billions of dollars of hard earned taxpayers' money on rather than focusing on the bigger picture and finding real solutions like energy efficient vehicles or renewable power that eliminates the need for oil? This waste of government resources and time will continue over the next ten years until the U.S. government and the American people are broke; except for the few elitists at the top.

These conditions are necessary so that the once proud American people will allow themselves to be controlled by foreign leaders without much of a fight. This is happening in our financial markets right now. An economic war is being fought and America is losing. Foreigners are manipulating the price of oil and the value of the U.S. dollar to break America's dominance. Plans are being formulated to replace the U.S. dollar as the currency of the world economy. This double barrel attack is causing the prices of everything in the U.S. to skyrocket at a time when jobs are being moved overseas and common folks have little means to afford them. Financial institutions are failing and some are being kept afloat with loans based on money that has no intrinsic value.

This kind of "bookkeeping" and poor fiscal responsibility by our government will lead to a hyperinflation economic situation where the value of money is effectively worthless. This is a serious condition that has occurred in other countries, most recently in Zimbabwe, Africa, and is caused by not controlling the money supply and printing money to repay debt. Can you print money to pay back your debt? No. It would be worthless and so is the American government's money. It is only a matter of time before foreign countries stop accepting U.S. currency as repayment for the loans they have given us. Notice how Zimbabwe went from one of the richest countries in Africa to one of the poorest (in about twenty years) because of government mismanagement, greed, and poor fiscal restraint—the same problem running rampant in the once great country of the United States.

> "Gideon Gono, the central bank governor, said he was no longer prepared to print notes of ever higher denominations. "Ten billion dollars today, will as from August 1 be revalued to one zimdollar dollar," he said earlier this week. The move follows an increase in Zimbabwean inflation to 2.2m%, and represents an attempt to get to grips with the economic crisis in what was once one of sub-Saharan Africa's richest countries. The $100bn note came into circulation only 10 days ago.

Gono's move follows the example of the president of the Reichsbank in the 1920s, Hjalmar Schacht, who helped end Weimar Germany's financial crisis by striking nine zeros from the currency and turning 10bn old marks into one new rentenmark."

Larry Elliott, economics editor
The Guardian, Friday August 1, 2008

This combination of higher prices and few jobs, along with the debt accumulation that many Americans took upon themselves to keep up with their neighbors and their own covetousness desires, has enslaved us to the elitists and foreign governments. In addition to our individual borrowing, our government's borrowing is at the end of its rope as well. Denial of these facts is denial of the truth. Listen; there is a universal rule that applies to this situation... "He who has the gold makes the rules." In other words, those who have the money and resources are in control. That used to be the United States and now the scales have tipped to foreigners without a means to rebalance the scales. Here is the lesson from God the Israelites were supposed to learn that applies to us as well.

For the LORD your God will bless you as he has promised, and you will lend to many nations but will borrow from none. You will rule over many nations but none will rule over you.

Deuteronomy 15:6 (TNIV)

The LORD will open the heavens, the storehouse of his bounty, to send rain on your land in season and to bless all the work of your hands. You will lend to many nations but will borrow from none. The LORD will make you the head, not the tail. If you pay attention to the commands of the LORD your God that I give you this day and carefully follow them, you will always be at the top, never at the bottom. Do not turn aside from any of the commands I give you today, to the right or to the left, following other gods and serving them.

Deuteronomy 28:12-14 (TNIV)

The foreigners who reside among you will rise above you higher and higher, but you will sink lower and lower. They will lend to you, but you will not lend to them. They will be the head, but you will be the tail. All these curses will come on you. They will pursue you and overtake you until you are destroyed, because you did not obey the LORD your God and observe the commands and decrees he gave you. They will be a sign and a wonder to you and your descendants forever. Because you did not serve the LORD your God joyfully and gladly in the time of prosperity, therefore in hunger and thirst, in nakedness and dire poverty, you will serve the enemies the LORD sends against you. He will put an iron yoke on your neck until he has destroyed you.

Deuteronomy 28:43-48 (TNIV)

From the first two sections of scripture we see that those who lend (have the money) and follow God's commands will be above those who borrow (don't have money) and don't follow God's commands. From the last text in Deuteronomy we find the consequences of not obeying the Lord's commands.

Crime will continue to increase as things continue to deteriorate. Break-ins, thefts, cheating and swindling, identity and computer financial thefts, murders, adultery, sodomy, etc. are all commonplace as the love of our fellow man grows cold. Years ago people were shocked to hear of a woman killing her children or being a mass murderer, but now days we witness these events on the news without batting an eyelash.

It wasn't that long ago when criminals wanted to steal money, they had to break into your house, break into the bank, or personally attack you to get it. Today someone anywhere in the world can do it with just a few clicks of a computer keyboard and take a lifetime of savings just because of the computer systems that are interconnected.

Not only that, but our government is passing laws to make these crimes even easier! In 2009 a new law will go into effect that is essentially a national ID act. It will tie together people's financial information with law enforcement information so that authorities in any state can access it. What this really means is more people can access "all" your information, without you knowing about it, at a time when more people cannot be trusted! What a great combination to ensure your personal information is safe and protected.

Increased wars and tensions can be expected over the next ten years because of all these converging trends causing uncontrollable stress. Presently wars are being fought in Iraq and Afghanistan. Smaller conflicts and civil wars litter the landscape of the African continent. China is fighting with Tibet in a civil war that is escalating, and is also conducting economic war with Taiwan (Republic of China). Armed conflict affects the northern countries of South America and of course the constant military conflicts between Israel and its Arab neighbors cannot be left unmentioned.

Conflict between India and Pakistan continue along with civil unrest in eastern European countries. Russia and the United States are butting heads because of U. S. policies to place missile systems in Eastern European countries that were once part of the old U.S.S.R. It is only a matter of time before the winds of war make their way to the unaffected areas of the globe like the United States. Prophecies are already being made about invasion from Russia, China, and Cuba.[b] Jesus said,

> *Watch out that no one deceives you. For many will come in my name, claiming, 'I am the Christ', and will deceive many. You will hear of wars and rumors of wars, but see to it that you are not alarmed. Such things must happen, but the end is still to come. Nation will rise against nation, and kingdom against kingdom. There will be famines and earthquakes in various places. All these are the beginning of birth pains.*
>
> Matthew 24:4-8 (TNIV)

We see from these verses that many conflicts will be going on near the end, but that they are not the end so do not become alarmed. The end is still to come… later on… after more things that need to occur take place i.e. the prophesied things that will occur in the last ten years before Christ's second advent.

But these wars are one of the key signs to watch for. One of these wars was forecasted in the book of Ezekiel 38-39. This war is known as the War of Gog and is a large scale war fought by many nations against

Israel and led by a nation far to the north of Israel… Russia. This war is projected to fall near the end of these next ten years.

Daniel says of this war,

> *He will be succeeded by a contemptible person [antichrist] who has not been given the honor of royalty. He will invade the kingdom when its people feel secure, and he will seize it through intrigue [politics]. Then an overwhelming army will be swept away before him [Gog's armies]; both it and a prince of the covenant will be destroyed. After coming to an agreement with him, he will act deceitfully, and with only a few people [ten] he will rise to power.*
>
> Daniel 11:21-23 (NIV)

We learn from these verses and the War of Gog being the next of Ezekiel's prophecies to be fulfilled, that the antichrist will use this war as a political means to gain power. These two prophecies converge at the end of this ten-year period and will usher in the last ten years of the Tribulation.

What we do know is things will continue to get worse and there are no reasons to believe otherwise. Planning for these trends will make these troubles more bearable, but knowing Christ as your personal Savior brings the peace of mind that will get Christians through these tough times and the even tougher times that lie further down the road when the antichrist is in power and God's wrath is poured out upon the world.

Chapter 8

Deciphering the Prophecies

T O BEGIN UNDERSTANDING THE BIBLE prophecies, so that we can know what events to expect during the tribulation period, we need to first realize there are methods, techniques, and rules to follow. We cannot just read these prophecies and interpret them as we wish, although insight from the Holy Spirit should never be resisted. Just understand that this inspiration is never at odds with what is already written in the Word and if it is, then this inspiration is not from God.

We need to know what the guidelines are in deciphering God's prophetic words if we are ever to learn what the future holds before it happens. This is an extremely difficult task to undertake, and may even be impractical, given that most prophecies have been impossible to understand until after they have occurred. Since we are at the very end of the Times of the Gentiles, we have an advantage over all those generations that have come before us. We know many of the prophecies must be fulfilled in the next twenty years and so we can narrow our center of attention.

Let's focus on the book of Revelation, since this book is the major prophetic book that covers the last days, to highlight what these methods of interpretation are. You should note that other books of the Bible have end-time prophecies in them, and the same rules we uncover will apply to them as well.

Redundancy

People who study the book of Revelation can be divided into two main groups. The first group sees this book as basically one sequential prophecy while the rest view the book as multiple prophecies covering the same series of events (the tribulation period) much like the book of Daniel is written. When I first began studying this book written by St. John many years ago, I used to be in the first group, but after continued in-depth study of Revelation and other prophetic books of the Bible, I have switched my position to the "multiple-prophecy" side.

When reading other books of the Bible written by the prophets of God, one realizes that none of them have just one prophecy within them, but many. Daniel is the closest book to Revelation as they share some of the same symbolism and end-time prophecies. The book of Daniel has seven major prophecies within and all these prophecies describe various details of human history from Daniel's time until the end of human history. Some aspects of these prophecies are redundant, providing more understanding and clarity, while other parts of each prophecy are unique and add missing details to the entire timeline of historical events. I believe the book of Revelation is much of the same in this regard in that it too has several separately identifiable prophecies.

Also, within each prophecy, the sequence of events should always be taken in the order given. Having said this, this requirement is not as easy to apply as it seems. First, one must be able to recognize where a prophecy stops and the next one begins. Next, all students of eschatology must realize that everything written within a specific prophecy is not always sequential even though the events are in order. Some events may be concurrent or what is written later is but a clarification of an event previously mentioned.

For example, in Revelation 12 we find a symbolic description of the conflict between the "Woman and the Dragon" that lasts for twelve hundred and sixty days. Let's see what this text says:

> A great and wondrous sign appeared in heaven: a woman clothed with the sun, with the moon under her feet and a crown of twelve stars on her head. She was pregnant and cried out in pain as she was about to give birth. Then another sign appeared in heaven: an enormous red dragon with seven heads and ten horns and seven crowns on its heads. Its tail swept a third of the stars out of the sky and flung them to the earth. The dragon stood in front of the woman who was about to give birth, so that it might devour her child the moment he was born. She gave birth to a son, a male child, who "will rule all the nations with an iron scepter." And her child was snatched up to God and to his throne. The woman fled into the wilderness to a place prepared for her by God, where she might be taken care of for 1,260 days.
>
> *Revelation 12:1-6* (TNIV)

In the remaining verses of Revelation twelve, be aware this same prophecy is reiterated using more literal language. Let's read,

And there was war in heaven. Michael and his angels fought against the dragon, and the dragon and his angels fought back. But he was not strong enough, and they lost their place in heaven. The great dragon was hurled down—that ancient serpent called the devil, or Satan, who leads the whole world astray. He was hurled to the earth, and his angels with him...

When the dragon saw that he had been hurled to the earth, he pursued the woman who had given birth to the male child. The woman was given the two wings of a great eagle, so that she might fly to the place prepared for her in the wilderness, where she would be taken care of for a time, times and half a time, out of the serpent's reach. Then from his mouth the serpent spewed water like a river, to overtake the woman and sweep her away with the torrent. But the earth helped the woman by opening its mouth and swallowing the river that the dragon had spewed out of his mouth. Then the dragon was enraged at the woman and went off to make war against the rest of her offspring—those who keep God's commands and hold fast their testimony about Jesus.

Revelation 12:7-9, 13-17 (TNIV)

This last text cannot all be taken literally, nevertheless some information provided should be. We are informed the dragon is in fact Satan, but still not told, with certainty, whom the woman symbolizes. In this further explanation, the amount of time this clash will take to complete is "time, times, and half a time."

Understanding that these verses are not "new" events that follow events covered in the previous section of scripture, but are the same events described with additional information detailed in other ways, allows us to grasp (in this instance) that the symbolic phrase "time, times, and half a time" means twelve hundred and sixty days! Furthermore, we know from our previous studies that these are prophetic times and that the specific term "time" used in this context is equal to three hundred and sixty P-days or one P-year.[60]

A careful analysis is required of each prophecy to ensure that the proper placement of end-time events is achieved without adding the same event more than once into the timeline. Failing to recognize that what appears to be a description of a new event, but in actuality has been previously covered using different imagery, will cause problems with the prophetic timing and sequence of events.

Subtitle Warning

When examining some Bible translations, one notices the translators have added subtitles to describe blocks of scripture with certain themes or ideas. These titles are helpful in identifying the accompanying text, while other times they can be a deterrent to grasping the biblical truths because they lead the reader down the wrong path… the path the translator "thought" was the true path.

An example of this problem is in the New International Version's (NIV) translation of Daniel chapter eleven. At verse thirty-six a subtitle was inserted that reads, "The King Who Exalts Himself." This added title seems appropriate considering the text that follows, but it draws the reader's attention to the notion that this is a logical break from the previous text. The scholars from the NIVSB (Study Bible) even bring up this fact in their commentary, mentioning from this point forward to the end of the chapter are details of the end-time antichrist. If the NIV translation had not added this particular heading where it did, the reader would have been left to their own conclusions without having them drawn subconsciously for them. When you read all of chapter eleven you will find a more logical break at verse twenty-one and the "person" who is the focus of the text in verse twenty-one, is the same person all the way through the end of the chapter. There appears to be no logical breakpoint at all in these verses. Insertion of a text label at a wrong spot can lead most Bible students astray. There are times when this method is very good at helping to comprehend sections of scripture, while other times it is not.

For the most part, the book of Daniel is broken up logically in a way that allows us to see the different prophecies clearly. Other Old Testament prophetic books are not divided as cleanly as Daniel, nor is the book of Revelation. The reader is left to determine the breakpoints for themselves (if there are any) in many translations and if these headings are added, like the NIV does, remember that they sometimes aid understanding while hindering it at other times. Keep this in mind when studying prophecies with Bible translations using this technique.

History Repeats Itself

The goal of eschatology students is to learn what the future holds. They usually don't know what compels them to find this knowledge, but most are driven nevertheless. Some of these students and professed experts who write on these matters study the sequence of prophetic events, but lack an understanding on when to expect them, what to do about them, and why these events need to happen in the order they do. In lacking knowledge of these events, people are unable to see or consistently interpret the symbolic language used in many of these prophecies.

Some people see key prophetic events as having occurred in the past, while others view them as unfulfilled future events. What they fail to realize is the old axiom that history is always repeating itself… that man never learns from his mistakes… that each generation needs to learn first hand the lessons of the previous generations. Some events foretold in scripture have not occurred, even though in the past similar things have.

A case in point is many preterism theologians believe the "abomination of desolation" spoken of by the prophet Daniel was fulfilled by Antiochus Epiphanes around 168 B.C. when he set up an idol of Zeus in the Jewish temple and sacrificed pigs on the altar. Although this was an abomination in God's eyes, these abominations have happened many times in the past. Chapter eight of Ezekiel is another good example of this. But Jesus himself mentions that the "specific" abomination of Daniel was unfulfilled even in his time and stated it would be a key sign of his return to watch for.[a]

You see, God is consistent and fair when dealing with His children.

He is better than the best earthly parents. When children act up or misbehave, good parents meter out punishment to fit the offense. God does the same. When I was a child and I said a "swear" word my grandmother would use a bar of soap to wash out my mouth. If this behavior continued even after the punishment, then sterner punishments were forthcoming. This is how the criminal justice system works and this is how God works as well. Punishments for the offenses of each generation are used consistently by God just as they are by parents and governments; history repeating itself over and over.

When Sodom and Gomorrah became so wicked that their "cup was full" of abominations and could hold no more, God sent down burning hail (severe meteorite shower) that destroyed those cities forever. We will see this same punishment used during the tribulation period on a grander scale.

The point is, that when we read about future events described with symbolism in prophecies, we can look to the past for clues as to what they might actually mean. It's unlikely the Lord is going to begin using a "new" punishment or penalty system that He hasn't used before. This awareness allows eschatology students to view future events using eyes toward the past. Let's see what the Wisdom of Solomon says on this subject.

What has been will be again, what has been done will be done again; there is nothing new under the sun. Is there anything of which one can say,

"Look! This is something new"?
It was here already, long ago: it was here before our time. There is no remembrance of men of old, and even those who are yet to come will not be remembered by those who follow.
Ecclesiastes 1:9-11 (TNIV)

We learn from these passages there is nothing new and if we apply this wisdom of the ages to prophecies of the Bible, then we will realize that during the tribulation God will not use some new punishment that

He hasn't used before. Although it may be grander in scale, it may not be new. We need only to look back in history to find the correct prophetic interpretation of what the future holds.

Those end-time experts who come up with wild, supernatural, interpretations to justify literal translations of symbolic language lack the wisdom that Solomon provides. Do not look for these paranormal events lest the real signs will pass under your watchful eyes undetected.

Scripture Interprets Scripture

I have read many commentaries and books on "end-time" prophecies. Most of which I find very troubling because many of the experts who write these commentaries lead their unsuspecting readers astray. They make their interpretation sound so convincing and even use other Bible verses to support their claims that theirs is the true interpretation. But they are like many Bible teachers who explain scripture every week in Bible studies around the world. They jump around the Bible to make their point by quoting a verse from this book or that book, many times taking the "supporting" verses out of context to prove the point they are trying to make in the lesson they are teaching.

Now it may sound like I am chastising this technique, but I am not. This in actuality is the foremost method to use to clarify biblical principles. Letting scripture interpret scripture. The point I am trying to make is that it must be done carefully and not haphazardly. The verses used to support a position must not be taken out of context when making a point. This method is the first study technique that should be explored in finding understanding of end-time prophecies.

Furthermore, all who study the end-time prophecies of the book of Revelation can be further separated into two additional categories. Some believe what is written is literal, while the others believe the language is mostly symbolic in nature. Now the experts in both groups do not hold tightly to one philosophy, but jump back and forth between these two viewpoints depending on their understanding of what they are reading. Again, I can find no fault with this approach because both styles of writing can be found in the book. The difficulty is in determining when to switch. When you read many of these commentaries you quickly realize that those who do a better job of distinguishing between when something is literal and when something is symbolic, write "better" commentaries. Better as in more believable and more accurate as to what will take place during the end-times. Those who do a poor job with switching, present the most off the wall, most unbelievable, and most likely false interpretations of all. I will provide an example from Revelation to make the point.

> *These have power to shut heaven, that it rain not in the days of their prophecy: and have power over waters to turn them to blood, and to smite the earth with all plagues, as often as they will.*
> Revelation 11:6 (KJV)

Some eschatology writers in the "literal camp" have argued that during the tribulation period God will send two prophets who have the ability to stop rain and turn water into real blood. I have found a few individuals on the Internet that go even further and claim that during this time (one thousand two hundred and sixty P-days... three and a half years) no rain will fall on the earth (anywhere on the earth) and that the oceans and fresh water will be turned to blood so that there will be no drinking water. Their position is that God has the power to do anything and He has done similar things in the past (Moses in Egypt).

Now let's be realistic. I too believe God has the power to do whatever He chooses, but I do not believe in this kind of blind ignorance. They will be waiting a long time for these signs to be fulfilled. Their faith prevents them from seeing the real truth of Christ and God's Plan. I know not why, but I sometimes wish I had that kind of faith (when properly applied). This is the kind of faith that will be needed during the tribulation period; when properly used with guidance from the Holy Spirit.

Now if we understand this to be symbolic language we can come up with equally interesting interpretations. Some experts in this group have developed a wide array of answers for what this verse might mean. Even though this group has more leeway with their interpretations, they actually come up with more reasonable answers as to what Christ's revelation to St. John really means for us. These theologians are more thorough and study the word in depth to make sure nothing is missed. Nevertheless, they are not perfect either. The first group, however, has to do no studying or research. Their interpretations just require them to believe.

My interpretation to this verse goes something like this. This three and a half year period spoken of in the complete prophecy this text was taken from, is a time when the Christians will be persecuted by the governments of the world controlled and influenced by the antichrist. Would God stop all the rain and turn all the water to blood, which would punish indiscriminately all mankind (including Christians)? I don't think so. Does God use nature (natural disasters) to get people's attention to turn them back towards Him? Yes.[b] So the rain will not stop everywhere on the earth for three and a half years, but you can expect severe droughts in many parts of the world as evidence of God's power and our living in the end-times. A punishment God has used before.

Listen: if God stopped the rain for three and a half years and turned all the waters of the earth to blood, would not the entire population of the whole world witness a miracle that would bring almost all to Christ? I believe it would. But God's plan for your salvation is not based on "might" or "unbelievable signs," but on faith. Jesus said no signs (magical, mystical, unbelievable, etc.) would be given to prove he is the messiah except one... his resurrection.[c] These kinds of signs should not be expected from God during the end-times either. Any unexplainable signs that are witnessed will be in accordance with how Satan, the antichrist, and the false prophet work.[d] So if we are to believe a literal translation for this text then we will need to ignore other parts of scripture, prophecies, God's method's, and His plan. Those things indicate a "literal" translation for this prophetic verse is not appropriate.

Painting The Same Picture

We already know some scholars teach the "tribulation period" is three and a half years long while others claim it is seven years in length.

However, the prophecies that speak about the tribulation timing and end-time events need to "paint the same picture." For instance, any commentaries on Revelation that render prophecies in Daniel or other biblical prophecies false are incorrect interpretations. The converse is also true. We will use this rule, along with the others we discussed, to find the true meanings for the end-time prophecies, which unlock the tribulation's secrets.

Many scholars who provide commentaries on Revelation must be unaware of what is written in the other biblical books because they break this rule all the time when they come up with their "private" interpretations. I have read many of their commentaries on Revelation and Daniel and have found all deficient which has prompted me to write my own.

I am confident that those who have written their own commentaries wrote them for similar reasons or were inspired by the Holy Spirit to write them. The problem in this day and age is there are so many end-time commentaries available. Anyone trying to uncover the truth finds himself or herself searching for a needle in a haystack: exercises that require patience, perseverance, knowledge, and a keen eye to reach either objective (finding a needle in a haystack or finding the truth of the Lord). Unfortunately, many Christians give up long before reaching the goal, just as they do in their personal lives serving the Lord.

Motivations and Baggage

Let us briefly revisit how writers and believers who support a rapture of the saints before the tribulation period act. Many of these Christians search for the signs of this impending doom, but to what end? If they truly believed in a pre-tribulation rapture, they will never witness many of these signs, experience the wrath of God, or even suffer under the antichrist's rule.

What is their real purpose for finding this information? Selling books? Self gratification? Their pre-tribulation rapture ideals only require them to believe Christ is their savior and to be ready at all times because the rapture could happen at any moment. Are they studying these things so that non-believers who remain on earth after the rapture can learn the truth from their work, even though the people who remain in all likelihood will not be actively searching for answers? Do they search for this knowledge so they can use this information to convince their unbelieving friends and relatives to repent and come to the Lord for true forgiveness? A few might, but I find most do not. Since they cannot tell their loved ones when this period is with any certainty, most do not find the urgency to spread the word when they are unconcerned about any tribulation judgments befalling them personally.

I find those who believe in a pre-tribulation rapture and want to know what the future holds, have a thirst within them that compels them to search for this knowledge. They just do not have the compulsion to act on this information as the Holy Spirit expects—to go out and testify to the world that Jesus is the only way to eternal life. When they are shown the truth of these cataclysms, most just thank the Lord they will not experience His wrath and rarely tell anyone else. Are you one of these Christians?

Strongly believing in wrong doctrines clouds ones judgment in uncovering the correct interpretation of God's prophecies. Catholics, Lutherans, and other mainstream Christian denominations teach Christ can come at anytime and so there will be no rapture. They believe all the signs have been fulfilled in the past and so it can be reasoned there can be no end-time signs (clues) to predict Christ's return. This kind of logic, doctrine, leads them to not teach on the prophecies of the Bible. They find no urgency to proclaim the coming of Christ, to actively watch for his return, or to recognize that we are the last generation spoken of by Christ in Matthew 24. These beliefs lead them to incorrect interpretations of the prophecies and the idea of not needing to study them at all.

Baptists and other evangelical Christian denominations however believe in a pre-tribulation rapture and so they wear the pre-tribulation rapture "glasses" when they behold the end-time paintings painted with the brushes of Bible prophecy. These "glasses" also cloud their ability to see the true picture of these future events. The point is, all the prophecies of the end-time events need to support each other to be the correct interpretation, and the baggage of one's Christian belief system gets in the way of coming to the truth.

Realizing the motivations and doctrines of others influence their interpretations of prophecies, we should guard against this bias when studying eschatology matters. This warning applies to my analyses and interpretations you have read and will read as well. Also, our own Christian upbringing and motivations can affect our ability to see the truth. Remember, those who are not Christian will never know the truth and conversely those who study and follow God's Word will come to the truth. Lukewarm Christians who practice only on Sunday and rarely or never study the Bible, fall in between these extremes and can expect varying degrees of success in finding the truth.

When trying to understand prophecies we need to use all the tools God has provided us: His Word, logical deductive thinking, past history, understanding God's methods and plan, prayer, and listening to the Holy Spirit. We need to guard against the possibility that doctrinal baggage (both ours as well as others), poor Bible translations and commentaries, and the motivations of those who write and teach on these subjects will impair our ability to understand the real truth of what God has planned for the time before Christ's return.

Chapter 9

Understanding Revelation

ALL WHO READ OR STUDY THE last book of the Bible struggle with The prophecies contained within. No one, including myself, completely understands all that is written in this book. What we do know is those who never read or study it will never understand.

> *Blessed is he who reads aloud the words of the prophecy, and blessed are those who hear, and who keep what is written therein; for the time is near.*
>
> Revelation 1:3 (RSV)

We find at the beginning of Revelation Christ gives a blessing to anyone who reads the book aloud. Not a blessing for those who read it silently to themselves, but to those who have the courage to read and teach it to others. Jesus also provides an extra blessing upon all those who hear the message and a warning that the time is near so take it to heart (actually believe it and live like you believe it).

Why does he bestow these blessings? Because Christ knows how hard it is to comprehend the prophecies provided within and knows most of those who study these revelations are searching for the wisdom of his return. Hoping for a better tomorrow in Christ Jesus and not the things of this world. Keeping watch as he has commanded and by understanding what is within the Bible we might escape the coming wrath (worldwide devastations) through faith in Jesus Christ.

Many who have read Revelation, give up trying to figure out its meaning instead of continually studying until the Holy Spirit reveals the knowledge they seek. Of the remaining Christians who study this book faithfully, we are all at different places along the road to complete understanding. Some calling back to others to aid their travel down the road, some stuck on the side of the road believing they have reached the end of their journey, and still others moving slowly down the road at their own pace without any aid but the Holy Spirit's.

Those who examine this book for the secrets of what the future holds, but who are not a Christian, will never be able to find the road. Many of them believe they have found the mysteries at the end of the road, just like the Christians who are stuck on the side of the Revelation road, but these people only fool themselves because they really are on the wrong path and don't realize it. The prophecies of the Lord are never unlocked for those people who are not His.

In the book of Daniel, there are seven prophecies and Revelation has several prophecies as well. Let me quickly highlight those that will aid our work and your individual studies.

Prophecy One (The Seven Churches)

To the angel of the church in Ephesus write: The words of him who holds the seven stars in his right hand, who walks among the seven golden lampstands. ...

... He who has an ear, let him hear what the Spirit says to the churches.

Revelation 2:1 - 3:22 (RSV)

Prophecy Two (The Seven Seals)

Now I saw when the Lamb opened one of the seven seals, and I heard one of the four living creatures say, as with a voice like thunder, "Come!" ...

... Then the angel took the censer and filled it with fire from the altar and threw it on the earth; and there were peals of thunder, voices, flashes of lightning, and an earthquake.

Revelation 6:1 - 8:5 (RSV)

Prophecy Three (The Seven Trumpets)

Now the seven angels who had the seven trumpets made ready to blow them. ...

... but that in the days of the trumpet call to be sounded by the seventh angel, the mystery of God, as he announced to his servants the prophets, should be fulfilled.

Revelation 8:5 - 10:7 (RSV)

Prophecy Four (The Two Witnesses)

Then the voice which I had heard from heaven spoke to me again, saying, "Go, take the scroll which is open in the hand of the angel who is standing on the sea and on the land." ...

... Then God's temple in heaven was opened, and the ark of his covenant was seen within his temple; and there were flashes of lightning, voices, peals of thunder, an earthquake, and heavy hail.

Revelation 10:8 - 11:19 (RSV)

Prophecy Five (The Woman and the Dragon)

And a great portend [sign] appeared in heaven, a woman clothed with the sun, with the moon under her feet, and on her head a crown of twelve stars. ...

... and the wine press was trodden outside the city, and blood flowed from the wine press, as high as a horse's bridle, for one thousand six hundred stadia [about 180 miles].

Revelation 12:1 - 14:20 (RSV)

Prophecy Six (The Seven Bowls of God's Wrath)

Then I saw another portent [sign] in heaven, great and wonderful, seven angels with seven plagues, which are the last, for with them the wrath of God is ended. ...

... and great hailstones, heavy as a hundred-weight [a hundred pounds], dropped on men from heaven, till men cursed God for the plague of the hail, so fearful was that plague.

Revelation 15:1 - 16:21 (RSV)

Prophecy Seven (Babylon & the Prostitute on the Beast)

Then one of the seven angels who had the seven bowls came and said to me, "Come, I will show you the judgment of the great harlot [prostitute], who is seated upon many waters. ...

... And the rest were slain by the sword of him who sits upon the horse, the sword that issues from his mouth; and all the birds were gorged with their flesh.

Revelation 17:1 - 19:21 (RSV)

These are the prophecies that speak about the end of this age. In Revelation another prophecy is given in chapter twenty about the age to come—the next age. This time deals with the Millennial Kingdom. Lastly, a final prophecy is given in Revelation chapters twenty-one and twenty-two about what will happen to mankind, after the Millennial Kingdom, when heaven and earth will be remade and all sin will be removed.

From the seven prophecies cited (along with other end-time secrets foretold through the prophets), we can learn many things about the last days. In this chapter I will address these prophecies that deal with the tribulation period. Focusing on when the beginning and the ending times for these prophecies will occur.

In addition, key points of these prophecies will be highlighted and in a few cases, a comprehensive analysis is provided to start our detailed investigation of the cataclysmic events foretold for the end of this age. In latter chapters, we will thoroughly examine many of the particulars that were skimmed over in these quick overviews.

The Seven Churches

After studying the letters to the seven churches, I get the sense that Christians who have written this is a prophecy of past and future church history, had spiritual insight. There appears to be strong evidence that these letters to early Christians in first-century Asia Minor portray the church age from Christ's death until his return. With this understanding, I would have to agree that the last letter to the church in Laodicea is an accurate portrayal of the condition of the overall church of today. It is neither cold nor hot, but "lukewarm" for Christ. This transition into the last phase of the Christian church began in the late 1950's, when Christian values began to decline and materialistic ideas began to take hold worldwide, and covers the last seventy years prior to Jesus' second coming. This is the Christian church of the last generation before Christ—the last church of this age.

These letters not only describe the past two thousand years of the church age during the Times of the Gentiles, but are also a snapshot of the current condition of the overall Christian church. Today's church exhibits the characteristics of the Laodicean church as a whole, but there are pockets of Christianity within the church that exhibit traits from the other six first-century churches.

Some churches are doing the right things and living for the Lord just as the churches in Philadelphia and Smyrna did in the first century, while there are other congregations who are lax in various aspects of their Christian lives and practice questionable doctrine like the churches in Ephesus, Pergamum, Thyatira, and Sardis.

Lastly, this prophecy of the churches not only speaks about the entire church age on the deepest level of understanding and the qualities of individual Christian churches of every era on another level, but speaks also on the most basic level to the individual Christian. As a Christian, if we search our hearts we can find that one or more of these letters applies to us personally. When we discover this truth, we will find Christ's words of wisdom for what is needed to synchronize our walk with him.

Should we choose not to follow these words of instruction, Jesus also provides words of warning. Take heed, for some professing Christians will not enter the Kingdom of God, but will reside with the unbelievers in the Lake of Fire.[a]

The Seven Seals

The prophecy of the "Seven Seals" covers the time from the antichrist's entrance to the end of the age. On our timelines the duration of this period is ten years (three thousand six hundred P-days). This prophecy is also the most famous of all the prophecies contained in the Apocalypse of St. John (Revelation) and is labeled "The Four Horsemen of the Apocalypse" by many theologians. However, the four horsemen are only a reference to the opening of the first four seals of God's scroll (plan) of end-time events. This prophecy covers much more than their ride of terror and worldwide destruction culminating with Christ's return at the climatic conclusion of the War of Armageddon. It provides details of other key events when studied along side other end-time prophecies.

During the early years of this prophecy there will be great distress and many people will envision they are living through the tribulation, but these early troubles will be unlike anything imaginable in the latter years when the real wrath of God is poured out.

> *I watched as he opened the sixth seal. There was a great earthquake. The sun turned black like sackcloth made of goat hair, the whole moon turned blood red, and the stars in the sky fell to earth, as figs drop from a fig tree when shaken by a strong wind. The sky receded like a scroll, rolling up, and every mountain and island was removed from its place. Then the kings of the earth, the princes, the generals, the rich, the mighty, and everyone else, both slave and free, hid in caves and among the rocks of the mountains. They called to the mountains and the rocks, "Fall on us and hide us from the face of him who sits on the throne and from the wrath of the Lamb! For the great day of their wrath has come, and who can withstand it?*
>
> Revelation 6:12-17 (TNIV)

This earthquake is the same quake in Revelation 11:13, which not only rocks Jerusalem, but other areas as well. This large tremor signals the end of the five-day window on the tribulation timeline. This is God's answer to the antichrist's blasphemies, and the arresting and killing of all who refuse the mark of the beast. Why is there an earthquake? Since this is the very instant when the antichrist proclaims he is the messiah, God has had enough. Remember, the antichrist will mimic Christ's plan—God's plan. When Christ returns there will even be a larger earthquake. If we think about this from the perspective of the "fake" plan, we notice that when the antichrist claims he is God (the abomination of desolation) there will be an earthquake and when Christ comes and proves he is God, there will also be an earthquake.

In addition, this event provides a signal for weak Christians. For obedient Christians this is no sign at all because they will know long before what is going on, but for the "lukewarm," fence sitting, materialistic weak Christian this will be their last warning—their last call. If you are one of these, one only has to watch for earthquakes in Jerusalem to determine if we are in the tribulation, but by then it will be too late to avoid God's wrath if you have accepted the mark of the beast to participate in the world economy. This earthquake signals the official point of God's wrath. "For the great day of *their* wrath has come, and who can withstand it?" Notice that the word "their" is used instead of "his" in this sentence. This is really God's planned wrath foretold in the Old Testament scriptures, but it is Christ who will administer and deliver that wrath.[b] It is *their* wrath indeed!

The prophesied events in Revelation 7 that follow these verses will occur during the thirty P-day window on our timeline. We know from our timeline that at the end of the thirty P-days God's sanctuary will be reconsecrated. This spiritual reconsecration is really the sealing of the 144,000 (first fruits of the harvest) and the rapture of faithful Christians. This is God's plan for the commencement of His wrath and the point when the mystery of God will be revealed to all. Since the Jews prior to this had welcomed the antichrist as the foretold messiah, most will not be taken in the rapture. The purpose of the 144,000's "sealing" is to protect them from God's planned desolations and the antichrist's anger while they spread the true Word of Christ to those who were not ready and remained on earth.

If all faithful Christians are removed from the earth, then there must be some way, some method, left to preach the Word during the last half of the tribulation to find every last convert and servant. Christ said that heaven and earth would pass away, but never his words. Someone must teach the truth of the Bible even during these times or else the Lord's comments would be untrue.

Unfortunately, the last converts to Christianity will have to suffer right along with those destined for a lifetime in the Lake of Fire. This is why God said in Daniel 12, "blessed are those who wait for and reach the end of the 1,335 P-days." He knew that for a Christian who came to the truth during the tribulation, if they reached the start of the Millennial Kingdom alive they would be deemed worthy enough to enter it. They would have demonstrated their faith in Christ with actions and deeds beyond all odds. These Christians, many Jews, will be truly special to have survived both the antichrist's anger and God's planned desolations.

After preparing the 144,000 for their last days' ministry, the church will be raptured. This is the "great multitude in white robes" that no one could count their number.[c] We know God can count to at least two hundred million because He does so in Revelation 9:16 and so this number must be significantly greater.

Those who cry Christians are not "good enough" and therefore not many will make it to heaven or that not many will be raptured are not very good with math or understanding that we are saved by grace, through faith, and not because of what we do or don't do.

Lastly, there is a phrase that comes up repeatedly in Revelation and does so in this prophecy.

> *Another angel, who had a golden censer, came and stood at the altar. He was given much incense to offer, with the prayers of all the saints, on the golden altar before the throne. The smoke of the incense, together with the prayers of the saints people, went up before God from the angel's hand. Then the angel took the censer, filled it with fire from the altar, and hurled it on the earth;* **and there came peals of thunder, rumblings, flashes of lightning and an earthquake.**
>
> *Revelation 8:3-5* (NIV)

This expression is highlighted and is really a description of God's wrath, the starting point of His planned desolations, and the return of Christ; a condensed description of the last forty-two months of the tribulation period. So anytime you come across similar phrases in Revelation you should realize it as a signal for the end of that particular prophecy and the start of a new prophecy.

The Seven Trumpets

The prophecy of the "Seven Trumpets" coincides with the time-frame of God's wrath. This is the period between when the antichrist sets up the abomination that causes desolation and Christ's return.

The Trumpet judgments are the product of God's patience finally ending. They are the result of Satan's revolt manifested through the antichrist's and false prophet's actions and mankind's rebellion to the truth of Christ. They describe natural and manmade disasters on a global scale that climax with the War of Armageddon. Let's zoom in and look at the events of this prophecy in depth so that we can begin understanding God's judgments, methods, and the future devastations that await us.

> Now the seven angels who had the seven trumpets made ready to blow them. The first angel blew his trumpet, and there followed hail and fire, mixed with blood, which fell on the earth; and *a third* of the earth was burnt up, *a third* of the trees were burnt up, *and all* green grass was burnt up.
>
> Revelation 8:6-7 (RSV)

The verses in this prophecy are clearly symbolic in nature. The first trumpet appears to reference meteorite fragments or bombs dropping from the sky. This disaster appears to affect the global environment burning up a third of all living vegetation. This could also be a reference to "selective" nuclear strikes whose radiation is so widespread it affects a third of the earth's surface. One should understand that as a large object from space enters near-earth orbit it begins to break apart. Smaller pieces (fragments) begin to splinter off and litter the sky before the main piece of the object finally enters and impacts the surface. For instance, the comet Shoemaker-Levy 9 collided with Jupiter in 1994 after passing too close to the planet and getting caught in its gravitational pull. Over a year's time that comet was pulled apart into several smaller pieces and all eventually impacted the planet over a five to six day window of devastation from July 16 through July 22 (Av 8 through Av 14 on the Jewish calendar).

We have all witnessed the space shuttle Columbia's reentry and destruction on February 1, 2003 as a result of damage to its heat resistant tiles. It exploded and disintegrated into thousands of burning fragments spread out over three states. One can easily imagine from these two examples how a comet, meteorite, or asteroid hitting the earth might play out—first burning fragments (hail mixed with fire) showering the earth followed by the main mass of the object striking the planet.

*The second angel blew his trumpet, and something like a great mountain, burning with fire, was thrown into the sea; and **a third** of the sea became blood, **a third** of the living creatures in the sea died, and **a third** of the ships were destroyed.*

Revelation 8:8-9 (RSV)

The term "sea" is interpreted in prophecy to mean all the nations and peoples of the earth[d] and it might apply here with that understanding, however the ships present a problem using this interpretation. Therefore, I will surmise a more literal explanation is in order and that it really does mean sea. So a "huge mountain all ablaze" (asteroid, meteor, comet) hits an ocean (one of the three major ones: Pacific, Atlantic, or Indian, but not the Artic as asteroids would not typically come in at that trajectory) and this calamity destroys every ship in that specific ocean. This explanation would cover approximately a third of all the water on the planet.

This unparalleled disaster in man's history would wipe out ships, kill ocean creatures, and destroy coastline cities with the resulting tidal waves. It could also raise the sea level around the world flooding the remaining coastlines of all countries that border oceans. If it hits in the Atlantic Ocean, New York City would be destroyed within an hour[e] and the financial/economic structure of the United States would collapse. The rest of the world's economies would suffer because of their financial links to this system. The coastlines of India, Pakistan, and Bangladesh may suffer severe flooding killing millions because of their low sea level. These are just the tip of the iceberg of the many problems that will be experienced worldwide.

*The third angel sounded his trumpet, and a great star, blazing like a torch, fell from the sky on **a third** of the rivers and on the springs of water—the name of the star is Wormwood. **A third** of the waters turned bitter, and many people died from the waters that had become bitter.*

Revelation 8:10-11 (TNIV)

This could be another description of an asteroid hit that poisons fresh water sources and many end-time scholars believe it is, but I don't. Just as the "sea" represents nations, ethnic peoples, etc. in prophecy, the term "star" usually is reserved for "angel." So using this interpretation we find a powerful angel whose symbolic name is Wormwood has fallen from heaven. Daniel says in 11:22 that a "prince of the covenant" would be destroyed during the tribulation time (although earlier on the timeline than this event). When the book of Daniel speaks about "princes" in prophecy it is more often than not referring to powerful, high-ranking angels.[f]

So if this is a reference to a fallen angel and not another large asteroid fragment hitting the earth, how does Satan destroy a powerful angel since they are immortal? By misleading and convincing the angel to join him— just as he deceives us! Basically this angel has switched sides. A third of the angels were cast out of heaven when Satan fell and maybe there are a few stragglers.[g]

Since the word "sea" in prophecy represents all the nations of the earth, then waters, springs, rivers, etc. represent smaller pockets of people… single nations. Subsequently a third of the nations on earth became bitter (their attitudes) about the things taking place under Christ's administering of God's judgments and the antichrist's perceived earthly leadership.

Are these the same nations that were affected by the asteroid hit? Almost certainly. Look, if New York City were destroyed completely along with many of the large coastal cities of the United States and other countries, wouldn't the remaining people in these countries be "bitter." Although many people inland will survive, their lives will be in ruins. Family members dead, the financial system collapsed, and many other unsolvable problems will now control every second of their lives—total anarchy.

These just punishments from God will cause civil unrest in those nations trying to cope with the basic necessities of life. Many people will be killed by governments, other neighbors, friends, and even family members who are just trying to manage their personal situations by doing what they think they must to stay alive. During this time it really will become "survival of the fittest." Unbelievers will finally be right about Darwin's ideology, but only for a time, times, and maybe half a time.

> *The fourth angel blew his trumpet, and **a third** of the sun was struck, and **a third** of the moon, and **a third** of the stars, so that **a third** of their light was darkened; **a third** of the day was kept from shining, and likewise **a third** of the night.*
>
> Revelation 8:12 (RSV)

It seems when God rolls the dice of wrath, they always come up threes... "a third." Have you noticed? Why? Remember God's number is 333.333 in its completed form and .333 (a third) in its most basic form! Is this another coincidence that the time period of God's wrath, which is the last 3.333 years plus another one hundred P-days of the tribulation (three times 33.333 P-days), matches perfectly with the methodology of God's judgments? You can make up your own mind. I believe this to be overwhelming proof our numerical analyses of the tribulation's timing were dead on and that the sounding of the trumpets coincides with the last third of the timeline.

So what is going on during the blowing of the fourth trumpet? This is one of the most perplexing of the trumpet desolations. I can think of no possible way in astronomy for this to take place exactly as described. If a huge "physical mass" strikes the planet, then depending on its trajectory, this could speed up the earth's rotation or even slow it down. To reduce the daylight and the moonlight by a third would require the earth to speed up by a third so that a day would be shortened to sixteen hours in length instead of twenty-four hours.

This seems unimaginable. This object would have to be massive! This would not only affect time as we measure it, but also cause all sorts of problems like enormous stresses on the earth's core. This could possibly shorten the timing of Christ's return by a third from when this event has been calculated to occur, but mankind would never survive such a huge cataclysm. So, what else might this mean since it is not this? Well we know a third of the stars (angels) were cast down from heaven with Satan during the war in heaven.[h] This symbolically could be talking about that struggle, but it is at the wrong place on the tribulation timeline.

Another possibility is it might be a reference to particles thrown in the air from natural disasters, armed conflicts, and man made pollutants that screen out a third of the sunlight, moonlight, and starlight from reaching the earth. It was reported that when the island volcano Krakatoa erupted and was destroyed in 1883 A.D., it changed the earth's environment for five years.[i] The volcanic ash spewed into the atmosphere transformed the sun and moon's light into various colors and the resulting tidal waves destroyed hundreds of coastal villages in that part of the world. This judgment may be similar, but much larger in scope.

However, the most logical answer is nothing so grandiose, but something more practical. You will find out later, as things come together, that the "Bowl" judgments are nothing more than a reiteration of the Trumpet judgments using different prophetic language so that Christians may glean a fuller picture of the last days. The fourth trumpet is the same as the fifth Bowl! Let's see what this scripture says,

The fifth angel poured out his bowl on the throne of the beast, and its kingdom was in darkness; ...

Revelation 16:10 (RSV)

A careful examination of the language from this passage reveals there is very little symbolism being employed when compared to the Trumpet prophecy. If we understand that God's anger will be more focused at those who support the beast (the antichrist and his kingdom), all we have to do is visualize the beast will control roughly a third of the world. Since his kingdom will be the European Union and its supporters, all you need to understand is that the electrical power grid will shut down. This event will satisfy all the symbolic language employed at the blowing of the fourth trumpet.

There will be no lights (electricity) during the day (sun), no lights at night (moon and the stars), for a third of the population of the earth (the antichrist's kingdom). Furthermore, if you comprehend that the preceding Bowl judgment (the fourth one) calls for very hot days for a long period of time, then you can without difficulty foresee air conditioning loads overwhelming the electrical grid—trends that are already in place. Widespread power outages have occurred in the United States in the past. On August 14, 2003 an extensive power disruption occurred that affected the Northeast and Midwest regions of the United States along with the Canadian province of Ontario. Ten million people in Canada and forty million people in the U.S. were affected.

Other major power outages affected Europe that same year in the United Kingdom (August 28th) and Italy (September 28th). It is easy to anticipate how a prolonged heat wave will be able to overtax the power-generating infrastructure beyond its design limits and bring it to its knees.

The Three Woes of World War III

Then I looked, and I heard an eagle crying with a loud voice, as it flew in midheaven, "Woe, woe, woe to those who dwell on the earth, at the blasts of the other trumpets which the three angels are about to blow!"

Revelation 8:13 (RSV)

This passage is fascinating because an eagle is shouting the warning of the next punishments that are forthcoming from heaven. Punishments that appear to be even more severe than those already described that have affected a third of just about everything imaginable. In all the other prophecies an angel announces the warnings, but here it is an eagle. Why? The "eagle" symbol has been used before in other prophecies,[i] but I have been unable to determine its meaning or significance in any prophecy with reasonable satisfaction so no interpretation is provided.

Furthermore, these three woes (the last three trumpets) all pertain to the War of Armageddon—World War III. And since they are more severe than the previous punishments we can deduce that if a large space rock hits the earth, it will cause less damage than a worldwide nuclear war. Just more evidence to consider as we unroll more of the scroll of God's plan for the end-times and continue to study its secrets.

And the fifth angel blew his trumpet, and I saw a star fallen from heaven to earth, and he was given the key of the shaft of the bottomless pit [Abyss];

Revelation 9:1 (RSV)

Again we see the use of the word "star" and this time it clearly signifies it is an angel by the context of its use. However, unlike the angel Wormwood, this powerful angel (demon) "had" fallen in bygone ages. This demon appears to be doing Satan's work by opening the Abyss, but in reality he is doing God's bidding because he was "given" the key to the Abyss—the key to hell. Why? Because it is being readied for an influx of men and women who have been deceived from the work of Satan's minions.

Maybe this disgraced angel and Satan believe they have outwitted God by obtaining the key to the Abyss, but he was allowed to take it so that he could indirectly carry out God's plans.

When he opened the Abyss, smoke rose from it like the smoke from a gigantic furnace. The sun and sky were darkened by the smoke from the Abyss.

Revelation 9:2 (TNIV)

We learn the sun and the sky were darkened by the "smoke" (ash rising into the air from natural disaster(s) or from the winds of war) rising from the abyss. This text provides us with a clue that this may be what was blocking out a third of the sun, moon, and stars in our previous interpretations, but I believe this is referencing something else entirely. Not a rehashing of previous punishments, but a new one and hence a new trumpet to signify the new event.

And out of the smoke locusts came down on the earth and were given power like that of scorpions of the earth. They were told not to harm the grass of the earth or any plant or tree, but only those people who did not have the seal of God on their foreheads. They were not given power to kill them, but only to torture them for five months. And the agony they suffered was like that of the sting of a scorpion when it strikes a man. During those days men will seek death, but will not find it; they will long to die, but death will elude them.

Revelation 9:3-6 (NIV)

These verses really are the key to the whole prophecy of the Trumpets. The "locusts" appear to be a plague or symbolically germ warfare like anthrax, but later on it says the three plagues were "fire, smoke, and sulfur" and not some infectious disease… but the "spreading plague" of warfare!

The best interpretation of this is a description of the beginning of World War III. The locusts appear to be a figurative description of many "military helicopters" attacking people, perhaps with chemicals, as they try to control battlefields, civil unrest, and riots.

The locusts looked like horses prepared for battle. On their heads they wore something like crowns of gold [they are not kings, but have the power of kings], and their faces resembled human faces. Their hair was like women's hair, and their teeth were like lions' teeth. They had breastplates like breastplates of iron [symbol for Rome in prophecy], and the sound of their wings was like the thundering of many horses and chariots rushing into battle. They had tails with stingers, like scorpions, and in their tails they had power to torment people for five months. They had as king over them the angel of the Abyss, whose name in Hebrew is Abaddon and in Greek is Apollyon [that is, Destroyer].

Revelation 9:7-11 (TNIV)

These military attack helicopters are in the beginning of a conventional war. This phase of the mankind's final war will last five months. The fallen angel of war "Abaddon" is influencing this "spirit" of war. Additionally, Revelation 9:7-11 is a rehashing (a clarification) of the previous verses of 9:3-6 providing more detail.

Some eschatology writers claim the angel Abaddon is none other than the antichrist because he rises from the abyss and the "beast" also rises from the abyss. However, at this point in the tribulation, the antichrist has been on the last stage of the Gentile Age for many years. Furthermore, we know it will be possible to recognize the antichrist by the number of man "666" and it is unfeasible to transform the name Abaddon or Apollyon into this infamous number. Lastly, we are informed this is a fallen angel from hell. Although the antichrist will be the evilest person alive, he is still a man and men and angels are separate beings.

*The first woe is past; two other woes are yet to come. The sixth angel sounded his trumpet, and I heard a voice coming from the four horns of the golden altar that is before God. It said to the sixth angel who had the trumpet, "Release the four angels who are bound at the great river Euphrates." And the four angels who had been kept ready for this very hour and day and month and year were released to kill **a third** of the world's people. The number of the mounted troops was two hundred million. I heard their number.*

Revelation 9:12-16 (TNIV)

This scripture is significant because it provides proof that God has a plan—a plan from the beginning. Moreover, the four angels have to be more demonic angels because they were chained for this task, for this very hour, day, month, and year—an "appointed" time on God's tribulation calendar. God does plan as we see more confirmation of this fact from these verses. Knowing there is a point on the tribulation timeline that is "locked and unchangeable" (planned to the very hour), should make all those Christians who believe Christ can come back at anytime stop and ponder. By locking down just one event during the tribulation period to a very specific date and hour of the day in future history, the end result is the whole ten-year timeline has been cemented in stone as well! There can be no floating of this time period or else the Word of God would be untrue.

Hence, if you are a Christian who believes (or was taught) Christ can return at "any time of the Father's choosing," realize the Father has already chosen centuries ago and it can't be altered. Our only choice is but to hurry up and wade into the battle; to start fighting—to start bringing our friends and family over to Christ's side. There can be no Switzerlands in this war. Neutrality is not an option, and the sooner Christians realize this, the better off they will be.

So these fallen angels, in all probability, envision they have "broken free" and are furthering Satan's goals, but in truth they are carrying out God's plans for punishing unbelievers. You see, how many Christians will be enlisted in the armies of the world and fighting wars at this time? None! By this time the church will have been raptured, and you can bet all those who fight during these last days will have the "mark of the beast" because they will not be allowed in the military without it! Some might think they are Christians, but Satan has deceived them because they haven't studied their Bibles. Are you one of these Christians? Do some of your Christian friends and family fit this category? Many of mine do and I find it troubling they are blinded to the realities and unwilling to study the Word of God so they might learn the truth before it is too late.

Studying later verses in Revelation we find out the job of these angels is to "dry-up" the Euphrates River so that the Kings of the East, with their two-hundred million man army, can cross easily into the Middle East battlefield. Without this provision, it would be impossible to deploy this many men from the orient into fighting positions for the final battle. This task will be simple to accomplish because of the dams already constructed on the upper tributaries of the Euphrates in Turkey and Iraq. Closing down the water flow gates completely, along with drought weather conditions from the heat waves discussed earlier, should dry-up enough of the upstream tributaries to allow crossing by the massive infantry forces from China and its allies. This is because the nation of China is the only country in the world that has the resources and manpower to have a military force this large.

Now if this interpretation is true, then we have additional information of interest hidden within the sounding of the sixth trumpet. I wrote in my first book, after a thorough analysis, that the antichrist would come from the country of Greece, but that there was a small possibility that he might come from Turkey. Now if Turkey is aiding the hordes from the East by shutting off all water from the dams on the headwaters of the Euphrates River, we can now be absolutely sure the antichrist will not come from this country because these armies will oppose the antichrist and his quest for world domination.

*The horses and riders [armies] I saw in my vision looked like this: Their breastplates were fiery red, dark blue, and yellow as sulfur. The heads of the horses resembled the heads of lions, and out of their mouths came fire, smoke and sulfur. A **third** of mankind was killed by the **three** plagues of fire [red], smoke [blue] and sulfur [yellow] that came out of their mouths. The power of the horses was in their mouths and in their tails; for their tails were like snakes* [rotating tank turrets], *having heads with which they inflict injury.*

Revelation 9:17-19 (NIV)

We discover from these three verses who the major combatants of this war will be. The horses (tanks) will have breastplates of fiery red, dark blue, and sulfur yellow. The "red" signifies the Chinese led two-hundred million man army. The "red" will also include Russia because we know the armies of the East and North[(k)] will oppose the antichrist and the European Union (dark blue). These understandings are based on the national colors of these nations' flags.

In August 2008, Russia invaded the country of Georgia and after a period of short intense fighting, a ceasefire was agreed to. Here is an excerpt from an article pertaining to this ceasefire and Russian troop withdraws.

"A handful of Russian military trucks stood ready to remove the remaining troops, and four European Union monitors stood by a pair of **blue** EU light-armored vehicles."

By MATT SIEGEL, Associated Press Writer, Wed Oct 8, 2008

Notice the color blue associated with the EU's vehicles? This is just further evidence that God's plan is moving forward and the European Union is the "blue" army of the last war.

Also, from Ezekiel 38, the War of Gog (Russian led invasion of Israel) will have already occurred and you can bet Russia will be looking for ways to get even. But who is the sulfur "yellow" army? Well the most logical interpretation is: the Muslim nations united under a "yellow" flag with a crescent moon and star.

A symbol that is used throughout the Islamic countries and is purely speculation based on current trends. Presently the Arab League sports a predominantly green flag so this is a stretch, but not a big one since these countries are the only players left that have a stake in a war with Israel on the plains of Megiddo (the War of Armageddon). The Arab nations have the only armies that can be interpreted as led by the King of the South spoken of by the prophet Daniel.

We now know that a third of remaining inhabitants of the earth will be killed by war. The only question is whether this third of the world's population is from the beginning of God's wrath or from the opening of the sixth seal? I would like to think it is a third of the world's citizens from the start of God's judgments—one third, in the end, will die from natural disasters, famine, etc. (acts of God) and in due course another third by the fires of war, civil unrest, and at Christ's return. This leaves a remnant third to ultimately be saved... one-third Christians (raptured and newly converted brought through the refining fires of the end-times). This is roughly the amount of people in the world who profess to be Christians at the present. It is also the exact percentage breakdown God informs Zechariah[1] will take place at the end of the age with the Jews.

> *The rest of mankind, who were not killed by these plagues, did not repent of the works of their hands nor give up worshiping demons and idols of gold and silver and bronze and stone and wood, which cannot either see or hear or walk; nor did they repent of their murders or their sorceries or their immorality or their thefts.*
>
> Revelation 9:20-21 (RSV)

The "rest of the people" are those who stayed home (women, children, old men, etc.) and didn't go to war. They still placed unwavering hope in the lies their leaders (antichrist, false prophet, other national leaders, etc.) were selling them—that things would get better and so they continued to blame God instead of looking in the mirror and repenting. Many will be addicted to their lifestyle before all these "problems" arise and wish only to go back to the previous way of life. They will do anything to make that a reality—apart from accepting Jesus.

Let's jump ahead to the last trumpet. To do this we need to skip forward to Revelation 11:14-19. Why? Because the verses in between cover the prophecy of the "Two Witnesses" which provides more details of earlier events on the tribulation timeline. This distinctive prophecy is shoehorned here between the sixth trumpet (second woe) and the seventh and final trumpet (third woe). It will be discussed briefly at the end of the Trumpets interpretation and again at great length in the next chapter. But first, the seventh trumpet…

The second woe has passed; the third woe is coming soon. The seventh angel sounded his trumpet, and there were loud voices in heaven, which said:

"The kingdom of the world has become the kingdom of our Lord and of his Messiah, and he will reign for ever and ever."

And the twenty-four elders, who were seated on their thrones before God, fell on their faces and worshiped God, saying:

"We give thanks to you, Lord God Almighty, the One who is and who was, because you have taken your great power and have begun to reign. The nations were angry, and your wrath has come. The time has come for judging the dead, and for rewarding your servants the prophets and your people who revere your name, both great and small—and for destroying those who destroy the earth."

Then God's temple in heaven was opened, and within his temple was seen the ark of his covenant. And there came flashes of lightning, rumblings, peals of thunder, an earthquake and a great hailstorm.

Revelation 11:14-19 (TNIV)

We find as the last trumpet sounds, that all those who are in heaven give praise to God and of His Messiah—Jesus. It is also the time for judging the dead, rewarding the faithful, and killing those who remain who are destroying the earth with their final acts of rebellion and will not repent. Additionally, we see at the final trumpet that the "kingdom of the world has become the kingdom of the Lord and of His Messiah"… Their kingdom. At the end of this prophecy we find the phrase *"flashes of lightning, rumblings, peals of thunder, etc."* that signals its conclusion and the return of Christ where he physically begins ruling the nations of the earth as opposed to ruling from heaven the previous three and a half years.

The Two Witnesses

The prophecy of the "Two Witnesses" begins at chapter eleven in Revelation, but before examining its revelations there are some "transition" verses worth investigating.

> *Then I saw another mighty angel coming down from heaven, wrapped in a cloud, with a rainbow over his head, and his face was like the sun, and his legs like pillars of fire. He had a little scroll open in his hand. And he set his right foot on the sea, and his left foot on the land, and called out with a loud voice, like a lion roaring; when he called out, the seven thunders sounded. And when the seven thunders had sounded, I was about to write, but I heard a voice from heaven saying, "Seal up what the seven thunders have said, and do not write it down."*
>
> Revelation 10:1-4 (RSV)

We can see from the verses above this is a description of Christ and not just any angel. Why? The adjectives and symbolism can only be reserved for Christ. God said in Psalms 110, "I will make your enemies a footstool" and we see the land and sea are indeed under his feet.

The words "sea and land" are used compared to "sea and earth." Usually when the word "earth" is used symbolically it doesn't mean land, but something else entirely. But their use here means that all nations, peoples, tribes, and the land of the entire earth are under Christ's authority.

Secondly, we'll see in the verses that follow that this "Angel" swears an oath using God's name. This is forbidden and no angel that follows the Lord would do this. This is reserved for God himself… Christ.[(m)] When he shouted, the voices of the seven thunders spoke. What angel commands the seven thunders? None, but the Angel of the Lord; Christ. Just further evidence this is our Lord and Savior Jesus Christ.

I don't know what the seven thunders are or what they revealed, but Christ commands them and John did hear what they said. But God the Father instructs St. John not to write down what was uttered either because it was not the right time in 95 A.D. for this mystery to be discovered, it had already been addressed in other prophecies, or it would be revealed to a modern day prophet during the last days when the Lord pours out His spirit.

We also notice this "Angel" is holding a little scroll which is opened—the scroll that Christ was given[n] from which the seals have already been removed.

Then the angel I had seen standing on the sea and on the land raised his right hand to heaven. And he swore by him who lives for ever and ever, who created the heavens and all that is in them, the earth and all that is in it, and the sea and all that is in it, and said, "There will be no more delay! But in the days when the seventh angel is about to sound his trumpet, the mystery of God will be accomplished, just as he announced to his servants the prophets."

Revelation 10:5-7 (NIV)

We see that with the blowing of the last trumpet at Christ's return, the mystery of God will be accomplished. God's plan for the end-times will be completed. There will be no more delay. Why was there a delay? The delay was to make sure that everyone had made their decision on whether to accept Christ as their personal savior or to reject him.

God searches the hearts of all people and the point when everyone has made their decision, Christ will return. This is so no one will be able to claim they weren't given enough time and information to decide, or be allowed to change their minds at the last minute when they witness Christ descending from heaven and realize they have made the wrong choice. All will be given ample opportunities to open the door of their heart and let Christ come in. Those who are remaining at Jesus' second advent and have received the mark of the beast will be "weeded out" and thrown into the Fires of Hell. But those who have come to know Christ as their personal savior after the rapture and have endured during the terrible tribulation period will enter the Millennial Kingdom.

The mystery of God will be accomplished at the last trumpet and we recognize something else that most overlook. That one of two things is true. Either the mystery was foretold ahead of time to the prophets and so it is really no mystery at all (if we can comprehend the prophets words) or simply, the prophets were told it would be revealed at the end of days—at the last trumpet call and nothing more. It is possible that both ideas could be true at the same time. That the mystery would be revealed to all at the end and was also foretold ahead of time, but misunderstood by most.

Then the voice that I had heard from heaven spoke to me once more: "Go, take the scroll that lies open in the hand of the angel who is standing on the sea and on the land." So I went to the angel and asked him to give me the little scroll. He said to me, "Take it and eat it. It will turn your stomach sour, but 'in your mouth it will be as sweet as honey'." I took the little scroll from the angel's hand and ate it. It tasted as sweet as honey in my mouth, but when I had eaten it, my stomach turned sour.

Revelation 10:8-10 (NIV)

This scroll at first tastes like honey… the good news, but afterwards turns sour in the stomach after digesting what is really contained in it… the bad news. This good news is Christ died for each of our sins and we will be saved for all eternity because of this. But the bad news is all the devastations that are to come upon the world to those people who rejected the free gift of salvation—those who rebel against the Word of God and want nothing to do with Christ. They love their lives and the things (idols) of this world so strongly they are blinded to the truth. They are destined to an eternity of torture away from the love of God. They are our friends and family!

Lastly, we find that at the opening of the seventh seal and the revealing of the mystery of God, that the scroll of prophecy will no longer be needed (because it was eaten).[o]

And I was told, "You must again prophesy about many peoples and nations and tongues and kings."

Revelation 10:11 (RSV)

We now stumble onto irrefutable evidence that Revelation is more than just one continuous prophecy (*you must again prophesy*) and that it is made up of multiple prophecies. This is a logical break point from the Trumpet prophecy, which was discussed at great length and not simply a continuation of it. As it turns out, the prophecy of the Two Witnesses is just a reiteration of previous prophecies providing more information about the tribulation timing and sequence of events.

This prophecy covers the time period from the beginning of Daniel's last 'week' when the covenant is confirmed and concludes at the end of the five days when the abomination that causes desolation is set up. This insightful prophecy provides us with times and information on how Christians are to react to living under the antichrist's rule. Things will be particularly bad for Christians, but the awareness of knowing Christ will return soon brings hope for a better tomorrow.

We know from our study of the tribulation timeline this prophecy will span twelve hundred and sixty P-days. Details of this key prophecy of the end-times will be examined in the next chapter.

The Woman and the Dragon

This prophecy also covers the time between the confirmation of the covenant and the setting up of the abomination of desolation. It gives us a look at what the antichrist will be doing during this time. By understanding the events that will take place, Christians will be better able to cope with the injustices that will overwhelm them.

Many of the details of this prophecy provide key signs to indicate just where we are on the tribulation timeline. These future events will be examined verse-by-verse in a later chapter so that you may learn about what is to take place and see further evidence of how prophecies are decoded just as we did when we looked at the prophecy of the Trumpets.

The Seven Bowls of God's Wrath

Then I heard a loud voice from the temple telling the seven angels, "Go and pour out on the earth the seven bowls of the wrath of God."

Revelation 16-1 (RSV)

It has begun. The antichrist's rule is officially over and Christ now reigns. The punishments begin for the antichrist, the false prophet, and all the people who followed them and persecuted Christians during the first three and a half years (2021 to 2025 A.D.) of the classical tribulation period. It's payback time!

This prophecy of the Seven Bowl judgments starts after the implementation of the mark of the beast and consequently is spread out over the last three and a half years prior to Christ's return. It runs concurrently with the Seven Trumpet pronouncements discussed in detail earlier. Although God is employing different imagery, these two prophecies together provide a sharper picture of the devastations following the abomination that causes desolation, the persecution of Christians, and the rapture of the church. Let's examine and compare these punishments to what we have learned so far.

So the first angel went and poured out his bowl on the earth, and foul and evil sores came upon the men who bore the mark of the beast and worshiped its image.

Revelation 16:2 (RSV)

Here we discover the evidence that the very first bowl judgment is poured out "after" the implementation of the mark of the beast. What about the "festering sores?" This might mean a couple of things. First, that some type of supernatural disease is unleashed that affects only those who have the "mark." Just like some of the plagues Moses unleashed on the Egyptians didn't affect the Israelites. However, I sense this verse still has some symbolism in it and cannot be interpreted literally from the elucidation I have provided. I feel another explanation is more plausible when you consider all the prophecies as a whole. Why was this "plague" only affecting those who had the mark? Because many who refused the mark had been martyred or raptured prior to this and so the vast majority of people who remain have already accepted the mark! The others, who will learn of the truth of Christ late and have not accepted the mark, will now be in hiding in the mountains and countryside away from centers of modern civilization.[p]

This may also be a description of radiation poisoning—fallout that may have drifted around the world. A large mushroom cloud from a nuclear explosion(s) could reach the upper atmosphere very easily and carry around the world. There may have even been multiple detonations since this event occurs very near the time for the destruction of "Babylon the Great (New York City)."

Now I don't really know the wind patterns over the Atlantic, but I do know they typically blow from west to east over the United States. If a bomb were detonated on New York City, the good news (if there really could be any) would be the radiation would most likely be carried out over the ocean toward Europe where the antichrist will be doing most of his work.

At the moment, the problem with this understanding is that it doesn't appear to match in any way with the interpretation of the first trumpet. The first trumpet seems to suggest falling space rocks (which should cause no physical ailments to humans other than death and destruction), while the first bowl indicates an event that triggers sickness in humans. Both are occurring near the same point on the tribulation timeline. For both to be describing the same event or different details of closely related events, a nuclear strike(s) explanation might be the better mutual interpretation.

The second angel poured out his bowl into the sea, and it became like the blood of a dead man, and every living thing died that was in the sea [that the meteor hit].

Revelation 16:3 (RSV)

This could be from the radiation as mentioned above. I'm confident the sea won't turn to real blood, but it may be possible that low levels of radiation are enough to kill sea animals (I don't know this for sure either) and pollute surface drinking waters. On the other hand, since it is a separate Bowl judgment we can infer it is a distinct event as opposed to a further clarification of a previous event.

Remember the second trumpet judgment? A large mountain all ablaze hits the sea and kills a third of the creatures and ships. Once you read these two verses side-by-side, many can easily grasp that they are talking about the same event. However, this problem is just the opposite of the one I mention for the first Trumpet and Bowl disaster.

When both the first Trumpet and Bowl judgments are taken together a nuclear bomb explanation seems most applicable, whereas the second Trumpet and Bowl sentences are in closer agreement with a comet, meteorite, or asteroid striking the earth.

The third angel poured out his bowl on the rivers and springs of water, and they became blood. Then I heard the angel in charge of the waters say: "You are just in these judgments, you who are and who were, the Holy One, because you have so judged; for they have shed the blood of your saints and prophets, and you have given them blood to drink as they deserve.

Revelation 16:4-6 (NIV)

This vision matches very closely with the third Trumpet. It talks about smaller bodies of water like rivers and springs just as the third Trumpet does. So we know this means the killing of people on smaller scales, not battlefields where nations collide, but internal civil unrest, riots, etc. because of the use of the words "rivers and springs" rather than the word "sea" which is a substitute for all the nations of the earth (mankind).

We see from this judgment that God's punishment is like a mirror... I will do unto others as they have done unto you. The unbelievers killed Christ's followers because they would not willingly receive the mark and participate in the new economy, which will be an "image" of the present system—but without physical money! A cashless economy with electronic "scorekeeping" of your possessions and wealth that are accessed with a personal ID that allows governmental (big brother) tracking of all your actions and whereabouts. But now God has turned the tables on them. They are killing each other because of the overwhelming problems piling-up from God's judgments.

I do not believe this verse really means they will be given actual blood to drink, but rather acid rain, tainted water, and water poisoned from the earlier desolations will add to their miseries. This imagery parallels the judgment given to Israel in 587 B.C. as prophesied by the prophet Jeremiah.

Therefore, this is what the LORD says: You have not proclaimed freedom for your fellow countrymen. So I now proclaim 'freedom' for you, declares the LORD— 'freedom' to fall by the sword, plague and famine.

Jeremiah 34:17 (TNIV)

This is not real 'freedom' that God proclaims, but death just as it will not be real 'blood' to drink, but death proclaimed again.

And I heard the altar respond:

"Yes, Lord God Almighty, true and just are your judgments."

The fourth angel poured out his bowl on the sun, and the sun was given power to scorch people with fire. They were seared by the intense heat and they cursed the name of God, who had control over these plagues, but they refused to repent and glorify him.

Revelation 16:7-9 (NIV)

This is interesting because almost all natural disasters (volcanic eruptions e.g.) and even nuclear war spew particulates into the atmosphere that block the sun's radiation and usually cause the temperature to drop. This appears to suggest at some point the skies will be extremely clear (maybe the ozone layer will open up) and the world is in for some hot days. It may also be a metaphor for selective nuclear warfare... seared by the intense heat that came from above (sun)... bombs and missiles.

But I believe it just signals many hot days during this time with very little rain—severe drought conditions in many parts of the world which along with major disruptions in the transportation of food and materials (remember a third of the ships on the oceans have been destroyed, a third of the fish in the sea are dead, and major rioting and anarchy prevail) will lead to widespread famine conditions worldwide.

The fifth angel poured out his bowl on the throne of the beast, and its kingdom was in darkness; ...

Revelation 16:10 (RSV)

Here we learn the European Union (the beast's kingdom) will be plunged into darkness while this verse implies the other nations of the world remain ok. This darkness is directed at the beast's kingdom and most likely is the result of large electrical power outages continent wide. These power outages could be the result of military strikes in retaliation to the things the antichrist has done to other nations as his rage grows from the knowledge that his time is short.[q] Realize that the first targets, during the beginning phase of war, are designed to disable the enemy's infrastructure, communications, and power distribution systems, which are at the very top of the list of things to disrupt.

> *... men gnawed their tongues in anguish and cursed the God of heaven for their pains and their sores, and did not repent of their deeds.*
>
> *Revelation 16:10-11* (RSV)

We see people are in misery and their sores are still present four Bowl judgments later. Instead of praying to God for help when things get bad as Christians do, unbelievers blame and curse God—they do just the opposite. These symptoms may be the accumulated results of thirst from the hot days, the scarcity of acceptable drinking water, lack of food and daily supplies, death everywhere, civil unrest, the lack of electricity that provides the conveniences of modern life, and many other unforeseen problems. They may also be from low level radiation poisoning circling the earth.

> *The sixth angel poured his bowl on the great river Euphrates, and its water was dried up, to prepare the way for the kings from the east.*
>
> *Revelation 16:12* (RSV)

We have seen this judgment before. The sixth Bowl is equivalent to the sixth Trumpet! That interpretation is further expounded on as the hot days of the tribulation drag on. These long drought weather conditions will begin to dry-up the shallowest upstream portions of the river. The symbolic language of both prophecies offer additional support to shutting down the flow of water from the Mosul and the Haditha Dams on the upper tributaries of the Tigris and Euphrates rivers so that numerous ground forces (two hundred million) can cross into Israel from the East.

Moving this many troops can only be done by land and funneling them across bridges would not be feasible. But shutting off the water sources to these rivers and allowing them to run dry in the narrow, shallow areas along the rivers would permit such a large invasion.

> *And I saw, issuing from the mouth of the dragon and from the mouth of the beast and from mouth of the false prophet, three foul spirits like frogs; for they are demonic spirits, performing signs, who go abroad to the kings of the whole world, to assemble them for battle on the great day of God the Almighty [the day Christ returns].*
>
> *Revelation 16:13-14* (RSV)

These three evil spirits will influence all kinds of things to rile up the nations to the point they will all advance on Israel for the last battle. Who knows exactly what the spirits controlled by Satan, the antichrist, and the false prophet will say or do—or what miraculous signs they will perform to provoke the mighty nations of the world to converge on the Middle East (either as allies of the beast or as enemies), but they will come.

Realize that many unbelievers will not support Satan and his plans anymore than they will believe in Christ's authority. True atheism, skepticism, and disbelief in a supreme creator accepts no gods and this includes Satan. The evidence for this is that the kings of the North and the East foretold by Daniel will oppose the antichrist during this period and these kings can only be explained as the nations of Russia and China. We know this because China has the only military that matches the additional information of the army from the east provided in Revelation. In addition, the northern army must be Russia because there are only two formidable military factions north of Israel. Since one is led by the antichrist, it can't be his own army that Daniel says will oppose him and therefore must be the other... Russia. The Islamic nations (South), European Union (the other northern army), and the Jewish forces will also be assembled for the last battle.

It is unclear where the United States interests will fall in this war… on the European Union's side or as Israel's protector. I believe there is evidence they will oppose the E.U. earlier than this and may continue to do so at this time, but we can't be sure. The U.S. is presently Europe's ally and I am sad to write there are no trends presently that would suggest a departure from this position. However, if New York City is bombed, it will be by the Europeans and if this can be proven, then this might be the event that changes the status quo. But, the U.S. will have limited capacity due to God's judgments, the rapture, and may be of no help to Israel requiring the Jews to turn back to God in the end for divine help.

> *"Look, I come like a thief! Blessed are those who stay awake and keep their clothes on, so that they may not go naked and be shamefully exposed."*
>
> Revelation 16:15 (TNIV)

Here we see Jesus' warning reissued from Matthew 24:17-19. Christ's return will be unexpected to most. Why? Because the Christians who were raptured took that knowledge with them (the prophecies of the Bible) and many of the remaining people will be ignorant—blinded so they would believe the lies.[r] But there will be many (the 144,000 e.g.) who will learn the truth and become believers. They will need to remain diligent until the end, or until they die, before they can receive everlasting salvation.

> *Then they [Satan, the antichrist, and the false prophet] gathered the kings together to the place that in Hebrew is called Armageddon. The seventh angel poured out his bowl into the air, and out of the temple came a loud voice from the throne, saying, "It is done!"* ***Then there came flashes of lightning, rumblings, peals of thunder and a severe earthquake.*** *…*
>
> Revelation 16:16-18 (NIV)

This is just more information on gathering the world's armies for World War III. Although these armies gather to fight, most are not expecting to fight Christ and his heavenly armies (except the antichrist who knows the truth). The expected battle will be between the antichrist (European Union) and his allies and the armies who oppose him (China, Russia, the Jews, and the Arab nations).

This is interesting because we see here that the Jews and the Arabs will actually have a common enemy after hundreds of years of conflict. The old axiom applies here—the enemy of my enemy is my friend. But this epic battle will not last long as Christ returns and those combatants, join together to fight against Christ and his host.

This text also informs us of the place where this final battle will take place. Furthermore we see our familiar closing sentence, "flashes of lightning, rumblings, peals of thunder and a severe earthquake." This is a phrase you will find in the book of Revelation seven times.[s] It is the expression used to encompass God and the Lamb's wrath, either in its entirety for some Revelation prophecies or as a "substitute" for the particulars that remain hidden when various details of God's wrath are identified (as they are in the Trumpet and Bowl prophecies). In this circumstance, it is substituting for the very last details of Christ returning with his obedient servants and the angelic forces of heaven.

> *... No earthquake like it has ever occurred since man has been on earth, so tremendous was the quake. The great city split into **three** parts, and the cities of the nations collapsed. God remembered Babylon the Great and gave her the cup filled with the wine of the fury of his wrath. Every island fled away and the mountains could not be found.*
>
> Revelation 16:18-20 (NIV)

This earthquake will rock the whole earth. This event occurs when Jesus plants his foot on the Mount of Olives and splits the mountain in half.[t] This may also be the result of another comet or meteor fragment hitting simultaneously with Christ's return. Regardless of the cause of the earthquake, this incident will cause Jerusalem to split into three parts and many cities in other nations will suffer one last devastation.

This is a final description of everything that happened during the last three and a half years of the tribulation period. Notice the past tense "remembered." The Lord remembered all of her abominations and after they had poured out Their wrath, nothing was left untouched during this earthquake. This is a statement of finality that things are coming to a conclusion.

From the sky huge hailstones, each weighing about a hundred pounds, fell on people. And they cursed God on account of the plague of hail, because the plague was so terrible.

Revelation 16:21 (TNIV)

We see at this time large hailstones fall from the sky. Maybe the water that had been building up in the skies from the long drought has let loose or perhaps these are fragments from one last meteorite hit that causes the massive earthquake. Regardless, it is the last of God's earth-wide judgments and many will still refuse to repent as they continue cursing God right up to the very end. When Christ returns the only judgments left to meter out will be personal in nature. This will be the "weeding" of the earth's fields of unbelievers when each receives their individual sentence of an eternity in fire.

We have now taken our first steps in laying down the foundation for the events of the last days. We have made cursory probes into all the prophecies of the Apocalypse of St. John and painstaking verse-by-verse analyses of a few key prophecies. The in-depth investigations of the Revelation prophecies quickly skimmed over, still await us in later chapters. They will provide more information and evidence of what to expect during the entire tribulation period, as well as allow you to continue growing in your abilities to interpret Bible prophecy.

Let's move along and study the prophecy of the "Two Witnesses of God that was wedged in between the "blowing" of the sixth and seventh trumpets of God. This prophecy covers some of the early years of the tribulation timeframe (prior to God's wrath), whereas the prophecies we have already investigated in detail have dealt with God and the Lamb's wrath during the last third of the ten-year understanding of the tribulation.

Chapter 10

God's Two Witnesses

1 HAVE READ MANY COMMENTARIES ON chapter eleven in the book of Revelation. Let's explore the problems with some of these interpretations by analyzing in detail the prophecy of the "Two Witnesses." We have briefly addressed the timing of this well documented prophecy that will span the three and a half years prior to the abomination that causes desolation.

Knowing whether the "two witnesses," or the "two prophets" as they are later labeled in many Bible translations, are literal or symbolic is crucial in understanding several of the events that will unfold during the last days. If these witnesses are literal, as some eschatology teachers believe, then the start of their preaching would be a visible sign that would allow a person to know where we are in God's plan! If they are purely symbolic, then they would provide no outward sign for which to gauge the progress of the classical seven-year tribulation period and other signs must be relied upon.

Let's begin our detailed analysis of this prophecy so that you are able to see what difficulties exist with various interpretations. Prophecy experts who interpret these verses rarely inform you of the troubles with their explanations, either because they are unaware of them or because they don't want you to know so you will believe their version of end-time events. Both of these reasons should concern you if you are trying to get to the absolute truth.

I was given a reed like a measuring rod and was told, "Go and measure the temple of God and the altar, with its worshipers. But exclude the outer court; do not measure it, because it has been given to the Gentiles. They will trample on the holy city for 42 months. And I will appoint my two witnesses, and they will prophesy for 1,260 [P-]days, clothed in sackcloth." They are "the

*two olive trees" and the two lampstands, and "they stand before
the Lord of the earth.*

Revelation 11:1-4 (TNIV)

Many interpreters believe the forty-two months prophesied are the
same twelve hundred and sixty P-days mentioned, but they are not. The
prophecies in Revelation need to support the Old Testament prophecies
of Daniel, Zechariah, Isaiah, Jeremiah, etc. and vice-versa. They cannot
contradict each other. Daniel indicates there is a seven prophetic year
period between when the antichrist confirms the covenant and the
return of Christ. Therefore, Revelation interpretations must also support
this timing. When you read the first verse carefully you realize the
apostle John is told to measure the "temple of God" and the worshipers
there. What answer did he get when he did this? Not forty-two months,
because he was told in the very next verse to *exclude* the outer court of
the Gentiles. The answer he received was the twelve hundred and sixty
P-days given in the next sentence.

One should realize the twelve hundred and sixty P-days is not the
timeframe for the "outer court," but the timeframe for the "worshipers
in the temple." Now it just so happens that twelve hundred and sixty P-
days are exactly forty-two months if you use "prophetic" accounting
methods. But just because these are equivalent numbers doesn't imply
they cover the same time period. We know they don't because John was
told to *exclude* the outer court—to leave out the forty-two months!
Therefore, we can deduce this is another equal period of time that needs
to be accounted for in the final interpretation.

Many experts use this verse to support that a temple will be rebuilt
in Jerusalem before Christ returns. This is just an example of taking a
verse out of context to support their position! Now I am not saying a
temple will not be rebuilt, because I believe it will, I am only saying this
verse speaks to something else entirely and therefore cannot be used as
proof of this claim. This verse is clearly symbolical and cannot be taken
literally. Why do I say this? Because when you measure a physical
structure with a measuring rod (stick), you get an answer that comes out
in units of length (feet, meter, cubits, yards, etc.) and not units of time!

Furthermore, God instructs John to take the measuring rod and measure the "number" of worshipers in the temple. You can't measure (count people) worshipers with a measuring rod! So we know the Lord is using symbolic language and anyone who tries to use literal interpretations on these few verses (like supporting the rebuilding of the temple) is unaware of what Christ really wants us to understand with these passages. So we need to continue searching for a deeper meaning.

When you understand the timing of Daniel's prophecies, one realizes that the twelve hundred and sixty P-days applies to the first half of Daniel's last 'week' (the classical understanding of the tribulation period instead of the ten-year version we have studied), while the forty-two months are designated for the second half. Both are equal amounts of time, but not the same period of time.

Next, what is the "temple of God" in this context? There is no physical temple presently and what is the purpose for "measuring it" and the altar's worshipers? Well, in this context God is waiting for the "right" amount (the complete number needed to fulfill His plan) to be reached before He unleashes His wrath. This number of righteous believers (Christians not Jews) will be reached, fulfilled, after 1,279 days.[38]

Not only is God telling us there are a predetermined number of people who will be saved, but there are also a predetermined number of martyrs (at the altar) as well.

> *When he opened the fifth seal, I saw under the altar the souls of those who had been slain because of the word of God [oil] and the testimony [lampstands] they had maintained. They called out in a loud voice, "How long, Sovereign Lord, holy and true, until you judge the inhabitants of the earth and avenge our blood?" Then each of them was given a white robe, and they were told to wait a little longer,* **until the number** *of their fellow servants and brothers who were to be killed as they had been was completed.*
>
> *Revelation 6:9-11* (NIV)

We see that the martyrs are the worshipers at God's altar and they must wait "until" their number is full... complete... a set number predetermined and known only in heaven. So John is ultimately measuring the number of martyrs and the "right" amount will be reached in 1,279 days from the confirming of the covenant mentioned in Daniel.

Also, in this context, the "temple of God" are those individuals who worship Christ. Jesus is the true "temple of God" and those who worship at this temple are Christians and new believers who will come to the faith during the last days before Christ returns. These believers are those who have the Holy Spirit (lampstand) and the Word of God (oil) written on their hearts and foreheads. The temple is not some physical building in this context or symbolically a reference to the Jews as some might argue, but the Christian church. Still, John does prophesy about the Jewish nation for some greater purpose of God's plan in other verses of Revelation we will explore next. Surprising to some, that specific prophecy will go on simultaneously and will be completed during the same 1,279 days.

What about the "gentiles" in the outer court? In this circumstance they represent all non-believers. They were excluded from entering the temple in the olden days and will not be found in the temple described here because their names are not written in the Lamb's book of life. Their time will be forty-two months long (actually 1,279 days too) and follow the time of the Two Witnesses when it is completed. We know this because very near the end of the first three and a half years of the antichrist's reign he will enter Jerusalem, set up the abomination that causes desolation, and the gentile armies (non-believers) from that point forward will gather in the Middle East (trample the holy city—Jerusalem) to ultimately fight the War of Armageddon.[a] The first three and a half years the Jews will be left alone, protected, while the antichrist wages war on the Christians. Here is what the Bible says on this point.

> *and the woman [Israel] fled into the wilderness, where she has a place prepared by God, in which to be nourished [protected] for one thousand two hundred and sixty [1,279] days.*
>
> *Revelation 12:6* (RSV)

> *He will confirm a covenant with many [Israel and her enemies] for one 'seven'. In the middle of the 'seven' [three and a half*

years—1,279 days] *he will put an end to sacrifice and offering* [break the covenant]. *And on the wing of the temple* [offshoot of Judaism—Christianity] *he will set up an abomination that causes desolation, until the end that is decreed* [planned] *is poured out on him.*

Daniel 9:27 (TNIV)

Let's further examine the "two witnesses" account that was provided earlier. During the first half of the classical tribulation, God will appoint them to prophesy for 1,279 "real" days during the same time the Jews are protected from the antichrist with the covenant they signed with him. These witnesses are further described using the symbolic language—"the two olive trees" and the "two lampstands" that stand before the Lord of the earth (God).

Here is the next area where prophecy scholars diverge. This text can be taken two different ways depending on which group they belong to: the literal group or the symbolic group. The literal group looks at the remaining text of chapter eleven and sees much of the language as literal. They argue that since the "two witnesses" later on in the chapter are referred to as the "two prophets," they will be real people who will be resurrected just as Christ was resurrected. Newer translations are lax with their use of these terms and carelessly substitute the words "prophets" or "men" in places where the original manuscripts do not use these words. By doing this, the translators interject their own beliefs and provide support to the idea that these are real people. Their biases can unintentionally affect the true meaning of scriptures and lead Christians astray. In this regard, the KJV is a better translation than some of the newer Bible translations.

Additionally, the symbolic language employed that describes them as "lampstands" and "olive trees" is nothing more than a description of their character. Just as Christ's character is described many ways (eyes like blazing fire, words like a double-edge sword, face like the sun, etc.) in the beginning chapters of Revelation (one through three), so too are the witnesses described here. These are persuasive arguments for their position.

Those who view this language as primarily symbolic, argue that when you look to the rest of scripture for understanding you find these symbols used in a prophecy given in Zechariah chapter four. This prophecy was to rebuild the physical temple after its destruction by the Babylonians in 587 B.C. Symbolic language that is translated for Zechariah into something other than real prophets who perform bona fide miracles! God informs Zechariah, "not by might nor power, but by my Spirit will my plans be accomplished." So using scripture to interpret scripture suggests these two witnesses are not actual prophets, but the text has a deeper spiritual significance. These experts also make good points.

So who is right? I don't think it's possible to answer this question just yet. Some of the prophecy experts who preach a literal interpretation go as far as to even suggest whom these two powerful witnesses (prophets) are. Some say they are Moses and Elijah because Jesus was seen talking with them before his death.

> *After six days Jesus took with him Peter, James and John the brother of James, and led them up a high mountain by themselves. There he was transfigured before them. His face shone like the sun, and his clothes became as white as the light. Just then there appeared before them Moses and Elijah, talking with Jesus.*
>
> *Peter said to Jesus, "Lord, it is good for us to be here. If you wish, I will put up three shelters—one for you, one for Moses and one for Elijah."*
>
> *While he was still speaking, a bright cloud covered them, and a voice from the cloud said, "This is my Son, whom I love; with him I am well pleased. Listen to him!"*
>
> *When the disciples heard this, they fell facedown to the ground, terrified. But Jesus came and touched them. "Get up," he said. "Don't be afraid." When they looked up, they saw no one except Jesus.*
>
> *As they were coming down the mountain, Jesus instructed them, "Don't tell anyone what you have seen, until the Son of Man has been raised from the dead."*

> *The disciples asked him, "Why then do the teachers of the law say that Elijah must come first?"*
>
> *Jesus replied, "To be sure, Elijah comes and will restore all things. But I tell you, Elijah has already come, and they did not recognize him, but have done to him everything they wished. In the same way the Son of Man is going to suffer at their hands." Then the disciples understood that he was talking to them about John the Baptist.*
>
> Matthew 17:1-13 (TNIV)

There are two things about these verses I would like to point out. Jesus says Elijah has already come. What the Old Testament teachers taught about Elijah returning before the "Day of the Lord," as foretold by the prophet Malachi[b] was not literal, but really symbolic. In the "spirit and power of Elijah," John the Baptist came whom Jesus said was even greater than Moses and Elijah.[c] Someone came to preach the word (oil) and to proclaim the spirit of testimony to Jesus' saving grace (lampstand). It just was not Elijah… but really the first Christian, John the Baptist who was born with the Holy Spirit.[d]

Christ also said,

"To be sure, Elijah comes and will restore all things."

Is it Elijah who will come and restore all things or is it Christ himself? Even if you still believe that Elijah is one of the two witnesses, you will see later the things the witnesses are going to do does not jive with the comment he "will restore all things." If you ponder the meaning of this sentence, one quickly realizes that Christ may be speaking symbolically again of his second coming because it can have no other meaning. He can't be talking about John the Baptist again because John has already died at this point in history and the word "comes" suggests a future event that will occur that will restore "all" things. Since Jesus' first coming did not accomplish this from an earthly perspective, then he must be speaking about his second advent.

The Bible says man is destined to die once (physically) and only once[e] and therefore Moses could not be one of these tribulation prophets because he has already died and the remaining verses, if taken literally as the prophets are, suggest they will die and be resurrected after three and a half days. Elijah could still be one since scripture indicates he did not die, but was taken to heaven on a fiery chariot in a whirl-wind.[f] Some prophecy experts who understand this biblical truth, get around this problem by claiming Moses is not one of the witnesses, but Enoch is. He will be the other prophet because he was taken to heaven and did not physically die either.[g]

Therefore, when these two "prophets" are killed after 1,279 days and resurrected three and a half P-days later, there will be nobody who has ever lived on earth that did not die physically before Christ returns. As if this rationale supports their argument? If you believe what Paul writes in First Thessalonians 4:13-18 (the rapture), there will be many who will not die at the end of the Gentile Age and so the requirement that all must die at least once is unfounded. This requirement cannot be the only support for their belief that the two end-time prophets will be Enoch and Elijah.

I have recently heard about a preacher who claims he is one of the two witnesses. His name is Ron Weinland; a leader in the Church of God and author of "2008 – God's Final Witness." Mr. Weinland claims he is one of the two witnesses (prophets) written about in Revelation 11. Hmm. Anyone who believes this should really pray to the Lord for guidance and understanding. This man obviously wants people to believe these Revelation verses are literal, but only certain parts of the text i.e. the part of the prophecy where he is one of the two witnesses. The rest he chooses to "spiritualize" away. He wants people to believe in a literal interpretation for this text and yet ignore a literal interpretation for the other text that goes along with this prophecy. This is exactly what I told you to watch out for in the chapter "Deciphering the Prophecies."

When we examine the entire prophecy later on, you will learn of all the things the "true" witnesses will do that Mr. Weinland does not. He doesn't wear sackcloth, he is not in Israel preaching, he says the timing will be thirteen hundred and thirty-five days long from the book of Daniel (not twelve hundred and sixty days from the pertinent Revela-tion text) until the Lord's return, he doesn't plan on dying and being resurrected, he didn't know who the other prophet was for a long time until April 17, 2008 when he claimed his wife was the other witness, he doesn't know who the antichrist is, he claimed there "may be" a nuclear attack in mid April 2008 on U.S. cities which didn't happen, and he believes Christ died on a Wednesday and rose on Saturday evening. He can't stop the rain from falling and he hasn't killed anyone with fire from his mouth, but he is prophesizing and deceiving many of his followers. It is now September 2008 A.D. as I write for the Lord and by the time this book is published, we may have learned he was just another end-time false prophet not in touch with the Holy Spirit.[h]

What does a person who selectively picks a piece of scripture to make a point while neglecting the text around it sound like? Satan the deceiver is the master of distorting the Word of God to achieve his objectives.[i] Does Mr. Weinland really think true Christians are going to believe that if God sends two prophets to preach at the end-times, that they wouldn't know whom the other one was? That God's witnesses wouldn't even know the real day of Christ's death or exactly when and where there will be a nuclear attack? Christ said that at the end of the age there will be false prophets and messiahs and do not be deceived by them. Do not get them confused with true Christians who watch for the signs of Christ's return. Clearly we are seeing evidence of these end-time signs already. Let's examine further the symbolic interpretation.

Zechariah was told that the "lampstands" and the "oil" are the "two who are anointed to serve the Lord of all the earth."[j] We see that this answer given in Zechariah is the same statement given in Revelation, "they stand before the Lord of the earth." I therefore believe whatever the point of Zechariah's prophecy was, the explanation given to him should probably be applied to this prophecy's symbolism.

If you read all of Zechariah's prophecy, you will realize these "two witnesses" are not actual people, but symbolism for God's methods—how He goes about accomplishing His plans. He uses the power of His Word and His will (Holy Spirit)… the power of the law (Word) and the prophets (testimony of Christ)… the power of the oil and the lampstands before the Lord of all the earth to achieve His purposes. These will also be used by the saints (Christians) in their spiritual war against Satan and the antichrist.

Fire will not literally come from their mouths anymore than it will come from Christ's at the end of the age when he returns. The "killing" fire that comes from their mouths is the truth—the Word of God and the testimony of Jesus. Remember, Zechariah is told not by might (force) will the Lord's purpose (plan) be accomplished, but by His Spirit and His Word… the two witnesses.[k]

When the antichrist and non-believers are persecuting the saints during this time, Christians testifying to the truth of Christ will pass judgment on those who hear their call to come to the Lord for salvation. Many outspoken Christians will save many non-believers and bring them to the truth of Christ with their actions, words, and deeds,[l] but just as it was at Ezekiel's time before the fall of Israel and the first temple, many will not heed the call—including friends and relatives. Those who will not accept Christ will die spiritually at the very moment they lockout the Word and the Holy Spirit when end-time Christians with humility (sackcloth) present them with the final ultimatum. Accept Jesus now and reject the antichrist and the ways of the world, or die and spend eternity in the Lake of Fire. They may not die bodily for several more years, but in God's eyes they have sealed their fate and they are already dead. The two witnesses (the Word and the spirit of testimony) have "killed" them. This is the "fire that comes from their mouths and devours their enemies."[m]

This spiritual battle between Christians and the antichrist for the souls of the undecided, fence sitting, masses will look a lot like the time during the first few centuries after Christ's death when Christianity grew in leaps and bounds. Why did it grow at such an accelerated rate? Because the Christians of those days really believed Christ would return at any moment and they believed strongly that Jesus was the Savior of the world. Nothing anyone did or threatened to do stopped them from spreading this message to everyone they met—even if it meant their death!

This lesson from history will be a repeat performance for Christians in the coming years and they need to make up their minds ahead of time on how they are going to handle it.

He who has an ear, let him hear. If anyone is to go into captivity, into captivity he will go. If anyone is to be killed with the sword, with the sword he will be killed. This calls for patient endurance and faithfulness on the part of the saints.

Revelation 13:9-10 (TNIV)

We know from these verses in Revelation that speak about the time when authority will be given to the beast that there is a plan for God's people. Unfortunately, it may not be a plan we will like. Some are destined to become martyrs, some will be put in prison, and many other tragedies will befall the saints and "test" their faith as Christians. Regrettably, many who reject Christ and turn against us in the last days will be people we know and love. Pay attention, listen, and don't deny Christ for any reason when put to the test during this time—or any time for that matter.

"Whoever publicly acknowledges me I will also acknowledge before my Father in heaven. But whoever publicly disowns me I will disown before my Father in heaven. "Do not suppose that I have come to bring peace to the earth. I did not come to bring peace, but a sword. For I have come to turn " 'a man against his father, a daughter against her mother, a daughter-in-law against her mother-in-law—your enemies will be the members of your own household.'

Matthew 10:32-36 (TNIV)

Just as Daniel was tested in the lion's den and Daniel's friends in the fiery furnace, all Christians' faith will be tested in the same way. These lessons from the book of Daniel were lessons of faith for first century Christians and for twenty-first century Christians as well. The antichrist will impose laws that forbid worship of anyone but him, just as Nebuchadnezzar's decrees did during Daniel's time.[n] He will pass laws that exclude true believer's in Christ from participating in the world economy by implementing the mark of the beast. Daniel's prophecies tell us much about the end-time events. Daniel's stories of faith, which are overlooked by eschatology experts, provide insights and clues into how we are expected to react when these statutes are enacted by the antichrist.

When Christians are being persecuted during the early days of the tribulation, the Word and the testimony of Christ will be their only protection and weapons for fighting the antichrist and the non-believers in authority.[o] Remember, Zechariah was told not by "might" will the temple of God be built, but with the two witnesses who are anointed to serve the Lord of all the earth! What temple was really being built? A physical building, as was the case with Zechariah, or the spiritual one John was told to measure?[p]

If we were to look back to page 193, we would notice that the time for the two witnesses is an exact number of days and the start of their work falls on the same day the covenant is confirmed by the antichrist. This is an odd convergence. Understand what this really means. If there are two real prophets, then Christians will not only have the covenant signing with the Jews as a key sign to watch for, but two real people who will start proclaiming to the world (at the very same time) the true identity of the antichrist and the coming wrath of God that will assist in the recognition of that important sign.

For instance, in the case of our false prophet Ron Weinland when he announced he was one of the Revelation prophets and months later announced his wife was the other prophet, the timing for both of them was not exactly the same! Their announcements did not occur at the same time nor did either of them occur on a day an agreement was signed with Israel. In other words, from the very start they opened their mouths it could be determined they were false prophets because the facts did not match the Word of God. Remember these truths, if you believe in a literal fulfillment of the two witnesses, when you are trying to discern whether someone is one of the two last-days prophets of Revelation.

I don't believe there will be two real prophets at the end and I wouldn't rely solely on this sign, if I were you. When I earnestly begin watching for confirmation we have reached the beginning of the tribulation, if these prophets really come—great! God's people will need all the help they can get and having two powerful witnesses spreading the truth will inspire both faithful servants and unbelievers who become Christians to persevere. However, their arrival will not relieve you of your responsibilities at this time.

As a side note, I waste very little time trying to identify the antichrist as some zealous Christians do. The time is indeed short and I am sure the reason for most of them hunting the man of perdition is not just for glory, but to sound the alarm for Christians who are asleep in the fire that surrounds them. Since we have a very good approximation of when these events will occur, I believe the present time is better spent spreading the good news of Christ to friends and relatives and trying to tell others they should be doing the same. Now is the time for planning and preparing for war. Going into battle without preparation is foolish and waiting until the last moment when the signs appear because we were searching far too early is a waste of precious time.

Considering everything, I feel this prophecy is by and large symbolic and that Zechariah's explanation is the most plausible choice. The real reason why the two witnesses' ministry will commence at the same time the antichrist signs his agreement with Israel is because many knowledgeable Christians are keeping watch (and more will be in the future) just as you should be looking for this key requirement of Daniel's prophecy when the time is ripe. As soon as they witness this event, there will be an outcry from these Christian sentries that will be heard around the world. Hence, the starting time for both of these prophecies will be the same (this line of reasoning could be applied to two literal prophets as well).

This grassroots Christian movement will accelerate over the twelve hundred and seventy-nine days and cause mounting problems for the antichrist as he tries to hide his true intentions, but the truth of his identity will be proclaimed with increasing frequency and boldness. This effect will drive the man of perdition to begin a campaign (war) to shut-up Christians who oppose him and reward many who aid him.[q] His efforts will escalate in severity as the number of Christians who realize the truth (wake up and resist him) grows in size.

The Rest of the Prophecy

I have acknowledged that I sense this prophecy taken as a whole is symbolic, unfortunately the later verses lead to a conflicting conclusion with this position. It would be so much easier if all the text came to a clear understanding. Thus we will investigate the rest of this prophecy for the additional problems that might be revealed so that you can make up your own mind.

The first four verses we have looked at were clearly symbolic in nature, but is it safe to assume the rest of the prophecy is also symbolic? Not in this case. Let us not forget that, although Revelation uses symbolic writing to convey future events, we know there are parts within that are literal and recognizing when to switch is the skill needed to uncovering more of what God wants us to know in the judgment years ahead.

For example, even though the first four verses are figurative as a whole, the timing within is not! That is to say it can be taken literally (after converting to "real" time) as opposed to rendering a symbolic translation for it so that it means something else entirely.

> *If anyone tries to harm them, fire comes from their mouths and devours their enemies. This is how anyone who wants to harm them must die.*
>
> Revelation 11:5 (TNIV)

Regardless of whether you support the witnesses are real prophets or a symbolic representation of all Christians' end-time responsibilities, this text is clearly figurative. Actual fire will not come from anyone's mouth. So what does it mean? Understand that the witnesses are God's people whether there are just two of them or millions of them, so those who wish to do harm to them are people who oppose the truth of their message. These people have hearts that have been hardened from their addictions to the things of this world and they refuse to give them up for any reason. They are the enemies spoken of in the text. Simply speaking, these are neighbors, friends, close relatives, and people in far away places who have tired of listening to God's Word being preached loudly, firmly, and without reservation. They are cynical unbelievers living through trying times who have made up their minds and have accepted the mark of the beast. In so doing these enemies have sealed their fate to eternal damnation.

So the death that comes from their mouth is the truth of Christ's return and authority. This "fire" is the only weapon the saints will have to fight the antichrist and unbelievers alike and will not come from their mouths anymore than it will come from Jesus' mouth when he returns. This condemning fire that comes from Christ's mouth at his return is the truth of his authority and the realization to the unbeliever that he or she has made the wrong choice. So also for those individuals who make the wrong choice, after hearing God's two witnesses proclaim the truth, prior to the Messiah's return.

> *These have power to shut heaven, that it rain not in the days of their prophecy: and have power over waters to turn them to blood, and to smite the earth with all plagues, as often as they will.*
> *Revelation 11:6* (KJV)

This verse is the hardest to explain using the 'symbolic' interpretation I favor. Why? Because I am unaware of any literal translation for interpreting "the stopping of rain" if this is symbolic. Thus, I am left with a literal understanding that it may not rain for the entire twelve hundred and seventy-nine days. Recognizing the unlikeliness of a symbolic understanding, I find it even harder to imagine that all Christians collectively would have the power to do the things written in this section of scripture. However, the Bible records some of God's prophets in years gone by did have these abilities. This would suggest two real-life prophets at the end of days would arise. Having just said that, I cannot fathom the whole earth not having rain for three and a half years. This idea I just cannot support because all plants would die long before God's judgments would begin and this event would be more severe than any of God's upcoming desolations. This idea would also suggest it would stop snowing everywhere on the planet for forty-two months. Something else to consider when pondering what is the correct interpretation.

A careful inspection of this text reveals that only the two witnesses will have the power to do these things, not necessarily that they will use the full power of their authority given to them. Subsequently we might infer that vast areas of drought conditions can be anticipated (especially the place at which they are prophesying), just not everywhere on earth at the same time. This also applies to the plagues' reference recognizing that when God uses the word plague it is not in the same context as we use it. He applies it to the horrors of war, famines, emotions, and not just diseases. These problems will all be progressing in severity at this time, for these trends are already in place.

As for the turning of waters into blood, this is something we have discussed before. There will be no changing of waters into real blood (no literal fulfillment of this text) and this is only the Lord's way of saying; with the preaching of the Bible, conflicts will arise in many nations (waters) between those who believe and those who do not. These conflicts will reach into every home and divide families—families of strangers, your family, and even my own family. These arguments will escalate and there will be much death because of it. Much blood will be spilled from the civil unrest of evil people angry with Christians for telling them the world is about to end and the lives they are addicted to are about to end as well.

Let's look closely at what is the "killing fire" that is being preached… the truth being prophesied during this time by God's witnesses. You should now know the time of Christ's return. These witnesses will be telling everyone the time is near… the tribulation time has begun… the end of the age is at hand. It further says in *Revelation 19:10* (RSV),

> …*Worship God!" For the testimony of Jesus is the Spirit of prophecy."*

In other words, prophesying is nothing more than testifying to everyone that Christ is the true savior and the only way to everlasting salvation and that following the antichrist and the ways of this world is the wrong path. Believers are to set high the light of Christ for all to see by what they say and more importantly, what they do—even if it means dying for Christ. We have discussed the bloody water, and the lack of rainfall, but what about the frequent plagues? It may be possible that some powerful Christians who are completely filled with the Holy Spirit will have the ability to do miracles just as the apostles did at Christ's time, but this text does seem to support a more literal interpretation even if I do not believe there will only be "two" specific individuals.

This confrontation between interpretations is a burdensome struggle. Zechariah was told not by might or power will the Lord's plans be accomplished (his temple built), yet if these two are real prophets, they are doing nothing that would be classified as anything less than miraculous!

The dilemmas of this scripture are left for the reader to wrestle with. Some of the consequences of people's rebellion to the truth appear to be clearly symbolic and are used by God in other prophecies, while others appear to have no known symbolic translation and must be taken more literally. Knowing how God works, it would be better suited if all three: the rain, the bloody waters, and the plagues were either literal or symbolic, but they appear to be a mixture. In summary, either they all are symbolic as the prophecy has been up to this point, and I am ignorant to what they mean, or they are all literal, which I find impractical to believe, or maybe they are a combination of both literal and figurative and I have interpreted them as best that can be expected. Again, you will have to be the judge.

Let's travel further on down the Revelation road of understanding.

Now when they have finished their testimony [after 1,260 P-days], the beast that comes up from the Abyss will attack them, and overpower and kill them. Their bodies will lie in the public square of the great city, which is figuratively called Sodom and Egypt, where also their Lord was crucified.

Revelation 11:7-8 (TNIV)

This is the Beast of the Sea, which we will learn about shortly. He is the antichrist born in Greece and may be resurrected in spirit as the second coming of Alexander the Great. Why? Well this is a complicated matter and covered in detail in my first book, but the short answer is the man of perdition came once before the time of Christ[r] and given all the facts of history and Bible prophecy the most logical conclusion is that he was Alexander the Great at his first coming.

As for other details given in these passages, we see symbolic language is being used and yet, the interpretation is also given. This city spoken of as "Sodom and Egypt" is Jerusalem because that was the city Christ was crucified in. Why is it figuratively called this and why was symbolic language even mentioned by St. John when clearly he could have just wrote Jerusalem since he knew where our Lord was crucified? I have no idea for I have never heard it called this, except in this passage, and have not uncovered any evidence to this hidden mystery other than it is not known symbolically as Babylon the Great, which is significant. But the real importance of knowing, without a doubt, the name of this city has been provided within the text by the Lord.

So far we know the two witnesses will be overpowered and killed by the antichrist in Jerusalem after twelve hundred and sixty prophetic days—a statement that by in large is literal. This would suggest the two witnesses are also literal for how does the leader of the European Union kill the Word and the Testimony of Christ by Christians? I have many thoughts in this regard, but before moving on realize that if you believe there will be actual end-time prophets, then there is not much more to add. According to this text, the antichrist will kill them in a public place in Jerusalem after one thousand two hundred and seventy-nine days from the signing of a covenant.

Now I have searched for this public place since there must be a public square if this is all literal... a town square in Jerusalem where this future event will take place. Having never visited Jerusalem or Israel for that matter, I have been limited to surfing the Internet to find such a place. As best I can determine, there is not a great square as there are in many international cities. The closest match I could find to this biblical reference would be Paris Square in downtown Jerusalem. Many civic protests are conducted there and in 2008 A.D. France donated a new fountain to commemorate the sixtieth anniversary of Israel becoming a nation again. This square is within view of many government buildings and intuitively fits the profile since the antichrist will set up his last days government in Jerusalem.[s]

Another possibility is "Solomon's Porch" a place for preaching to the Gentiles in the outer court of the second Jewish temple where first century Christians gathered.[t] This place was destroyed in 70 A.D. and does not currently exist, but the now vacant area is located near the Dome of the Rock Mosque. If this mosque is destroyed in the future and a temple rebuilt in its place, this might be the "public square" once again where two real prophets might proclaim the truth and then meet their death at the orders of the antichrist.

As for the symbolic understanding, we only need to comprehend what will be going on at this very moment. The antichrist will be setting up the abomination that causes desolation and attacking Christians because of their failure to renounce Christ and embrace the new economic system. He will be instituting laws that will take effect and become punishable even unto death. These enforcement activities will continue for the thirty to thirty-five days of conflict we discussed in chapter six. One can visualize that making laws which will punish anyone who preaches the Word of God or testifies to the saving grace of Christ Jesus will constitute the symbolic "killing of the two witnesses."

Now a most troubling quandary becomes apparent. It has to do with the timing of events around this time. This prophecy said the two witnesses would prophesy for an exact amount of time—twelve hundred and sixty P-days. It cannot be more or less than this amount or else the Word of God would be a lie. Therefore, it must be an exact amount (within twenty-four hours). In the next verse we will learn these two witnesses will be dead for three and a half P-days and then be resurrected so the three and a half P-days cannot be accounted for in the twelve hundred and sixty P-days total because they can't do their work while they are dead.

What does this really mean? Well in chapter five's timelines we took this requirement into account, but in chapter six it was not portrayed that way. So if there are literally two real prophets, then the timeline in chapter five appears to be the correct choice. However, if these two witnesses are only symbolic, then we have to conclude the resurrection of the two has to be symbolic as well. Consequently, if the two witnesses really mean Christians are proclaiming the Word and testifying about Christ and they are the ones dying, then resurrecting them may be nothing more than a description of the rapture or the setting up of the 144,000 who will spread the Gospel to the world during the last three and a half years of the tribulation.

Following this train of thought to its conclusion, one arrives at a deserted station because we know the rapture will happen at the end of the thirty-five days when the *"sanctuary is reconsecrated"* and not three and a half days after the abomination that causes desecration is set up.

And he saith unto me, Till evening – morning two thousand and three hundred, then is the holy place declared right."

Daniel 8:14 (YLT)

The Young's Literal Translation Bible says instead the *"holy place will be declared right"* rather than the *"sanctuary will be reconsecrated."* God will declare… deem, decree that the holy place (Christians) will be set right.[u] That all the ungodly things the antichrist will be doing will be cut short[v] of the required time with the rapture of the church. This timeline of events is not consistent with a literal two prophets under-standing because the three and a half days does not fall at the end of the thirty-five days.

This is confusing, but what is really being said is if there are literally two real prophets, then the three and a half P-days are also literal and cannot be accounted for in the twelve hundred and sixty P-days. Consequently, the timeline in chapter five has to be the most accurate one. And if the two witnesses are not real, but only figurative, then the three and a half days must also be figurative as well because they cannot be accounted for within the thirty-five P-days of conflict in any logical or mathematical way given the sequence of events. This divergence of interpretations will continue through the rest of the prophecy.

For three and a half days many from every people, tribe, language and nation will gaze on their bodies and refuse them burial. The inhabitants of the earth will gloat over them and will celebrate by sending each other gifts, because these two prophets had tormented those who live on the earth.

Revelation 11:9-10 (TNIV)

These verses are self-explanatory given a literal interpretation. The two prophets will be killed in Jerusalem and their bodies will lie in the public square while the whole world watches (probably on the news networks around the globe) and celebrates because they had caused so much pain with the things they were saying and doing.

As we continue forward through the rest of the prophecy, a symbolic interpretation gets harder to support, but I will put forth an interpretation for consideration. Since the three and a half days cannot be literal for a symbolic understanding because of timing problems, then it must mean something else. If we apply the rule of "a year for a day" to this time we would see that the time could now be viewed as three and a half years— the same as it is for the length of the whole prophecy. This verse then would be saying that during the time the two witnesses are testifying (all Christians), many will be persecuted and killed and the unbelievers (enemies) around the world will support these activities and celebrate with their other unbelieving friends the rounding up and killing of Christians. Applying this understanding goes against the grain of every other Revelation prophecy interpretation because all the times in those prophecies were taken literally (after converting to real time) and none required applying the "day for a year" rule! The time manipulation in this instance troubles me. Nevertheless, I provide an interpretation because there are literal things that bother me as well. There just are not any within these verses.

But after the three and a half days a breath of life from God entered them, and they stood up on their feet, and great fear fell on those who saw them. Then they heard a loud voice from heaven saying to them, "Come up hither!" And in the sight of their foes they went up to heaven in a cloud.

Revelation 11:11-12 (RSV)

Again a literal application is easy to comprehend. The two prophets will be resurrected after three and a half P-days and ascend into heaven just as Christ did. But the real question is why? What is God's purpose for doing this? These three and a half P-days coincide with the abomination being set up, but are roughly a month before the rapture of the faithful. Would not all who witnessed this event around the world repent immediately and in so doing be included in the rapture? All were in terror when they witnessed the resurrection and ascension. The nations of the world were amazed and should now recognize the truth. It seems irrational to think anyone would not believe the truth after viewing something like this… and yet billions will not.

Why resurrect only two individuals and then thirty days later resurrect millions in the rapture? Hmm. It just doesn't make sense knowing the true timing of events. A symbolic translation is equally easy to comprehend because the wording is almost a definition of the rapture of the saints. But for this to be correct, the three and a half days has to be either three and a half years as we discussed or mean thirty-five days (ten times the amount… a higher level within God's plan by multiplying by the number of completion) and this idea just does not ring true to me either. Although God uses this method, applying it in this instance to force a symbolic interpretation is probably unwise. Let's examine the last verse so we can tie all this together.

And at that hour there was a great earthquake, and a tenth of the city fell; seven thousand people were killed in the earthquake, and the rest were terrified and gave glory to the God of heaven.

Revelation 11:13 (RSV)

We now learn a severe earthquake will shake Jerusalem and the Middle East within minutes of the ascension of the two witnesses. Earthquakes are always to be taken literally in prophecies because there is no biblical evidence to the contrary. One tenth (a tithe and the number of completion) of the city will collapse and seven thousand people will die from it. A most curious number that I also believe is literal. The *"survivors were terrified and gave glory to God."* This Bible translation is problematic. "The survivors" is really a poor choice of words.

We just saw in the previous text that all were terrified, but had been partying and celebrating the death of the two witnesses. This verse implies most people in Jerusalem repented and began giving glory to God. But that is not what will happen. The unbelievers will be blaspheming God at this time including many of those in Jerusalem. Why? Because as the abomination of desolation is set up, which is occurring concurrently with the death, resurrection, and ascension of the two witnesses and this earthquake, Christ's instructions are for all in Judea (Jerusalem) to flee to the safety of the mountains—to get out of town as fast as possible. Will there be Christians still here? Not likely. How about Jews? Yes and some will now realize they have made a mistake trusting the antichrist. They are the ones giving glory and not all the survivors, which you can bet will include foreigners who have come with the arrival of the antichrist.

The King James Bible translates this way, *"and the remnant were affrighted, and gave glory to the God of heaven."* This is a better translation for we see it in not all the survivors who give praise, but the "remnant" who are the ones. The "remnant;" a most interesting choice of words. If we were to study the use of this word in the Bible, we would find many occurrences where God saves a "portion", a "remnant" of the servants, He deems needs further refining to carry out His remaining plans.

When Jerusalem and the first temple were destroyed in 587 B.C. at the hands of the Babylonians, a remnant of Jews were saved (a third)[w] that would later be used to rebuild the temple and the city after seventy years in exile. Those selected suffered God's wrath right along side the wicked God was punishing. So too in the last days as Zechariah prophesies in verses 13:8-14:2 will a remnant be saved.

Furthermore, this earthquake is the same one described by the opening of the sixth seal. If these are real prophets, expect considerable Christian bloodsheding on a massive scale, for if we know when the rapture is likely to occur, then so does the antichrist. If there are no real prophets, but just Christians preaching the word and testifying to Christ's return, then this is nothing more than a description of the rapture and the signaling that the last three and a half P-years of the tribulation have begun.

Many Jews who remain in Jerusalem after witnessing this event will now realize they were wrong in believing the antichrist was the prophesied Messiah and will begin resisting just as the Christians had been doing. Countless numbers will realize the error of their ways and will come to the true Christ and spend the last three and a half years trying to convert any last minute Jews and Gentiles (stragglers; new tribulation saints who will repopulate the earth—a remnant brought through the tribulation fires) during the wrath years. This will be the time of God's wrath, as we already know, but also the time of the 144,000's message.[x] The antichrist will be furious with the turn of events and will begin lashing out at them because there are no easily identifiable Christians left to pick on and he knows his time is short. Christ's servants have either been raptured or gone into hiding.

This ends the prophecy of the Two Witnesses. A prophecy that is easy to understand, but has many problems, if you take everything literally. The same can be said for a symbolic interpretation. You will have to make up your own mind, but I leave you with these last words of wisdom. Plan for the worst and hope for the best. That is to say, prepare to fight for the Lord without their help, but hope they do come for they will provide not only spiritual guidance and inspiration, but also a sign to point to for your unbelieving friends and relatives—your loved ones who need convincing. For if you can show the Bible predicted their coming thousands of years in advance, then maybe those you love can believe all that has been written and everything they (and you) have been saying.

Chapter 11

The Woman and the Dragon

WE HAVE INVESTIGATED IN DETAIL roughly half of the prophecies that deal with the end-times in the book of Revelation. During that work we have established a solid foundation of the judgments that will befall mankind during the latter days of the tribulation. However, there is still much to be learned for thoroughness will only aid in preparing us for what to expect and how to handle what is to come. These prophecies not only provide insight into God's judgments, but details of Satan's plans and key events that will occur around the time of the abomination of desolation.

Now most people view prophecies as future events planned by God since the Creator gave them to His prophets. I know I did until just this very moment when I realized that although we are given details of God's planned wrath, everything that occurs throughout history is not God's plan, but includes Satan's work, and man's own actions. But the Lord is omniscient and not only provides details in prophecy that are plans of His own making, but also details of what will happen due to Satan's work and our own free will. For instance, do you really suppose the abomination that causes desolation is one of God's ideas or is the Maker just reporting to you Satan's plans in advance? "See, I have told you ahead of time" so that you will not be unaware of these facts.[a] All three of these forces are intertwined and affect human history.

In this collection of prophecies, "the Woman and the Dragon," I chose to group as one because they deal with the details of the first half of the classical seven-year tribulation period, we will find information pertaining to the counterfeit trinity. That is to say, particulars of Satan's (father) rebellion, the antichrist's (son) actions, and the false prophet's (Holy Spirit) work as well as God's plans and the consequences for going along with and foolishly following these three—whether willfully or from ignorance.

Let's continue our studies with an in-depth look at the prophecies pertaining to Satan's plans: truly a noteworthy time to pay careful attention to in man's future.

The Dragon: Satan

A great and wondrous sign appeared in heaven: a woman clothed with the sun, with the moon under her feet and a crown of twelve stars on her head.

Revelation 12:1 (TNIV)

This text is widely recognized as a description of the Israelites and Judaism. The people the Creator chose to bring salvation to the entire world through the atoning blood of Jesus. The twelve stars signify the twelve tribes of the Jews. The woman is clothed with God's light of truth (sun) and Satan (moon) is under her feet. This heavenly sign also has a deeper meaning with respect to astrology, but that interpretation was covered in my first book and will not be reviewed here. What is of further notice is there are twelve stars and, by an odd twist of fate, the European Union's flag has twelve stars. Is this another attempt by Satan to mimic God's plans? When discussing how the E.U.'s flag was developed, here is the answer from the European website "http://europa.eu:"

"It is the symbol not only of the European Union but also of Europe's unity and identity in a wider sense. The circle of gold stars represents solidarity and harmony between the peoples of Europe. The number of stars has nothing to do with the number of Member States. There are twelve stars because the number twelve is traditionally the symbol of perfection, completeness and unity. The flag therefore remains unchanged regardless of EU enlargements."

So we see an answer that says the number of stars is based on numerology: the number twelve to signify perfection and completeness. Is this really the number of perfection or completeness? Not the way God uses numbers, but it is a number the Lord uses for other purposes and I think its use here is nothing more than Satan's attempt to usurp God's future kingdom with one of his own. Satan will use the E.U. to achieve his purposes just as God uses the Jews to fulfill his plans. The flag's symbols are not based on earthly things as many country's flags are, but admittedly on a spiritual mysticism based on numbers (twelve) and symbols (stars and circle). Fascinating.

> *she was with child and she cried out in her pangs of birth, in anguish for delivery.*
>
> *Revelation 12:2* (RSV)

The nation of Israel gave birth to the Messiah. Consequently, through his birth, the religion of Judaism (based on law) painfully gave birth to a religion based on forgiveness: Christianity.

> *And there appeared another wonder in heaven; and behold a great red dragon, having seven heads and ten horns, and seven crowns upon his heads. And his tail drew the **third part** of the stars of heaven, and did cast them to the earth: ...*
>
> *Revelation 12:3-4* (KJV)

We see here Lucifer (red dragon) corrupted or will corrupt a third of the angels (stars) in heaven and they were or will be thrown out of heaven. How do we know this? As the prophecy continues, we will observe confirmation of who the dragon really is in plain terms. We also know from our previous studies that "stars" in some cases can represent angels. Now if twelve is the number of perfection or completeness as the European Union asserts, then why does Satan have "seven" heads with "seven" crowns and "ten" horns? He has these numbers because they are the true numbers of perfection and completion and he aspires to be like the Creator and to mimic Him in as many ways as possible.

> *... and the dragon stood before the woman which was ready to be delivered, for to devour her child as soon as it was born. And she brought forth a man child, who was to rule all nations with a rod of iron: ...*
>
> *Revelation 12:4-5* (KJV)

This is the birth of Jesus that Satan tried to stop unsuccessfully by rousing Herod the Great to murder all male children under the age of two. Clearly the description of a male child who will rule all the nations with an iron scepter and was prophesied in many scriptures is Jesus.[b] Knowing this provides added support to the idea of whom the woman symbolizes.

> *... and her child was caught up unto God, and to his throne. And the woman fled into the wilderness, where she hath a place prepared of God, that they should feed her there a thousand two hundred and threescore days* [1,260 P-days].
>
> Revelation 12:5-6 (KJV)

In these verses we see the death, resurrection, and ascension of Christ. Next, we fast forward to the beginning of the tribulation period where we find out the Jews (the woman) will be protected for twelve hundred and sixty P-days. Why fast forward almost two thousand years to the end-times? Because if we were to apply a strict literal interpretation of forty-two months directly after Christ's ascension, it would be meaningless. Nothing happened in history that could be classified as "taking care of" the Jews.

Additionally, applying the prophecy rule a "year for a day" also is pointless because the Jews (woman) were not taken care of for one thousand twelve hundred and sixty years between 33 A.D. and 1293 A.D.! So this must mean something else and since Christ's revelation given to St. John was primarily about the end-times, then the logical answer is so is the timing.

Why are the Jewish people protected? Daniel implies the antichrist will confirm a covenant (more than likely a peace accord) with them and their enemies, which would place the nation under his protection for seven P-years. Afterward, the antichrist will break it half way through and invade Israel.[c]

This breaking of the agreement will coincide with the start of his downfall. But more importantly, this treaty will occur because Israel will be attacked by her Arab neighbors with Russian assistance sometime between 2018 A.D. and 2021 A.D. and the antichrist, the E.U., and its allies will come to her aid. Since Israel doesn't believe the messiah foretold in the Old Testament prophecies has already come, it consequently doesn't recognize Jesus as the promised one. They continuously hope for a powerful political leader to arise who will lead and protect them from the nations who persecute them—and the antichrist will meet their qualifications.

They will be easily deceived by the antichrist because they yearn so desperately for a messiah that they are open to his deceptions and this charismatic leader of Jewish descent will fit the qualification for the anointed one they long for. He will provide the means for them to fulfill a two thousand year old dream to rebuild a temple to God—a temple that will not be to God, but only for himself.[d]

*And there was war in heaven. Michael and his angels fought against the dragon, and the dragon and his angels fought back. But he was not strong enough, and they lost their place in heaven. The great dragon was hurled down–**that ancient serpent called the devil, or Satan**, who leads the whole world astray. He was hurled to the earth, and his angels with him.*

Revelation 12:7-9 (NIV)

These verses are nothing more than a literal description of verses three and four; a rehashing providing more clarity in terms that are easier to comprehend. Nothing new has been added. They just verify that the dragon is Satan, and the stars are angels!

And I heard a loud voice in heaven, saying,

"Now the salvation and the power and the kingdom of our God and the authority of his Christ have come, for the accuser of our brethren has been thrown down, who accuses them day and night before our God. And they have conquered him by the blood of the Lamb and by the word of their testimony, for they loved not their lives even unto death."

Revelation 12:10-11 (RSV)

217

Now has come the salvation, the power, and the kingdom of God and the authority of his Messiah—Jesus. When is this all going to take place? At the end of the twelve hundred and sixty P-days... at the midpoint of the seven-year tribulation! Proof that this is the time, the point on the tribulation timeline, where the Messiah (Jesus) is given authority to begin reigning from heaven. These verses say "now" right after talking about the twelve hundred and sixty P-days—not some later time, but now. Remember, verses seven through nine are just a clarification of the first verses of the prophecy i.e. they are redundant verses so verse ten really is following verse six where we are given the length of time for this prophecy.

Who are "they?" These are martyrs who triumphed over Satan and the antichrist by the blood of the Lamb (believing in Christ as their personal savior and therefore having the light of the Holy Spirit within them) and by the word of their testimony. They resisted the antichrist's tactics, the mark of the beast, and everyone who threatened them by speaking out and proclaiming the truth of Christ and the coming of the end of the age. They are strong Christians who will not cower from their responsibility to testify by acting on their faith—even to the point of being killed. Just as it was in the early years of Christianity, so shall it be at the end of the Times of the Gentiles. There will be millions of martyrs from many generations under God's altar in heaven.

Rejoice then, O heaven and you that dwell therein! But woe to you, O earth and sea, for the devil has come down to you in great wrath, because he knows that his time is short!

Revelation 12:12 (RSV)

All the loyal angels to God, the saints, and martyrs are the ones dwelling in heaven at this point and are rejoicing. Satan and his angels have been "kicked out." Interesting use of the word "woe" here because it signals the start of God's wrath. The sea is the remaining inhabitants of the world. And the earth... I am uncertain of its exact meaning because it is used interchangeably in various verses to designate different things; but most likely means "nature" in this context because of the natural disasters (desolations) decreed for the wrath years and the destruction that Satan, the antichrist, and the false prophet will inflict to the planet.

And when the dragon saw that he was cast unto the earth, he persecuted the woman which brought forth the man child. And to the woman were given two wings of a great eagle, that she might fly into the wilderness, into her place, where she is nourished for a time, and times, and half a time [three and a half years], from the face of the serpent.

Revelation 12:13-14 (KJV)

We find out here that Satan knows his time is short and his anger is first directed at the Jews who were being protected for three and a half years with the covenant signed by the antichrist. He will break this covenant and turn on them. We know this will occur when the abomination of desolation is set up, but we also know Jesus instructs all to flee Jerusalem (and Israel) to the mountains where they will be safe for the remaining three and a half years. Daniel provides the place. Words of wisdom for all Christians to follow during this time... to find safety in the countryside away from large cities.

He will also invade the Beautiful land [Israel]. Many countries will fall, but Edom, Moab and the leaders of Ammon will be delivered from his hand.

Daniel 11:41 (TNIV)

All will not flee, for all do not know the Word of the Lord... this instruction is given in the New Testament[e] which is not studied by many Jews and clearly not by secular Jews who have no religious beliefs. They will be left behind because they are not God's people and will not abandon all their worldly possessions or their rebellious ways. The area of Edom, Moab, and Ammon of bygone years is located along the mountainous border that separates Israel with Jordan.

Then from his mouth the serpent spewed water like a river, to overtake the woman and sweep her away with the torrent. But the earth helped the woman by opening its mouth and swallowing the river that the dragon had spewed out of his mouth. Then the dragon was enraged at the woman and went off to make war against the rest of her offspring–those who obey God's commandments and hold to the testimony of Jesus.

Revelation 12:15-17 (NIV)

"Then from his mouth the serpent spewed water like a river, to overtake the woman and sweep her away with the torrent" i.e. then the antichrist, inspired by Satan's anger, will set his armies (like a river) upon on the fleeing Jews in great force and try to overwhelm ("torrent") and kill them. This clearly is symbolic writing and therefore smaller bodies of water like rivers mean smaller groups of people… in this case an army.

I will gather all the nations to Jerusalem to fight against it; the city will be captured, the houses ransacked, and the women raped. Half the city will go into exile, but the rest of the people will not be taken from the city. Then the Lord will go out and fight against those nations, as he fights in the day of battle.

Zechariah 14:2-3 (TNIV)

This prophecy of Zechariah's pertains to the last days before Christ returns and gives us a glimpse of what will happen in Jerusalem prior to Armageddon when at that time the Lord will fight the same way as man does. We also learn that half the city will heed Christ's warning and flee while the other half will not obey and will be victimized by the wickedness of the antichrist and his armies.

"But the earth helped the woman by opening its mouth and swallowing the river that the dragon had spewed out of his mouth" i.e. but there is an earthquake at that very moment which disrupts the armies' actions and allows more of the remnant to evade the antichrist's efforts.[f]

"Then the dragon was enraged at the woman and went off to make war against the rest of her offspring—those who obey God's commandments and hold to the testimony of Jesus." Afterwards, the antichrist will step up his efforts to kill Christians worldwide and these serious endeavors will be curtailed with the rapture of the church about thirty days later.

The Sea Beast: Antichrist

> *The dragon stood on the shore of the sea. And I saw a beast coming out of the sea. It had ten horns and seven heads, with ten crowns on its horns, and on each head a blasphemous name. The beast I saw resembled a leopard, but had feet like those of a bear and a mouth like that of a lion. The dragon gave the beast his power and his throne and great authority. One of the heads of the beast seemed to have had a fatal wound, but the fatal wound had been healed. The whole world was filled with wonder and followed the beast. People worshiped the dragon because he had given authority to the beast, and they also worshiped the beast and asked, "Who is like the beast? Who can make war against it?*
>
> *Revelation 13:1-4* (TNIV)

This prophecy starts all over again from the beginning of the seven-year tribulation and we now learn about the antichrist and how people will be enamored with him; "Who is like the beast? Who can make war against it?" Chapter twelve of Revelation was showing us this time period from Satan's perspective. This imagery is a description of the European Union and its future leader during the same time. It appears he was given authority by Satan to rule, but that is not so. This authority given is only "allowed" by God so that the antichrist will fulfill God's plan so all that was foretold to the prophets would take place.

What else can we learn? Let me rewrite this scripture with plain English for you to contemplate:

And Lucifer(g) came to the inhabitants of the world.(h) And I saw a great nation, empire,(I) arise above the other nations of the world. This empire had ten leaders(j) and was the last in a series of seven world empires(k) throughout history (Egypt, Assyria, Babylon, Medo-Persia, Greece, Roman, and the European Union). And each of these empires was an abomination unto God and blasphemed and rejected His authority.(l) This empire I saw resembled the old Greek Empire(m) and its president will come from this country.(n) This nation will have the brute military strength of the old Medo-Persian Empire(o) and the economic ideals (idol worshiping) and cunning of the Babylonians(p) whose leader

221

Nebuchadnezzar considered himself above God[q]— just as this leader will.[r]

Satan gave this future leader/empire its great power.[s] One of the Empires (Rome) appeared to have died, but had been healed in the last days[t]—the Revived Roman Empire (1958): the European Union. The whole world was infatuated with this nation and its charismatic leader. Those who do not know the true God worshiped this powerful leader and indirectly Satan, for whom in the world is like this great leader and can oppose him?

> *The beast was given a mouth to utter proud words and blasphemies and to exercise its authority for forty-two months. It opened its mouth to blaspheme God, and to slander his name and his dwelling place and those who live in heaven.*
>
> *Revelation 13:5-6* (TNIV)

The antichrist will make war on Christians for forty-two months—the first half of Daniel's remaining seven prophetic years. During this time the leader of the European Union will utter appalling things about Christians, Christianity, and God himself. He will fan the flames of those who are not Christian into riotous acts against all believers and will institute laws that will discriminate towards them. The mark of the beast will be implemented during these years.

> *It was given power to make war against God's people and to conquer them. And it was given authority over every tribe, people, language and nation. All inhabitants of the earth will worship the beast—all whose names have not been written in the Lamb's book of life, the Lamb who was slain from the creation of the world.*
>
> *Whoever has ears, let them hear.*
>
> *"If anyone is to go into captivity, into captivity they will go. If anyone is to be killed with the sword, with the sword they will be killed."*
>
> *Revelation 13:7-10* (TNIV)

These are powerful verses and they are all literal! We know this because God chooses to use the phrase *"over every tribe, people, language and nation"* instead of its symbolic equivalent "sea." *"This calls for patient endurance and faithfulness on the part of God's people."* Those who believe in a pre-tribulation rapture are in for a rude awakening. They will be surprised when they find out they are still here and know who the antichrist is. They will be caught unprepared and some will even believe that God lied to them... but it was their "shepherds" on earth that led them astray. Those pastors and priests who knew not the Word of the Lord[u] and did not prepare them properly for the days to come.

This period of time will be symbolically like "wars of old" where foreign invaders came and began a siege on an ancient fortified city. The saints for all practical purposes are the people within the city under siege. As in those days, the purpose of the siege was to starve to death the people inside the city walls, to coerce them into surrendering, or to just simply slay them if the opportunity presented itself during a skirmish or attempted escape. These will be the effects the "mark of the beast" has on Christians!

Some passive Christians will starve to death because they are not allowed to buy food, some will be killed because of their bold testimony of Jesus, many will be attacked for the food they saved because they had prepared ahead of time for this eventuality, and still others will be coerced into accepting the mark. Regardless of the reason(s), Christians are told to patiently endure until the forty-two months are over. They are to survive as best they can, preparing their minds for these eventualities—to not shirk one's responsibility to preach to unbelievers even if it means death!

We are not to go and hide out for these three and a half P-years. The majority of those whose names are not written in the book of life will believe the antichrist is the greatest world leader of all time! They will "worship" him, trust in him, and will not understand why Christians hate him so much. You should now realize many will be your own family and friends! We must remain steadfast in ceaselessly refusing to take the mark of the beast for any reason prior to death, prior to the rapture, or prior to Jesus' return!

The Earth Beast: False Prophet

Then I saw another beast which rose out of the earth; it had two horns like a lamb and it spoke like a dragon.

Revelation 13:11 (RSV)

This "beast" is usually interpreted as a religious leader—a "false prophet" who will institute a counterfeit "one world" religion that requires worship to the antichrist. I do believe the antichrist will use the remnant of the Catholic Church at the time of the end to set up his false church because of a prophecy attributed to Saint Malachy around 1139 A.D. pertaining to the succession of popes. This prophecy is not from the Bible and is just a personal belief of mine. It foretells the sequence of future popes of which we are near the end. There is only one pope remaining from this list who will be known as "Peter the Roman."

However, I do not think this scripture supports the beast of the earth having anything to do with it. When we examine the entire text of this section as a whole we come across the key clue that rejects that idea. We find this is the "beast" who forces everyone to take a "mark" which will eventually be required to participate in the world economy to buy or sell. Clearly the idea of a "religious" false prophet having this much secular authority or even caring about world economics is ridiculous!

Now I'm not saying there won't be efforts to establish a one-world religious system (this trend is already in place), I'm only stating the beast of the earth is not the champion of this effort. Furthermore, this phased-in requirement to participate in the world economy will be fully implemented with strict governmental enforcement when the abomination that causes desolation is established. Afterward, half of those claiming to be Christians will need to worry about refusing the mark for only a short time (thirty P-days) since Christ will rescue his bride, the church, from the wrath years that follow.[v] Stumblers, those who realize the error of their ways and come to Christ after witnessing the rapture (like the Christians who were not ready and many Jews), can expect to suffer greatly during the last three and a half years of the tribulation and will now need to go into hiding from authorities. There is a time for fighting (the first three and a half P-years) and a time for running (the last three and a half P-years).

When examining the symbolism of this verse, we need to realize that "beasts" are interpreted for us in the book of Daniel and later on in Revelation chapters 17-19 as kingdoms, or more accurately empires. Horns are symbolically identified as kings or powerful leaders of a specific nation. The dragon is the devil and the Lamb is interpreted as Christ or a religious system. However, this beast is slightly different from all the other beasts prophesied by Daniel because all those beasts were beasts of the sea! They were secular governmental systems of control that provided administrative, political, and protection systems for the inhabitants of their kingdoms, countries, nations, or empires.

But this beast is a religious system based on economics—idol worshiping. Why? Well the term "earth" goes hand and hand with pagan religions. Pagan religions all seem to flow back to the idea of worshiping nature: the sun, the moon, the planting seasons (mother nature), idol worshiping, etc. Furthermore, the two horns are "like a lamb," indicates a false ("like a," but not the real deal) substitution for Christ and the one true religion.

So this beast, the "beast of the earth" is passive (two horns like a lamb), but has Satan's backing, lies, and deceptions just like the antichrist. And just as the sea beasts signify not only the governmental control of the empire and at times their specific leader, so does this duality exist with the earth beast. It has to signify both an individual leader (false prophet), as we are told in Revelation 19:20, as well as some kind of organization that affects the population as a whole. In this case a religious system based on idol worshiping… the capitalistic economic system the world has come to embrace. The purpose of this beast is to accomplish things; to implement things that will lead people to like, admire, and follow the antichrist—in other words, to worship the leader of the European Union and his nation.

It exercised all the authority of the first beast on its behalf, and made the earth and its inhabitants worship the first beast, whose fatal wound had been healed.

Revelation 13:12 (TNIV)

The first beast's fatal wound signifies this beast was in existence in years past and "seemed to have died," but now has come back into existence at the end of the age. This is the "new" Roman Empire—the European Union and it may also be even more specific and additionally mean that Alexander the Great will be reborn in spirit since he was the first antichrist as previously touched upon on page 206. A symbolic forgery of Christ's death and resurrection: the beast that once was, now is not, but will come again.[w] So the symbolism is associated with the nation itself, but may possibly be applied to the antichrist as well. The idea of Alexander cannot be known with certainty until we actually see who the antichrist is in the years between 2018 and 2025 A.D.: a time of recognition that depends on a Christian's ability to distinguish the signs.

> *And it performed great signs, even causing fire to come down from heaven to the earth in full view of everyone. Because of the signs it was given power to perform on behalf of the first beast, it deceived the inhabitants of the earth.*
>
> Revelation 13:13-14 (TNIV)

These and the next two verses are some of the hardest scripture in this prophecy to interpret! That is because of what the beast of the earth really represents. It's easy to envision how a "false religious prophet" could be allowed to perform miraculous signs to deceive the masses, but not why he would bother with economics and implement the mark of the beast.

The converse of this is also true for a symbolic interpretation of the beast of the earth. It is easy to see how the beast will institute the mark, but not how or why it will cause fire to fall from the sky! Let's logically think about this further. If the beast of the earth is a religious leader and he is able to cast down fire from the sky, would not the masses of people who witness this miracle want to worship this beast instead of the first beast i.e. the antichrist? What would be his purpose for raining fire from the sky? Just as a sign? Probably not: this awesome spectacle would cause destruction to whatever the fire struck—would it not?

So if this "sign" is mystical in nature, it would only bring praise to the beast of the earth instead of on behalf of the beast of the sea (the antichrist) as the text says. These miracles would have to be performed in such a way as to bring glory to the antichrist and not detract from him. Based on human nature, it seems so irrational to think a person with the capabilities to do miracles would not be held more highly and worshiped than someone who did not perform unexplainable wonders. And yet, we are left to consider this idea is true.

So what is this sign really? I am unsure because it appears to be literal and I cannot find a consistent interpretation for a symbolic meaning that fits with the rest of the prophecy as well as the other end-time prophecies of the Bible. Besides, coming up with a "normal" explanation to what the fire might mean negates the effect of being able to deceive men. If this "miracle" were something more explainable, then men would not be so easily awe inspired. So we are left with the idea of a person who has all the power of the antichrist, has the authority to pass laws that apply to economics, and performs miracles in full view of the world. What an interesting combination of skills—a set of skills that point to these views probably not being entirely true.

At this time, I do not have a good understanding of this miracle and so I leave you with only my thoughts and move on to an equally challenging section of scripture.

> ... It ordered them [those who are deceived] to set up an image in honor of the beast who was wounded by the sword and yet lived. It was given power to give breath to the image of the first beast, so that it could speak and cause all who refused to worship the image to be killed.
>
> Revelation 13:14-15 (TNIV)

Many eschatology scholars and students want you to believe this image is some supernatural statue that looks like the antichrist, but I believe they are living in a fairy tale world. They do make good points and if you read Daniel chapter three you can find evidence that leads one to this idea.

Nebuchadnezzar had an image made of gold that was sixty cubits high and six cubits wide (sixty by six; sixty-six) set up on the plains of Dura in the province of Babylon—a statue or monument that after it was finished required all to bow down and worship the image at the sound of any music that was played. A monument that by today's measurements would be almost one hundred feet tall! This is where these scholars get their idea. Now it may come to pass they are correct in their interpretation and the text is literal on whole, but it just doesn't ring true that anyone in today's world (let alone millions) would bow down and openly worship any inanimate object.

To get around this problem, the false prophet is allowed to perform another miracle and give "breath" to the image so that it could speak. This implies life, but does not say this (although some Bible translations use this word) and the image appears only to be able to speak. Interesting. It does not mean "life" as if this structure would get up and walk, but only the ability to speak. If the image is large, as was the case of Nebuchadnezzar's, can you imagine it moving around? No, this just isn't going to happen. So the image can only talk; but then so does my computer and many things in the high tech world of today and would you worship them?

These experts need only to believe what is written instead of constantly studying to see if the Lord's Word reveals another meaning—a symbolic interpretation. Listen: the beast of the earth will not have the power to bring to life a large statue. Satan does not have this power or he would just make as many men as he wanted to do his bidding i.e. to fill his own kingdom so they could worship him—to really be the creator of things from nothing. No, that power is reserved for the Creator and He did not give it to anyone. These ideas just don't make sense. They are literal interpretations for symbolic language.

So, what "kind" of image is it that will be set up by the inhabitants of the earth (unbelievers, idol worshipers) that will bring honor to the antichrist? We need to keep in mind that the first beast (the beast of the sea) not only represents the antichrist, but also the European Union! When reading the prophecies pertaining to the beast, at times they are talking about the antichrist, while other times the E.U.—and sometimes both. We need to analyze all three perspectives to see which one applies. In the interpretation provided from other end-time writers, the beast of the sea and the earth were understood to mean real people and thus the things written were taken literally. However, I believe these verses speak more to the governmental and economic ideals these beasts can also represent. Why, because if you read the prophecies of Daniel, even more specifically the prophecy of the Kings of the South and the North, and narrow in on the last kings of this prophecy we find details of the last years before the end of the age. We know the last King of the North will be the antichrist and if you were to read the events of his reign, one would find no instance of him being killed by a sword or in any other way for that matter.

Some eschatology scholars claim these verses support that the antichrist will be killed during an assassination attempt, but will survive. Others even go as far as to say he will die, but be resurrected just as Christ was—another forgery of Satan's and in so doing obviously bring much praise to the antichrist. Persuasive ideas I just don't buy. Why? Because if something this significant were true, those details would have been provided in the prophecy of the Kings of the South and the North, which they were not! There is no reference or text that could even remotely be understood to mean this and since that prophecy is really an interpretation that must be taken literally, there can be no misunderstandings with the language.

Therefore, we can conclude the beast of the sea that was "wounded by the sword and yet lived" or "seemed to have a fatal wound" is not the antichrist himself, but a description of the European Union. We must not mix up our symbolism. Remember it was one of the "heads" of the beast of the sea that had been wounded i.e. an empire and not a horn, which symbolizes actual leaders! Thus, the image that will be set up is an "image" of the European Union and not of the antichrist.

So what is this "image" or copy that will be set up? As of now, I would say it is a replication of the European economic model that will be set up worldwide. Let me explain. If we were to go back and examine the trends in place today one would find that the United States' leadership in world economics has been poor. The countries of the world have become dependent on the U.S. for controlling economic forces and its mismanagement and greed at the top levels of Wall Street, corporations, and government has become apparent. The 2008 A.D. crises in the financial markets has caused hardship not only in America, but also worldwide.

It's clear this system in its current form and leadership will not be acceptable in the future by these foreign nations. With the United States destined to go bankrupt as a nation in the not to distant future because of the policies and programs that are being put into place to mask the problems of greed, corruption, and negligence, it is apparent another commerce system in the future must take its place. The lack of trust and faith in the United States monetary policies and the devaluation or collapse of the worth of the U.S. dollar in the future will be the impetus for change.

This decline will create a power vacuum for the antichrist to step into. Now the false prophet (an economic leader in the EU) will be assigned by the antichrist to quickly set up a "new" economic system… an image that will ultimately bring honor to the antichrist by him becoming the savior of "all things lost." This system will be moneyless and require a "mark" to participate in. With the deterioration of the wealth of the United States, the EU will step into the top role as the nation(s) that will lead the world economies. The rest of the world will rejoice at this turn of events because the world now hates Americans almost as much as they hate Israel. This highlights another point: When a country is liked because of its policies (usually prosperity is wide-spread), regardless of how or why things happened, the leader at the time things are going well receives the praise and recognition for these good fortunes. History bears this out. During economic booms of the past, those presidents in office were considered good presidents while during economic busts; those leaders in office were considered poor presidents. George Bush will go down in history as a terrible president and is now hated by the majority of Americans and clearly by almost all foreigners. A president that after September 11, 2001 was regarded in very high esteem. He could do whatever he wanted without question.

These days I have foreigners writing and telling me George Bush is the antichrist (even after leaving office)—that is how much he is now hated. The lesson to be learned: when a country and its people are hated, its leader will get the blame and conversely when a nation is liked, its leader at the time will get the accolades… the praise and honor… the worship! This is really the only way for the false prophet to do things that will bring honor to the antichrist and not to himself.

So this image is probably not a statue that will be worshiped be-cause that brings no honor to the antichrist anymore than worshiping a statue of Christ brings honor to him. This is really an "image of" the European Union's economic system at the time this prophecy will take place (about twelve years from now). The beast of the earth; the false prophet who promises prosperity for all after the hardships that have been caused by the United States' leadership will set up (bring to life) an image—tie-in the other nations of the world economies to Europe and in so doing bring honor to the antichrist as the "world savior." Giving "breath to the image" which is just an expression for getting the system up and running.

This "false prophet," false leader of hope will most likely be the leader of the European Central Bank. And to ensure a quick implementation and full compliance, the European Union will do two things. First, they will make it mandatory by law to participate in this system. The use of money as we know it will be outlawed and secondly, they will bomb New York City shortly after this law goes into full effect to destroy the old system still hanging on and force those still using it to join the new one.

Now what did the false prophets of biblical times do? They were not necessarily religious, but many were advisors to kings. Daniel was a true prophet, but he wasn't a religious leader who went out and proclaimed the Word of the Lord. Yes, he did believe in God, set good examples by the life he led for all to witness, and yes he did testify to these facts when put to the test, but he was no John the Baptist (who was another type of prophet). He was an advisor to the king—to the leader of Babylon.

Most false prophets were in this category and what were the signs of a false prophet? They didn't always preach the true Word of the Lord and many tried to put a "good spin" on what the future held. They rarely told the king he was screwing up or that bad things were coming... those in power never want to hear these things. Furthermore, most feared for their lives if they spoke of these things. True prophets never worried about what was going to happen to them. They told it like it was (usually those were bad things) and let the chips fall where they may. Those who have ears let them hear. As bad as things are right now, they will be the best they are going to be in the years leading up to Christ's second advent. Those who say otherwise do not understand the Word of the Lord or they are just a false...

He also forced everyone, small and great, rich and poor, free and slave, to receive a mark on his right hand or on his forehead, so that no one could buy or sell unless he had the mark, which is the name of the beast or the number of his name. This calls for wisdom. If anyone has insight, let him calculate the number of the beast, for it is man's number. His number is 666.

Revelation 13:16-18 (NIV)

What is the infamous "mark of the beast?" Most people alive today have heard this term whether they are a Christian, belong to some other religion or are even an atheist. But regardless of your religious beliefs, most people have no clue what this mark is and therefore are unaware of what the consequences will be for them when the time comes for this system to be implemented.

The "mark of the beast" is a personal identifier like a tattoo or electronic chip that will allow a person to conduct financial transactions in the last years before Christ returns. At the same time it will allow a nation's government to track an individual's every move and action. Anyone who does not have this mark will not be able to buy or sell goods or services. They will not be able to participate in economic activities. This is the simple answer. But let's look at the deeper things the Holy Spirit reveals to see what else we might learn.

This term comes from the book of Revelation in the Bible. It is a warning; a sign, to Christians that Christ is about to return. That the end of the world, as some unknowledgeable people label it, is near. But this is really another signal for the end of the age and the spiritual rebirth of a new age—the Millennial Kingdom of Christ. This identification "mark" will be a separation tool used by God to distinguish those who believe Jesus is their savior from those who love the world and do not acknowledge him in that regard.

Those individuals, who really know Christ and study the Word of the Lord, will not accept the mark because in doing so they understand they will be condemned to eternal punishment. These are the true believers. They will resist acceptance even unto death. Some will resist openly and some passively, but all faithful servants will do their part and fight by testifying to Jesus' saving blood and proclaiming the Word of the Lord (the two witnesses) to the best of their abilities until the church is raptured.

Those Christians who accept the mark during these last days or rationalize taking the mark believing Christ will forgive them because they were only doing so for the "good of their family"—for the "good of their loved ones" are only deceiving themselves. They are Christians who only profess with their mouths, not with their hearts or actions and have not studied the Word of the Lord. They are of the synagogue of Satan. They are really lovers of the materialistic things of this world... the idols of olden days. God wants Christians during these times to stand up and fight Satan using His two witnesses as their guide so that as many as possible might ultimately be saved. What better witness is there for a non-believer than to have devoted Christians die right in front of their very eyes because they believed in Jesus so strongly. This will be a repeat performance of the beginnings of Christianity in the first century. This persecution is already occurring in other parts of the world like Iran where anyone who converts to Christianity does so under penalty of death.

Those unbelievers who cannot be convinced by their Christian friends and family to finally make the decision to accept Christ and who also take the mark will be lost forever. They will become the enemy of Christians and their trust cannot be relied upon. They will be under Satan's influence and will do things one would never expect i.e. like turning in believers to authorities for reward when the mark finally becomes mandatory.

Let's investigate the timing and the details of the well-known "mark of the beast" so that you can be prepared in advance of what to expect when this time comes. Most people today and this includes Christians, have no idea how close we are to this system. They speed through life, overwhelmed by the changes that are taking place. But the old adage "what we don't know about will not hurt us" is a false saying. It just is not so.

As of 2008 A.D. I find myself on the brink of a decision. Having to make a choice. My driver's license is up for renewal and I can't renew it by mail this year as I had in previous years. Why? Because there is a new federal law in the United States that has gone into effect: an "optional" law that will need to be complied with by 2009 A.D. This is the tagline as spewed-out by lying politicians. You see the new "optional" law that was instituted says all the states in the U.S. must conform to a national standard for identifying their residents or else they will suffer certain penalties; like forfeiture of federal funds for many programs. This "optional" law requires that the states have in place a computerized system that can be accessed by all the other states, their associated law enforcement agencies, and the federal government bureaucracies so that it is easy to track criminal activities. Individuals who choose not to participate will be unable to open bank accounts, get loans, and drive as in my case—not a big problem right? Hmm... Participation is optional?

So back to my driver's license dilemma: for me to renew my driver's license I must now provide a social security number or I won't be able to renew it and thus will not be able to drive. Currently my state does not have this piece of information on file for me. Why do I need to provide this information? So that my state can conform to the "optional" law and not be denied money from the federal government. Now I have had no problem driving for many years without this requirement, but for some reason a social security number is now needed to make me a better driver. Federal government officials claim this is a "security" thing, but in actuality it is "big brother" on its last leg to controlling and spying on every citizen's activities. The excuse for passing this law and others like it is so illegal aliens cannot get U.S. IDs that allow them to carry out some of the awful things that have been taking place.

The original social security law that was passed stated a social security number could only be used for the administration of retirement benefits and could not be used as an identification number. This was to appease the populous at that time and reduce their fears that a national ID system was being put into place. In other words, this lie was spread to get the government's foot in the door and sell the country on President Roosevelt's New Deal programs in the 1930's. This national ID effort began about seventy years ago.

In the years that followed, this number became a person's tax ID number. Next, another law was passed that expanded its use which claimed again that it was "optional." It stated that if you wanted to claim your children as dependents and get tax write-offs, your children must obtain a social security number. So those who were too young to work now needed a social security number that was required only for the working. Following that law, a few years later, another law was passed that lowered the age at which a child "had" to get a social security number to be claimed on a parents taxes to two years of age. Step-by-step the trap around closes in. Along the way, social security numbers began being used by businesses in the economic systems of America: banking, loan application, credit reporting, etc. as a way to identify and track a person's financial transactions.

How far we have been led down these dark stairs in the United States. We have almost reached that last step on this slippery staircase that was constructed with the enslavement of the elderly seventy years ago with promises of "free" money; money that would be paid to them at retirement, but earned on the backs of their children's labor. A great pyramid scheme that is also nearing its end.

This number that was suppose to be used only for administering retirement benefits has become a person's national ID number in just a couple of generations by gradually sliding in different requirements without the population ever realizing what has been happening or agreeing to it. I'm sure systems like this are now in place in every nation of the world with few exceptions—some countries are even farther down the slippery slope than the United States.

So as of 2009 my driver's license will effectively become my national ID card and any state or federal agency will now have instant access to everything about me—and probably you. Not to mention if they have access to this information, then other countries' law enforcement agencies have access too! Many nations already share this information freely between each other without your permission.

We are now near the end of our slimy rope and there is only one remaining unfinished item that will cause many to lose their grip and hit rock bottom… the bottom of the bottomless pit to hell. This piece will take about another fifteen years to sell people on and phase-in. This remaining item in the plan is to replace the license with a personal identification mark that can be used for everything and theoretically not be counterfeited. You see, the system that is being put into operation right now will only aid criminals. In the past if I lost my driver's license or it was stolen, the thief would have no way to tie this document to my bank accounts or social security number. It was only used by law enforcement agencies to keep track of when I broke the law.

Now that these two systems (the financial and law) have been linked together, it will be much easier for organized crime to counterfeit IDs, use stolen licenses, or worse yet, entice government employees to steal people's information and sell it to them so they can electronically "clean-out" people's accounts. The only method to stop this after the problem arises, and it will, is to convince ("sell") the unsuspecting masses that they need this mark because it can't be counterfeited. When you start hearing this greasing of the skids, this sale pitch for a new law to combat these computer crimes, we will have finally reached the end of the slide to hell.

So again I ask, what exactly is this mark? Some believe it will be an electronic chip implanted in the forehead or under the skin of a person's right hand (as the Bible says) that can be used to uniquely identify each one of us for conducting everything needed in everyday life. If you need to buy food, go to the grocery store and gather what you need. Scan your items purchased, along with your hand, using the electronic store scanner. The scanner will total your purchases, read your mark (microchip), and "walla" everything will be taken care of. No need for real currency to exchange hands. Everything will be handled electronically.

If governments need more money, they can just add a few zeros to the central bank computers and "walla" they have more money. Want a new television, some clothes, or to just go to the movies and dinner? No need to run to the ATM or bring your purse or wallet. The mark replaces everything. The convenience of the mark makes everything so much easier. No waiting for tickets at the theater or bills at the diner; just swipe your hand under the scanner as you enter the theater or leave the restaurant and the cost will be automatically deducted from your account. Sounds fantastic—just more grease that will be used to convince people to accept the mark.

Have you seen the Visa commercials where they make it seem easier to use their card than to pay cash? I find no system easier than to pay cash. How many eyes roll in the store checkout line when someone gets out his or her credit or debit card to pay—then they have to wait for a verification code to find out if the purchase was approved or denied. Next a signature and photo ID check. What a mess if the card is denied! Does this ever happen with cash? And yet people are swayed and deceived into believing using a credit or debit card is easier than paying cash.

This is just another indoctrination technique (television commercials) being used to pave the road to the future just like the acceptance training going on with young people and their use of ATM's. Many are being programmed to accept these electronic systems and soon they will wonder why there are no jobs left: when bank tellers are no longer needed. What a great system for the rich banker. His cost of doing business practically disappears. There will be no need for bank buildings, safes to store money, armor car services, armed guards, printing and minting of currency, etc. We are already seeing these advances with the invention of on-line banking and automated bill payment. There are really very few jobs that will be needed in the financial sector of an electronic cashless society. Just millions of dollars of profits waiting to be reaped for the mega-rich at the top.

What about these effects on retailers? Cashiers will go the way of the dinosaurs as well. Many large retailers are already installing automated machines to replace cashiers... the brainwashing and training continues as people grow accustom to their use and use them more frequently. I myself refuse to utilize these machines and would rather wait in a line than to aid in this transformation to a total electronic world that reduces job opportunities for people. Christians should not be using these systems. Where will people get jobs when these trends have fully matured? Resistance to the antichrist's plan ahead of time is "fighting the lies" ahead of time, although the end will still come at the designated time on God's calendar.

With all the jobs disappearing, getting a job as a cashier was one of the few that could be found since almost everyone is addicted to having as much material wealth as possible... to out duel the Jones next door... even if it requires borrowing well past the point of certain bankruptcy. This addiction is worse than any drug, pornography, sex, or gambling habit. People don't even realize they have it and the governments of the world, which are even worse addicts, are encouraging it. The United States is already at the mercy of foreign hostile governments and so are many third world countries at the mercy of the industrial nations who control the world's wealth... remember, he who has the gold makes the rules. These conditions will only deteriorate as the unaware U.S. population is deceived to the point of no return... actually many are already past this point setting up conditions that will further aid in the call for the mark of the beast.

The eschatology scholars who believe this will be a microchip, like the ID chips that veterinarians have been implanting into pets, may be in error. I too envision how easily this idea may be expanded to humans as you have read, but here are the problems I see with that system. Chips can be forged too. The sophisticated criminals will have access to the same technology as the governments and will be able to acquire these same chips and program them. If it is a simple matter of implanting one under the skin, like they do in animals, it will be a simple procedure to take one out and insert another one in. It is not a permanent thing. One could be taken out and another one put back in—a counterfeit one with another's identity. Maybe they wouldn't even need to do that since a live person will probably not operate the scanners. A person could just hold the fake ID chip in one's hand and walk right through or pass it right underneath the scanner.

Furthermore, no one will easily know whether you have accepted or rejected the chip (the mark of the beast) because it will be under the skin and not visible. The electronic chip could even be mounted into a ring worn on the hand so it would emit the electronic signature of the mark of the beast. Using this technology, there really will be no visible evidence, so a person could walk around town without one and nobody would be able to distinguish between them and someone who has already taken the mark. This would be a severe problem for enforcement efforts.

No, I believe this "mark" in whatever form it will ultimately take must be visible unlike the mark used by Christ to mark his servants, which is almost certainly invisible.[x] This way when a Christian walks down the street they will be easily spotted by his or her neighbors, friends, and all authorities and nonbelievers who are using the system. In this manner it will be easier to identify and eventually persecute those who are resisting the transition.

Let's suppose you are a policeman during this time and you already have your "mark" for you must have one to hold a job in the future. You are driving around on patrol and you spot a person without the mark. What are you going to think? You know they can't legally do anything (buy or sell) so clearly this individual must be up to no good or have something to hide. They may even be breaking the law if it has passed the phased-in time and the law has gone into full effect. Let's stop and question them—or possibly arrest them depending on the timing. Are the authorities going to believe your story on why you don't yet have a mark, when clearly they already have bought into the deceiver's lie? I think not. And to make matters worse, there will be some who will not get a mark and are not Christians[y]... they are people who are real criminals or who cherish their freedom. This will make a Christian's testimony to law enforcement agencies and neighbors even harder to swallow because those with the mark are already unable to distinguish the truth from lies.

The same could be said for neighbors who see you around in the neighborhood without a mark. Will they not call authorities and report you? What about your suspicious neighbors right now? Most people do not trust neighbors who don't conform to what is considered the norm.

Isn't this how you view neighbors and strangers who are a little differ-ent… with suspicion? This is how you will be viewed if you are a Christian and have not conformed to the norm i.e. have not accepted the mark.

Those who refuse the mark will become outcasts, outsiders, and re-bels who will be looked down on with mistrust from those people with marks. This practice will leave Christians open for all kinds of abuse since they will be in the minority. The tables will be turned and Chris-tians will be looked upon as criminals by the real laws break-ers—especially when this law becomes a mandatory requirement to "buy or sell or do anything" instead of "optional." The real persecutions of the last converts will start in earnest during the final three and a half P-years before Christ returns.

The KJV states the mark will be received "in" a person's right hand or forehead, which leads one to envision an electronic chip implant. The NIV uses the word "on" instead of the word "in" which implies a visible sign. So, more than likely, this mark will have to be visible using this translation. Clearly, this word is important, and the Greek and Hebrew manuscripts indicate the correct choice is "on" or "upon." A visible mark that is permanent suggests a "tattoo" of some sorts. Something the Bible forbids[z] and this may be the reason why. However, implementing this form of the mark throughout the world would be time consuming whereas the microchip would be easy to do. There is also a blur factor with tattoos, and the thin areas of skin between the black carbon pigment lines of the barcode will inevitably be blurred by the natural skin cell division causing ink particle movement. This will make scanning a tatooed barcode unreliable. Theses problem make certain that this will not be the system either—even though a personal barcode tattooed on one's hand fits all the requirements better than a microchip implant.

Regardless, each person's ID will have to be unique and this presents some problems for every "mark" being associated with the number 666. The best guess is it might work like a phone number. If you live in a certain area, then all those in that area have that same area code, but a different phone number. This same number and area code might be assigned to someone else in another country because each country has its own country code.

When all the countries are linked together then a world code is necessary and placed in front of each national code, which is in front of essary and placed in front of each national code, which is in front of each area code, which in turn is placed in front of each telephone number.

Applying this idea to personal IDs worldwide, the phone number analogy would be replaced with a person's social security number and tied in with similar numbering systems used by other nations to identify their citizens.

Combining different countries' IDs into one computer database by using country codes and possibly a universal code (666) that might be added for some unknown reason, would make each individual in the world unique. If one were to understand how barcodes were developed in the 1970's and are now used worldwide in thirty short years, one can understand how the number 666 could be uniquely "built into" each and every barcode. This may suggest that the number of the beast will not only be used by Christians to identify the antichrist, but if barcodes are used in some way for the mark of the beast why the number 666 would be included in each person's ID.

The 144,000: the Firstfruits

> Then I looked, and lo, on Mount Zion stood the Lamb, and with him a hundred and forty-four thousand who had his name and his Father's name written on their foreheads.
>
> *Revelation 14:1* (RSV)

Now this verse cannot mean that Jesus has returned to earth with 144,000. Why? Because when Christ returns he will stand on the Mount of Olives[aa] as the prophet Zechariah foretold and as the angels informed the apostles at Christ's ascension on the Mount of Olives.[ab] The Mount of Olives is not Mount Zion. This is symbolic language and cannot be taken literally.

These are God's people and they too have a mark on their forehead that separates them from the unbelievers who garnish the "mark of the beast" that brands them as Satan's property. However, this heavenly mark is invisible. These are men who keep God's laws and follow Christ. One cannot receive eternal salvation by believing in Christ, yet are always breaking the law, and conversely those who try to keep the law, but do not recognize Jesus as their Lord and Savior will not obtain everlasting life either.

> *Hear, O Israel: The LORD our God, the LORD is one. Love the LORD your God with all your heart and with all your soul and with all your strength. These commandments that I give you today are to be on your hearts. Impress them on your children. Talk about them when you sit at home and when you walk along the road, when you lie down and when you get up.* **Tie them as symbols on your hands and bind them on your fore-heads.** *Write them on the doorframes of your houses and on your gates.*
>
> Deuteronomy 6:4-9 (TNIV)

These laws are to be on your hearts. To be practiced at all times in all that we do. To be reflected in the works of our hands and the thoughts of our minds. God's people need no visible mark for they are known by the Lord and will be distinguished by their character and the mark they don't have. Nevertheless, our undetectable (lack of a) mark will be a 'highly visible' sign to unbelievers and fellow Christians alike in the last days—a beacon of light that will shine brightly for all to see. Simply put, those who have the mark are unbelievers and those who don't (with some exceptions) are God's people.

> *And I heard a voice from heaven like the sound of many waters* [raptured saints] *and like the sound of loud thunder; the voice I heard was like the sound of harpers playing on their harps, and they sing a new song before the throne and before the four living creatures and before the elders. No one could learn that song ex-cept the hundred and forty-four thousand who had been redeemed* [harvested] *from the earth. It is these who have not defiled them-selves with women, for they are chaste; it is these who follow the Lamb wherever he goes; these have been redeemed* [paid for by Christ's blood] *from mankind* [not just Jews] *as first fruits* [the

first ten percent of the harvest] *for God and the Lamb, and in their mouth no lie was found, for they are spotless.*

Revelation 14:2-5 (RSV)

These are the same 144,000 mentioned in Revelation chapter seven and just as it was in chapter seven they are mentioned before the great multitude (roar of rushing waters). We learn further information about them in these latter verses. They are "firstfruits" and they follow (report to) Christ. Why are "firstfruits" needed here at this point? Because this is midway during the seven-year classical tribulation period and this is where the antichrist's authority is taken away and Christ begins ruling the nations from heaven. This is the beginning of the harvest. The antichrist is weeding the world of believers just as it will be when Christ comes and he weeds the unbelievers from the world.

Furthermore, we learn that the 144,000 are not Jews, but are from the entire human race—that the 144,000 Jews in chapter seven are only symbolism for true Christians.

There is neither Jew nor Gentile, neither slave nor free, neither male nor female, for you are all one in Christ Jesus. If you belong to Christ, then you are Abraham's seed, and heirs according to the promise.

Galatians 3:28-29 (TNIV)

So these redeemed Christian "Jews" (Abraham's seed) will be used by the Lord to preach to the remaining world's population during the wrath years. Why? Because Christ will remove his "two witnesses (the Word of God and the testimony of Christ)" when the church is raptured. These symbolic 144,000 males will be used to spread the Gospel so that the last stragglers for Christ's sheep pen may be found. This is so that, when Jesus arrives at the end of the tribulation, all will have made their decision! This event occurs at the point where God begins His punishment of those who would not listen to Christian testimonials during the first three and a half P-years of the antichrist's rule.

Now let's put this timing in perspective by digging deeper into the sequence of events surrounding the 144,000. In Revelation chapters 6-8 we find the prophecy of the seven seals we have already discussed. But that earlier discussion did not address this topic. The sixth seal occurs when the abomination of desolation occurs. Right after this event we learn this is the time when God's wrath begins.

> *for the great day of their wrath has come, and who can stand be-*
> *fore it?*
>
> > *Revelation 6:17* (RSV)

However, there are a few things of importance worth noting. First, when the seventh seal was opened (this seal is the last seal and signifies God and Jesus' wrath) there was silence in heaven for about half an hour. We talked about this back at the end of chapter five and even tried to calculate how long this was using various understandings. However, the most likely understanding after taking everything into consideration is this means thirty days. Why? Although there is no basis in the Bible for what I am going to say, we have learned through our studies that not only does a day stand for a year and even a thousand years, but also other time analogies can be made i.e. it can stand for Jubilee years.

On page 110 we found in our analysis that a day may also be symbolically equal to a minute in God's eyes and in this case I believe it is. Since half of an hour is thirty minutes, we can see that this really could mean thirty days. We know that the sixth seal is at the end of the abomination of desolation and if the seventh seal is opened immediately afterward which it appears it will be, then the thirty days that follow this event would correspond exactly with the thirty minutes of silence in heaven.

> *When the Lamb opened the seventh seal, there was silence in*
> *heaven for about half an hour.*
>
> > *Revelation 8:1* (RSV)

After this silence we find God's planned wrath is unleashed. But what occurs between the sixth and the seventh seal during these thirty days? We know there will be a lull in natural disasters, one of the key methods God uses to punish mankind.

After this I saw four angels standing at the four corners of the earth, holding back the four winds of the earth to prevent any wind from blowing on the land or on the sea or on any tree. Then I saw another angel coming up from the east, having the seal of the living God. He called out in a loud voice to the four angels who had been given power to harm the land and the sea: "Do not harm the land or the sea or the trees until we put a seal on the foreheads of the servants of our God." Then I heard the number of those who were sealed: 144,000 from all the tribes of Israel [except the tribe of Dan].

Revelation 7:1-4 (NIV)

We notice from this text during this silence... this lull... that God is holding back His anger and preparing for the things to come. Envision this lull is like the eye of a hurricane. It falls exactly in the middle of the classical seven year tribulation period where troubles are building from the beginning of the antichrist's rise to the midway point and then there is a thirty P-day lull followed by the rest of the storm during the three and a half years of God's planned desolations. This really will not be a lull for Christians, but the worse time in the history of mankind for them.

During the silence Jesus is finding and sealing the 144,000. What else do we know about this time? It is the time the mark of the beast has been finalized and the time the antichrist has began killing Christians in earnest. Next, we find confirming evidence to verses 14:2-5 that the "roar of rushing waters (a great number of people)" that follows the sealing of the 144,000 is the "great multitude" in heaven that no one could count for it also follows the sealing of 144,000 in Revelation chapter seven.[ac]

Their Wrath Begins: God's Winepress

Then I saw another angel flying in midheaven, with an eternal gospel to proclaim to those who dwell on earth, to every nation and tribe and tongue and people; and he said with a loud voice, "Fear God and give him glory, for the hour of his judgment has

come; and worship him who made heaven and earth, the sea and
the fountains of water."

Revelation 14:6-7 (RSV)

This is the beginning of the 144,000's last day's ministry because those who knew the truth have been removed and yet there is still three and a half years remaining. This symbolism supports a time when the gospel is being preached with conviction. Many today believe the gospel of Jesus Christ has been preached to all parts of the world and in most places it has, but there must still be areas and people who have been hindered from hearing its full message.

Jesus provides the signs for the end of the age in Matthew 24. What most people do not realize is these signs are in sequential order! That's right, they mesh with the prophecies of Revelation. When does Christ mention that the gospel will be preached to the whole world, so that no one can claim they did not know? In verse fourteen, after prophesying of the persecution of Christians, the apostasy of the unbelieving masses, and immediately before the abomination of desolation. He says the truth of God's kingdom will be preached to everyone and then the end will come.

Most scholars and teachers believe this means that during the age of Christianity (the Times of the Gentiles) that the Gospel will be spread throughout the whole world and when this effort is completed by God's people, the end will come. This is most certainly true. But on another level, a more basic level, we should understand Jesus was not asked what would be the signs covering the entire age, but the signs to recognize the "end" of the age. If we apply his answer in the context it was given and recognize the relationship of the 144,000 on the tribulation timeline with other events, one can see that it is the 144,000 who are preaching the gospel. This is another interpretation, in addition to that which Christian leaders teach, which I feel is the best one that fits all the information provided.

Jesus said, heaven and earth will end, but never his words.[ad] There must always be some method to pass the Word on to those who are ignorant of the existence of God and Christ's authority. If the dutiful believers are taken at the midpoint of the tribulation, as I believe the prophecies support, then there must be a means to keep the light of truth alive during the remaining years of spiritual darkness.

245

A second angel followed and said, " 'Fallen! Fallen is Babylon the Great,' which made all the nations drink the maddening wine of her adulteries.

Revelation 14:8 (TNIV)

Again, this text is interesting in that it follows right after the sealing of the 144,000, the great multitude in heaven and the end of the thirty days—the end of the thirty minutes of silence in heaven. Babylon the Great's demise has been touched upon in earlier chapters for this event is one of God's first judgments and will be analyzed in thoroughness in the next chapter. This is the failure of the capitalistic systems we enjoy today. The most likely cause is a nuclear bomb that will destroy all of New York City at one time—in one hour. It is possible that something else might transpire that will accomplish the same things, but this is my best interpretation after studying the prophecies as a whole. This is because this catastrophe will be the destruction of a city that can be seen from ships at sea that has the power to disrupt the world economy and cause much of it to collapse.

A third angel followed them and said in a loud voice: "If anyone worships the beast and its image and receives its mark on their forehead or on their hand, they, too, will drink of the wine of God's fury, which has been poured full strength into the cup of his wrath. ...

Revelation 14:9-10 (TNIV)

We learn that in addition to God bringing down the worst religion of all time (idol worshiping in its purest sense—capitalism) God will punish forever those who worship the new system (have the mark of the beast and participate in this new cashless economic system—the image of the old system). They will receive His full wrath just as Babylon the Great will. We also learn from this warning that the mark of the beast and the image will be set up prior to God's wrath, for how can Christ pour out God's wrath on people who accept these things, if implementation occurs afterwards?

How will the rebels to the truth experience this wrath? Well if they accept the mark, they will miss the rapture boat and when that boat leaves, Christ will begin actively ruling from heaven by dispensing God's planned desolations (punishments)! More specifically, Christ will begin actively ruling thirty days after the abomination of desolation is set up. These deceived souls will then be forever lost to an eternity of love and peace and will spend perpetuity in everlasting damnation.

... He will be tormented with burning sulfur in the presence of the holy angels and of the Lamb. And the smoke of their torment will rise for ever and ever. There is no rest day or night for those who worship the beast and his image, or for anyone who receives the mark of his name." This calls for patient endurance on the part of the saints who obey God's commandments and remain faithful to Jesus.

Revelation 14:10-12 (NIV)

We see the final consequences for those who reject the truth is an existence in the Lake of Burning Sulfur forever and ever. All those who accept the mark sign their own death certificate. We also find out that all sins are not forgivable. He who has an ear Listen!—There is only one unforgivable sin… the rejection of the Holy Spirit and this mark is a witness to that rejection! No one will be forgiven who openly accepts the mark by choice or by coercion.

All who take the mark will not be allowed to change their minds and ask for forgiveness later after realizing they should have listened to their Christian friends and family instead of branding them outcasts, religious fanatics, and turning away from them.

And I heard a voice from heaven saying, "Write this: Blessed are the dead who die in the Lord henceforth." "Blessed indeed," says the Spirit, "that they may rest from their labors, for their deeds follow them!"

Revelation 14:13 (RSV)

This is a blessing from God. The very same blessing at the end of Daniel when Christ, prior to his incarnation, says, "blessed are those who reach the end of 1,335 P-days!" Why is this blessing given in both of these prophetic books at the very same point on the tribulation timeline? Because the mark of the beast has become mandatory and the church has just been raptured. All true believers have left for the wedding feast of the Lamb and have been spared the coming desolations of God's wrath. They were not spared the evil plans and tortures of Satan, the antichrist, and the false prophet, but spared the retributions that the Creator is about to meter out.

The majority of those left on earth are unbelievers destined to eternal punishment. But as I stated, there will be millions of stragglers and for them to survive to the end of the tribulation will take a heroic effort—basically an impossible feat to accomplish. They will need to survive without accepting the mark for several years instead of the thirty P-days of persecution that Christians who kept watch had to endure. A task most will die from trying to complete.

Wicked people with nowhere to turn will surround them on all sides. Many Christians during their trial and tribulations (prior to the rapture) could rely on fellow Christians to aid them through their tough times. These last converts' obedience to find and help the remaining stragglers like themselves will be overwhelming. They will have fewer supporters and God recognizes this and bestows upon them a special blessing. For those who are successful and live through these terrible times, they will enter the Millennial Kingdom and their deeds will be rewarded by Christ.

> *I looked, and there before me was a white cloud, and seated on the cloud was one" like a son of man [Jesus]" with a crown of gold on his head and a sharp sickle in his hand [to cut the wheat; harvest the saints]. Then another angel came out of the temple and called in a loud voice to him who was sitting on the cloud, "Take your sickle and reap, because the time to reap has come, for the harvest of the earth is ripe." So he who was seated on the cloud swung his sickle over the earth, and the earth was harvested [saints raptured].*
>
> Revelation 14:14-16 (NIV)

Now we back up slightly on the timeline. There is a lot going on around this time and it is impossible to capture simultaneous or closely spaced events in writing. This prophetic imagery is really a description of the rapture! Christ is told the harvest is ready so go down and gather the wheat... those who belong to him. Gather them for the wedding feast. We see this duty is not left up to the angels, but for Christ himself to do whereas all the other tasks are assigned to angels if you have not noticed! This is after the firstfruits were taken (prepared)—the 144,000. We also learn there is a scheduled time for this responsibility (*"the time to reap has come"*) and we learn that Christ does it from the air (clouds) and does not put his feet on the earth yet just as Paul prophesied in Thessalonians.

Also, this event is to occur before Armageddon. We learn this from the verses that follow. The believers are taken before the Great War of Armageddon. So the fields are reaped before Armageddon and not afterwards. The good crop is harvested and the remaining weeds are left behind to be burned. Thus, Christ's "official" return (second coming) is unmistakably after the armies of the world are gathered on the plains of Megiddo when he plants his foot on the Mount of Olives and a massive earthquake rocks the world.

Therefore, those who do not support the rapture or believe in a pre-tribulation rescue of the elect have more evidence to think about. Why is Christ told to harvest at this point... before the War of Armageddon and around the times of the events that are taking place in the middle of the classical tribulation period? Remember that prophecies are given in sequential order. If there is no rapture at this point in time, then who or what is Christ reaping?

Another angel came out of the temple in heaven, and he too had a sharp sickle. Still another angel, who had charge of the fire, came from the altar and called in a loud voice to him who had the sharp sickle, "Take your sharp sickle and gather the clusters of grapes from the earth's vine, because its grapes are ripe [the armies are gathered]." The angel swung his sickle on the earth, gathered its grapes [gathered the armies of the world to the Middle East] and threw them into the great winepress of God's wrath. They were trampled in the winepress outside [plains of Megiddo] the city [Jerusalem], and blood flowed out of the press [millions died], rising as high as the horses' bridles for a distance of 1,600 stadia [180 miles].

Revelation 14:17-20 (NIV)

Now we see angels are used to symbolically "gather the grapes for mashing into wine"—to pour out God's wrath. But we already know this, for these are the seven trumpet judgments and seven vial judgments previously discussed. The armies will be gathered in Israel for the battle of Armageddon (the great winepress of God's wrath) outside the city of Jerusalem and the killing will be so great that the bodies of the dead will reach the height of a horse's bridle for 1,600 stadia! This is the "winepress" Christ "treads of the fury of the wrath of God Almighty."[(ae)]

I noticed something odd about the number that is provided here. It is not a unit of measure that most people are familiar with and so I provided a modern equivalent… about one hundred and eighty miles or three hundred kilometers. I am unsure as to why this number was chosen by God at this time, but if one were to convert one hundred and eighty miles into feet there would be 950,400 feet.[(61)] This transformed number comes out to be 6.6 times 144,000 exactly: both numbers we know the Master uses. Man's number and the number of "sealing." This is man's number (6.6) without God in his life and those who are dying are the unbelievers who were left behind because they rejected the truth of Christ and didn't follow God's laws. Maybe, just maybe, we can envision God sealing their deaths with the use of this number?

Those people who believe there is no rapture really do not understand the sequence of events of this prophecy and how the other end-time prophecies are intertwined to paint only one portrait of the last days of this age. First the wheat will be harvested (Christians taken by Christ to ready their return), next the grapes will be gathered to be crushed in the last battle (World War III), and then finally Christ will return with the saints and his holy angels to judge those who survived and to set up his kingdom on earth!

We have uncovered more of God's plans for the last days including the details of the infamous mark of the beast. The work has been exhausting and requires extensive commitment from the eschatology student to learn all there is to know beforehand for the time is short. There is but one Revelation prophecy left to investigate. Although it is not as widely known as others, such as the Mark of the Beast or the Four Horseman of the Apocalypse, it is undoubtedly the most significant of all the prophecies you will learn about.

Chapter 12

Babylon the Great

THIS PROPHECY IS BY FAR THE MOST important of all the prophecies given in the book of Revelation and is the key to unlocking all their meanings. Why? Because it is not really a prophecy, but the interpretation of a prophecy. In other words, it is the answer from God to the question of what the vision of the "Prostitute who rides the Beast" represents. Therefore, we cannot interpret it as we wish, but only try to comprehend its meaning. Secondly, since several of the prophecies are redundant, we can use this interpretation as the backbone for interpreting the other prophecies we have investigated. In actuality, that is what we have been doing when the details of the earlier prophecies were discussed.

Many find reading the book of Revelation troubling. That's because they fail to understand what I just said: God didn't give us these prophecies without giving us their meaning! In the last chapters that deal with the final prophecy of this age, an angel is instructed to tell St. John what it means. Just as every prophecy in Daniel is interpreted, so too we can find the meanings of the prophecies in Revelation. The problem most Christians have when they study the symbolism within the prophecies is they are blinded to the idea that an interpretation is included because that interpretation also appears to be symbolic.

Let me give you some examples. In the book of Daniel, many times Daniel has visions of beasts and in every case an angel instructs Daniel that all these beasts represent countries and nations, but more specifically they signify empires. Some interpretations even tell us (by name) whom these beasts represent—like Babylon (lion), Medo-Persia (bear), and Greece (leopard).[a] However, the future beast that will arise in the last days is not named so clearly. But just because it is not named, does not mean we cannot know who it is or, even worse, that eschatology experts can interpret this beast as meaning something else entirely as some do. Make no mistake, the last beast will be a world dominating country… an empire and any interpretation to the contrary is in effect altering the interpretation already provided by one of God's messengers.

In the last chapters of Daniel, Daniel has a terrifying vision that we are not given the specifics of. That's right, the last chapters that deal with the Kings of the South and the North are not a vision, but the interpretation of Daniel's dream. So those who view this prophecy as symbolic language are missing the point. This is literal language that looks like symbolic language only because many do not understand its meaning.

The facts are that these chapters cover major historical events from the Persian Empire up through Christ's return. If you are unfamiliar with past history, then you will never understand these verses. Conversely, if you are a historian who is not a Christian, then you too will have problems with a total understanding. Only a Christian who knows world history can figure out this interpretation, and then only with the Holy Spirit's tutoring may one unlock its secrets.

I am reminded of another important point that flows from this line of reasoning. I find many Christians who hold firmly to the notion that no one can know when Christ will return. These are the same Christians if you were to ask, "on what date did Christ die or start his ministry," would have no clue. How are they to understand prophecies, which are only details of future history provided in advance, when they have no knowledge of past history—things that are knowable? If a person doesn't recognize or study events that are knowable, they can never expect to discover things about the future that are also knowable!

Let me provide one last example to make the point before beginning our study of Babylon the Great. It comes from the Kings of the South and the North prophecy. I read a posting on a Christian website, about the antichrist. You can find many "antichrist hunters" on these forums. The problem they all seem to have is their obsession with being the first one to find the antichrist. Many are educated in scripture, understand the time is short, and even keep watch for the signs. Those are the things that Christians are instructed to do—along with many others.

The problem, as I see it, is the determination used in pursuing their quest blinds them from finding the truth. I read a post from one such person who claimed the antichrist was the leader of Iran. Why? Because he appears to be the most vocal Muslim leader opposing Israel. I have read some of his other postings and he is knowledgeable on the prophecies of Daniel and Revelation. However, being well-informed doesn't necessarily mean he completely understands them, even though he claims he does and states other Christian's interpretations are incorrect. These limiting Christian attitudes block the Holy Spirit's efforts to correct any misbeliefs a person may have. It appears their obsession blinds them from seeing everything else that surrounds them and, in so doing, keeps them from learning the complete truth. Consequently, I wrote this man back and explained that the Kings of the South and the North cannot be interpreted as we wish, because it is an interpretation and not a vision! It is literal language and not symbolic and furthermore, this explanation is from God and we are informed the antichrist will be the King of the North. I told him that if the leader from Iran was the antichrist, then he must be the King of the North. Since this is true, I asked him who then is the King of the South?

Recognizing that since it is an interpretation, directions are literal and the King of the North must come from a country north of Israel and the King of the South from a country south of Israel. Iran is neither north nor south, but mostly east of Israel. If we are to understand that the King of the South will be a leader from the Islamic nations because the vast majority of these nations are south, then we might logically conclude he could be the last King of the South. But he can never be the King of the North and hence, he is not the antichrist. By not knowing all the prophecies, or accepting interpretations as actual prophecies and developing private understandings different from their intended meaning (because the language "appears" to be symbolic), a person will never find what they are searching for.

We too who study these matters, will never find the truth if we suppose the information is not provided. If you believe all the information contained in Revelation are symbolic prophecies, then you are limiting the Holy Spirit from showing you the truth. Blocking the Holy Spirit from aiding you in understanding the interpretations already given, will blind you to the truth, just like the gentleman who searches for the identity of the antichrist. I personally know Christians who consciously choose to live in darkness even when the Holy Spirit provides the light that leads to the truth of God's Word.

Babylon and the Prostitute on the Beast

One of the seven angels who had the seven bowls came and said to me, "Come, I will show you the punishment of the great prosti-

tute, who sits by many waters [many nations or countries]. *With her the kings of the earth committed adultery, and the inhabitants of the earth were intoxicated with the wine of her adulteries.*

Revelation 17:1-2 (TNIV)

The great prostitute who sits on many waters is another metaphor for the "Beast of the Earth (false prophet)" and also later, the "Woman who sits upon the Scarlet Beast." We know this because John was told to come and he would be shown its true meaning. The kings of the earth who are committing adultery are powerful leaders and wealthy men who have turned away from worshiping the Lord. Understand, Jesus is the groom and the Christian church is his bride. Many have turned away from the true God and sold their spiritual souls (prostituted) in the worship of idols—in the pursuit of the materialistic things of this world that has corrupted everything (the American dream). These people care for little else except how to acquire even more things and money, even if it requires them to lie, cheat, murder, or commit every sin imaginable to achieve their goals. Many sins they have no inkling they are committing—such as slandering a fellow coworker to get ahead. *The inhabitants of the earth are intoxicated* (caught up in this way of life) *with the wine* (the pleasures of a good life) *of her adulteries* (worshiping everything except God; notice this is plural), that they are unable to control their actions—much like an alcoholic at an open bar.

How many reality shows do you watch on television where it doesn't matter if contestants lie, cheat, strip naked, fornicate, entice, abuse, etc. just so they can be the last person left to win the prize? Most people, when given the opportunity to win large sums of money, would trade their morals and beliefs; their very souls to get money, power, and wealth and would see nothing wrong with it! Many of these people state publicly they are Christians, but they are blinded to the transgressions they commit. Powerful people with money are even worse than the ones who don't have any! Just turn on the business channel and you will see what I mean. Their greed is a fire that is all consuming.

And he [angel] *carried me away in the Spirit into a wilderness, and I saw a woman sitting on a scarlet beast which was full of blasphemous names, and it had seven heads and ten horns.*

Revelation 17:3 (RSV)

The woman sitting on the scarlet beast signifies she has control of this beast, but not total control. Imagine a rodeo rider atop a bull. The bull (beast) is controlled by the rider (woman) at times—like when the bull is in the chute just before the ride begins. Here the bull has trouble doing as it pleases because of the rider and other constraints. Other times the rider, in this case the whore, has trouble controlling the bull—like when it is let out of the chute to buck wildly. They share a common bond, just not a harmonious one. The scarlet beast described here is the "Beast of the Sea." This symbolic beast is the European Union and its leader—the antichrist, who we have already studied in other prophecies utilizing different imagery and will soon confirm once again.

The woman was dressed in purple and scarlet, and was glittering with gold, precious stones and pearls.

Revelation 17:4 (TNIV)

Many scholars want to point to this verse and claim the woman is a representation of the Christian Catholic Church or the Christian religion as a whole because of the purple and scarlet garments she is wearing. But this is not exactly true as they portray her. Theologians who hold to this interpretation are usually evangelical Christians or belong to fringe non-denominational churches. They interpret these symbols because they are unhappy with the teachings of Catholicism and major protestant denominations. They further imply that the catholic segment of the Christian Church needs to repent or they will not see the kingdom of heaven. But the reality of this bias clouds their ability to interpret these passages correctly. They see only the splinter in their brother's eye, but not the plank in their own. There are passages in Old Testament prophecies that lead one to conclude that a third of the world's population will be saved and two thirds will perish.[b] This idea is consistent with the way God uses numbers and other prophecies we have investigated. Currently about a third of the world's inhabitants profess to be Christians and at least half or more of those claim to be Catholic. Many of these Christians are weak in faith and rarely worship the Lord regularly, but other denominations have this problem as well. They only call on Him when times are bad and forget who He is when times are good (i.e. when they are committing adultery with idols). But the fact remains, no matter how many sins you commit or what those sins are, as long as you truly believe in your heart that Christ is your savior, you will be saved. The evidence provided within the prophecies[c] indicates untold numbers of Christians will ultimately be saved and for this to be true, Catholics must be included in the numbers contrary to what these splinter Christian church denominations assert.

Weak faith saves and so the evidence supports that many borderline Christians will be saved even when they have parts of their doctrine wrong and don't practice exactly what the Word of God says. Now there is a time and place for pointing out our brother's faults and calling for repentance, but never at the expense of ignoring unbelievers in Christ, which should be a Christian's first calling—especially now.

This description of what the woman is wearing can be more accurately interpreted by comparing it to what Daniel was promised by the Babylonians:

> The king called out for the enchanters, astrologers and the diviners to be brought and said to these wise men of Babylon, "whoever reads this writing and tells me what it means will be **clothed in purple and have a gold chain placed around his neck**, and he will be made the third highest ruler in the kingdom.
>
> Daniel 5:7 (TNIV)

What lessons can be gleaned from this passage? Quite a few. First this is not symbolical writing, but literal language that can be used to aid in determining who or what the woman represents. The king's wise men were in effect idol worshipers, just as they are today; powerful men and women in high places who use every means available (or taught to them) to counsel the top leaders of many countries. Many are academic elitists who feel they are so intelligent that most have little use for God.

What were these advisors promised as a reward in Daniel's account? They were promised power! The power to rule more of the empire; to control more of the riches and people of the empire (i.e. the economy; the day to day operations of the country just as Joseph did in Egypt) and everyone knew in those days that anyone wearing these symbols of power was in charge. So these symbols have nothing to do with religious beliefs, as most people visualize religion today, and everything to do with controlling economic power. They denote economic authority and we can infer that the woman is really a very powerful financial leader that has some autonomy, but is indirectly accountable to a higher authority (king, president, prime minister, Satan etc.).

An example of this relationship is the connection between the president of the United States and the chairman of the Federal Reserve. The actions of each affect the ability of the other to achieve their own directives: neither reporting to the other, and yet both responsible for the economic conditions of the United States.

She held a golden cup in her hand, filled with abominable things and the filth of her adulteries.

This cup represents the full measure of all the sins that were committed by people to achieve **power and wealth** *they were so deluded into doing.*

This title was written on her forehead:

MYSTERY
BABYLON THE GREAT
THE MOTHER OF PROSTITUTES
AND OF THE ABOMINATIONS OF THE EARTH."
Revelation 17:4-5 (TNIV)

Finally, we see why the woman is connected to Babylon when we study the Daniel account I already mentioned. This practice came to its first pinnacle under the Babylonian Empire. This mother (the beginning) of prostitutes (those who will sell their moral standards, sell their belief in the one true God, sell their very soul in some cases) who worship materialistic things—idols.

Have you ever seen the bumper sticker or sign that says, "he who dies with the most toys wins?" This is the modern motto of the Babylonian ideology as it is practiced today. This worshiping of idols is the abomination of the earth in God's eyes and even the antichrist will detest this practice ("he will hate the woman") because he does not receive the praise and worship either. Only Satan does! Satan enjoys everything the Beast of the Sea and the Beast of the Earth does. Those who claim this is symbolically tied to religion are partially right. It is the idol worshiping religion of ancient Babylon of which an unsuspecting Christian Church partakes regularly along side the unbelievers of the world.

> *I saw that the woman was drunk with the blood of the saints* [Christians who were killed because they refused the mark of the beast], *the blood of those who bore testimony to Jesus* [Christian witnesses]. *When I saw her, I was greatly astonished.*
>
> Revelation 17:6 (NIV)

The woman was drunk (overindulged) with the excesses of capitalism. This system will ultimately be the cause of physical death to many true believers of Christianity because the "mark of the beast" will be required to participate in the economy during the tribulation years. Many of those who refuse the mark will be hassled, persecuted, and even killed by governments or overzealous unbelievers during the riotous times ahead—especially those Christians who continuously testify loudly to the truth of the end-time judgments, the identity of the antichrist and false prophet, and the return of Jesus. Their constant cries will irritate those authorities who are in power and force them into action. Others will die from starvation because they were unable to purchase or obtain the essentials of living—shelter, food, and water. It would be far better to die from starvation at this time, than to accept the mark of the beast.

> *Then the angel said to me: "Why are you astonished? I will explain to you the mystery of the woman and of the beast she rides, which has the seven heads and ten horns.*
>
> Revelation 17:7 (TNIV)

Here is where we begin to get some clarity. The angel is going to furnish the explanation of the symbolism being employed in the earlier verses we have already been discussing. The angel asks, "Why are you astonished?" as if he is astonished himself that John would even consider all these visions would be shown to him without an explanation. Just as every one of Daniel's prophecies were explained in literal terms, so will this revelation of Christ's. Now that we know this foretelling is redundant with the other Revelation prophecies, we have the means to unlock them all.

This same question is for all to answer who study the book of Revelation: "why are you astonished… do you also not understand these prophecies because I have explained many of them to you?" Do not worry: Christians who keep watch, study the Word of the Lord, and have the Holy Spirit as their guide, will understand the prophecies with greater clarity as the time of the tribulation draws near.

> *The beast, which you saw, once was, now is not, and will come up out of the Abyss and go to its destruction. The inhabitants of the earth whose names have not been written in the book of life from the creation of the world will be astonished when they see the beast, because it once was, now is not, and yet will come.*
>
> *Revelation 17:8* (TNIV)

This is a reference to the scarlet beast (the beast of Revelation 12 with ten horns and ten crowns) and is not the woman. This text can only be referring to the antichrist. That is because it is talking about the fate of the antichrist with the comment of coming up from the Abyss and going to his destruction, which is a reference to other prophecies that he will end up in the "Fiery Lake."[d]

The inhabitants of the earth whose names have not been written in the book of life from the creation of the world will be mesmerized when they see the beast, because it once was, now is not, and yet will come. But Christians won't be. If you are star-struck by any leader, then look out for your eternal life hangs in the balance.

The inhabitants of the world who are not Christians will be amazed when they see the beast. Why? Because this beast once existed prior to the time John received his revelation, yet at the time of the vision (around 95 A.D.) this beast was not in existence. Many years later, at the end of this age, the beast will reappear once more. Will this fact openly play into his success at deceiving the nations of the world? I don't know the answer except to say, "beware of leaders the masses are captivated by." Let's examine the basis for this point further.

With the understanding that the beast can denote a leader (king) or an empire (kingdom), then it is either the antichrist or the Revived Roman Empire (European Union). Given that the Roman Empire was in control during John's time, that interpretation can be eliminated so that this text can be narrowed down to mean the antichrist himself. Bear in mind the devil is forever trying to copy Christ and God's plan and what this is really saying is the antichrist came once before at a time prior to the first century. Just as Christ came once before, in the "spirit" of the first antichrist will the future man of lawlessness come for a repeat performance. A second advent just like Christ! He will be able to deceive unbelievers (his people), but not God's people.

> *This calls for a mind with wisdom: the seven heads are seven mountains on which the woman is seated;*
>
> Revelation 17:9 (RSV)

Some writers claim this is a reference to the Vatican and more explicitly the Catholic Church. This is because the city of Rome was founded on seven hills. They say it is a representation of a corrupt Church in the end-times. These eschatology writers mix up their symbolism and fail to remember or grasp it is the "woman" who represents the religious system (idol worshiping) and not the "beast" that is the governmental system of authority and law. Satan is the prosecutor of the law and thereby lives under the law whereas Christ is the giver of grace and forgiveness. We can expect the antichrist (the beast with seven heads) to be a law enforcer as well—a conquering king and not a priestess of worldly pleasures.

Given that the seven heads are of the beast and not the woman, then the seven heads must represent the antichrist and his nation and cannot be a symbol for the Catholic Church or any religion for that matter. I remind the reader again; we cannot interpret or interchange haphazardly (as some writers do) the symbols as we wish because this is the explanation given by one of God's messengers for this prophecy. We are compelled to only try to understand its true meaning. Realizing this, long before the existence of the Vatican there was just plain old Rome and this is just another confirmation of many that this is the European Union i.e. the old Roman Empire reemerging.

*They are also seven kings. Five have fallen, **one is**, the other has not yet come; but when he does come, he must remain for a little while. The beast who once was, and now is not, is an eighth king. He belongs to the seven and is going to his destruction.*

Revelation 17:10-11 (TNIV)

Since the passage "it once was, now is not, and yet will come" is a reference to the antichrist, then this latest passage cannot be a reference to him because the "now is not" portion is contradictory to the comment "one is." Both cannot be logically true at the same time. Therefore this passage really is not about specific kings, but the extension of the kingdoms they rule i.e. their empires.

The use of the word "kings" can be understood to mean either actual kings or kingdoms in prophecy. It is always important to know the context of how it is being used in a particular piece of prophecy before deciding how it is to be applied. This is true because for every kingdom, there is only one king (one leader), and so in describing one of the two, you are by default always describing characteristics of other. This text switches from talking about the specifics of the end-time leader in the previous passages to details about the empire he will come from.

The five empires that existed prior to John's time were Egypt, Assyria, Babylonia, Medo-Persia, and Greece. The "one that is," was the Roman Empire, and the one that is to come is the Revived Roman Empire—the European Union and it must remain for a little while (seventy years – 1958 A.D. to 2028 A.D.) just as long as the Babylonian Empire (609 B.C. to 539 B.C.). The next sentence in this text we are examining switches back again to discussing the beast as a leader and not as an empire. The beast who once was, and now is not, is an eighth king (when he comes again). He belongs to the seven (the last world empire) and he is destined for destruction in the "Fiery Lake of Sulfur" as planned.[e]

"The ten horns you saw are ten kings who have not yet received a kingdom, but who for one hour will receive authority as kings along with the beast. They have one purpose and will give their power and authority to the beast."

Revelation 17:12-13 (TNIV)

Now some theologians have said the ten horns of Daniel's beast[f] were ten successive kings throughout history. Even Isaac Newton claimed these were ten kingdoms in Europe throughout history and the number ten was reached in 800 A.D. He later implied, that by adding twelve hundred and sixty years to this date, Christ would return in 2060 A.D. Many students of eschatology believe this "successive" elucidation for either kings or kingdoms to support their false end-time ideologies.

However, we see here in Revelation that this is the exact same beast Daniel spoke of. What additional details was St. John given about this empire? That the ten future kings "would not be kings or kingdoms over time (throughout history), but would arise at the same time! In "one hour" together they would receive authority along with the beast." Those who believe a "successive" interpretation of Daniel's ten-horned beast are blind to the clarity this verse provides. They are incorrectly applying symbolic language to justify their private interpretation when this in fact "is" the interpretation from heaven so it must be taken literally.

These "kings" will be ten leaders from the European Union who will pass a law in the European Council that will give dictator like authority to the antichrist to govern over all of the European Union nations. They will be motivated by God to accomplish His plans[g]—"they have but one purpose" i.e. to give their authority, allegiance, and power to the antichrist. Now is there anywhere on earth where ten leaders with such power take office at the same time? Only one place... Europe! This is just more proof that the beast is the European Union. They currently hold elections across all the member countries at the same time. The leaders of the European Council, which these ten powerful leaders will hail from, will rise to power at the same time.

Those who claim the beast from the sea is the United States and the antichrist is the president of the United States fail to understand the American election process and governmental structure. The president is not placed into office with the aid of other leaders, but is elected by the populace of the country. However, the President of the European Council is not elected by the people, but by "appointed" leaders from each of the member nations of the E.U.

> they [leaders blinded by God] *will make war on the Lamb, and*
> *the Lamb will conquer them, for he is Lord of lords and King of*

kings, and those with him are called and chosen and faithful.
Revelation 17:14 (RSV)

These select few will assist the antichrist in making "war" against the saints and will be in opposition to everything God and Christ stands for. When Christ returns, only seven will remain of the original ten.[h] They will be aiding the antichrist in the final battle planned for the end of this age—at the end of the tribulation when Jesus will have with him the raptured church i.e. all his called, chosen, and faithful followers.

Realize that if the church has not been raptured as many Christian teachers contend, then how does one explain the verse that his servants are accompanying him? Let's not forget this is the answer for this prophecy. Besides, those who claim the rapture will occur simultaneously with Christ's return, fail to appreciate that the elect will be given white robes in heaven and cleansed of their sins[i] prior to his arrival. For this reason, these events cannot logically be viewed as one in the same.

And he said to me, "The waters that you saw, where the harlot is seated, are peoples and multitudes and nations and tongues."
Revelation 17:15 (RSV)

This explanation is self-explanatory. Its meaning is to be applied to words like "seas, rivers, streams, waters, etc." when they are used in Revelation prophecies to denote various amounts of people. This is the evidence that informs eschatology students how parts of the prophecy interpretations discussed in previous chapters were actually developed.

The beast and the ten horns you saw will hate the prostitute. They will bring her to ruin and leave her naked; they will eat her flesh and burn her with fire.
Revelation 17:16 (TNIV)

The antichrist and his ten disciples (European Council leaders) will hate the prostitute (American controlled capitalism spread throughout the world as the American dream). This is the spot where we learn it is this group of ten presidents that plot and scheme to bring the economic system down. It will not be Arab terrorist groups who will bomb Babylon the Great as popularly believed. That is not to say they won't

263

continue their terrorist actions, it is only to point out that the nuclear bomb that will destroy New York City (NYC) will not be from their deeds.

Daniel says early in this contemptible leader's counterfeit reign,

...He will plot the overthrow of fortresses [large cities, maybe even entire countries]—*but only for a time.*

Daniel 11:24 (TNIV)

Why? Because one of the best ways to make the world believe you are its savior, and get them to follow (worship) you, is to destroy those things people depend, rely, and covet most: money and the materialistic things of this world. Then set up another system (image) that replaces the things they have lost. Thereby becoming a hero to them. In other words, create their crisis, blame it on someone or something else, and then solve their crisis thus indebting them to you. These are the methods of the powerful and deceitful.

How do the E.U. leaders and the antichrist accomplish this? By destroying NYC completely (not just a few buildings) with a massive nuclear blast! Burn her with fire.[j] What effect will this have? Well the stock exchange, the United Nations, and the largest U.S. banks, controlling trillions of dollars, are all headquartered there. Realize the stock market was shut down, for the first time ever in an unplanned way, when the World Trade Centers were destroyed in September 2001 A.D. (a sign, warning, or prelude to the coming destruction). This tragedy caused major disruptions in the worldwide economy, but all those financial problems could be fixed.

What will happen to the international banks headquartered in NYC and the New York Stock Exchange if total destruction were to occur to this great city? Nothing would be left. Many of the rich and powerful people of the world would be killed and all of the computers that keep track of the wealth of millions of citizens would be gone forever. "In just one hour" all would be lost without the hope of things ever returning to normal or the ability to recover that lost wealth.[k]

This incident would require a massive bomb (about fifty to one hundred megatons) and only Russia currently has this capacity. These European leaders may even plot to strike the capital of the United

States, Washington D.C., to further incapacitate the infrastructure of the U.S. and bring it to its knees. There is really no protection from this type of unexpected attack for any coastal city of which the United States has many. The reality is an enemy nuclear submarine could sit right off the coast and within an hour do all the necessary damage without any means of stopping it.

This event(s) would shift world power to the European Union and would immediately make them the most "powerful" group of nations on the face of the earth. Russia will already have lost ground in its capacity to influence world politics because of the severe losses from the War of Gog that will have occurred prior to this dreadful event taking place. The Russians will be looking to recover from this military disaster (just as they are now trying to do from the fall of the Iron Curtain during the 1980's) and just may be tricked into doing the dirty work for the E.U. and its devious leaders.

> *For God has put it into their hearts* [the ten E.U. presidents] *to accomplish his purpose* [plans] *by agreeing to give the beast* [anti-christ] *their power to rule, until God's words* [prophecies] *are fulfilled. The woman you saw is the great city* [NYC] *that rules over the kings of the earth.*
>
> Revelation 17:17-18 (TNIV)

Many eschatology experts and students believe this to be Vatican City, some a rebuilt Babylon city over the next twelve to twenty years, and others even claim it is a reference to Jerusalem. Let's investigate their false claims under the watchful light of the Holy Spirit. The reality is that most certainly none of them are this prophesied great city as we have already seen. This city is a coastal city. We will learn shortly that merchants will witness the burning of this city and lament its loss from their ships at sea.[1]

This added information rules out Babylon. Besides, it is unrealistic to believe this ancient city from biblical history will be rebuilt this fast in an area that is under constant fighting. Not to mention being able to wrestle control of the world economy to the degree of it "ruling over all the kings of the earth" is a ridiculous idea. Being an American, I can envision no scenario where the U.S. would ever give up this power willingly nor has any empire throughout history ever done so.

As for the Rome/Vatican City idea, these cities are roughly thirty-to-forty miles from the Mediterranean Sea. Should it be blown-up, it might be possible to see it burning from ships along the coast of Italy. Are the merchants of the world anchored off the coast of Italy? I think not.

Nuclear explosions can be seen from a very long distance especially a blast of this magnitude. However, the Catholic Church and the Pope have little influence over all the kings of the earth and definitely have no control of the economic fortunes of the world. Two thirds of the world's population are not Christian and could care less what the Pope and the Catholic Church does. Do you really think the antichrist is going to destroy a city in Europe when, more than likely, one of the ten leaders who helps him is Italian? Not to mention he may use the shell of the Christian Church (after the rapture) to implement his one world religion with him as God. This idea is not supported in Bible prophecy and is based upon the insights of others, current trends, and intuition.

What about Jerusalem? Well it too is about thirty-to-forty miles from the Mediterranean Sea and so the same logic applies to witnessing its destruction from sea. This city does "impact" the kings of the earth because they are always trying to figure out how to stop the conflicts there. But this city will be in existence when Christ returns and clearly God will not destroy it with a nuclear blast (fire)—a blast this great would leave nothing but a large hole contaminated by nuclear radiation. This city will also be used by the antichrist in the last days as his base of operations.[m] So what great city has control of the kings and they don't like it? New York City, of course, with all the economies of the world linked to it!

This really is the only city that logically fits with all the facts presented in Revelation and the Bible. This will probably not be the end of the United States, but it will be the end of its leadership role in the world forever. Nevertheless, there is more to this story. If we were to examine our timeline and place this event on it, which we will do in a later chapter with all the end-time events, we would discover this disaster occurs shortly after the thirty-five days of conflict and is the start of God's wrath. This would be the late spring of 2025 A.D. just after the abomination that causes desolation, the thirty days of intense

Christian persecution, and the rapture of the church.

However, the Bible provides details of the United States in Daniel chapter eleven. The King of the North (the antichrist) will attack the King of the South (the Islamic countries) and gain great wealth[n]… probably control of the world's oil fields under the guise of Israel's protector. Afterward, the antichrist will return to his country for a time, but then attack these countries again.[o] Why? The best guess is the Arabs will continue to resist Europe's control and break the peace treaty signed.[p] However, this time it will be different. A navy from the western coastlands[q] will come to their aid.

Now those with wisdom listen closely. There is only one navy west of Israel that has the power to resist the combined forces of the European Union armies and that is the United States. This Navy will oppose the antichrist and he will lose heart.[q] Then he will turn back and vent his anger against Christians.[q] These events take place just prior to the abomination of desolation, but after the confirmation of the covenant both spoken of by Daniel.[r] This would place this mini-war most likely in the year 2024 A.D. Why do I write these things? Though the majority of theologians claim these events were fulfilled during the reign of Antiochus IV Epiphanes (175 - 164 B.C.), they were not.

Jesus said to watch for the abomination that causes desolation, spoken of by the prophet Daniel, as a critical sign for Jews and Christians alike[s] to run and hide. In view of the fact that Daniel spoke of only one such atrocity, this prophecy could not have been fulfilled before the birth of Christ because it is a key sign for the end of this age. So these prophetic verses can only be applied to the end of the age and not arbitrarily to ancient historical events as misguided historians and theologians contend.

So why does the United States attack the European armies after years of alliance and come to the aid of the Arab nations? Prophecy is silent as to the events that will lead up to this confrontation, but we can deduce from current trends this friendship will continue to sour over the next fifteen years as crude oil disagreements, dealings with economic woes, and friction with how the war on Islamic terrorism is being mishandled will play a major role in that decline. Many already hate America as much as they do Israel.

This military defeat for the antichrist and his forces will not sit well with him and his supporters. It will only add to the fuel for revenge that will bring upon the nuclear bombing of New York City. After administering this punishment of God's we can't be sure where the United States will end up during the remaining years of God's wrath, except it appears its days as a superpower are over and it will probably have little to do in the War of Armageddon.

> *After this I saw another angel coming down from heaven, having great authority; and the earth was made bright with his splendor. And he called out with a mighty voice,*
>
> *"Fallen, fallen is Babylon the great! It has become a dwelling place of demons, a haunt of every foul spirit, a haunt of every foul and hateful bird;"*
>
> Revelation 18:1:2 (TNIV)

Now we see language being used in the past tense. As if the actions or deeds have already taken place. Babylon the Great "has" fallen. Some assert this is the United States. I guess you could look at it that way from a certain perspective. But I would like to think with all the analyses we have completed so far, that this is not the country of the United States in as much as it is the capitalistic structure that has spread to the whole world from the United States.

God is destroying that idol worshiping "religious" system and the best way (the fastest) to achieve His purposes will be to make New York City go away! When the time comes for this to happen, just after the rapture, everything will be electronic and depend on computers. How many times have you called a company or government agency to straighten out a matter and were told, "the computers are down?" How much more will mankind be relying on computers in another sixteen years when New York City is destroyed?

Even now, if the computers are not working, banks won't let you withdraw funds. What will happen when there are no computer records after they are destroyed? You won't have any money to take out is the answer, especially if the guys at the top who want to eliminate money and go to a cashless society succeed with their plans before this disaster occurs. Not to mention with bank consolidations continuing over the past twenty years and into the foreseeable future at an alarming rate, the wealth of the nation and its people are being consolidated in one place.

The nest eggs of Americans are being placed in one large basket that will be boiled all at once! What kinds of civil unrest will this calamity cause?

> For **all the nations** have drunk the maddening wine of her adulteries. The kings of the earth committed adultery with her, and the merchants of the earth grew rich from her excessive luxuries.
>
> Revelation 18:3 (TNIV)

Next we witness that not only are the kings of the earth controlled by the woman (Babylon the Great) who sits on the scarlet beast, but so are the merchants! If the harlot controls the leaders and merchants of the world, we can also conclude she will control just about everyone else in the world! As our studies move forward, we will continue to find mounting evidence that the woman is really about economics—the religion of material things and not a "spiritual religion" as most people perceive this term.

Forewarning to Flee Babylon's Judgment

> Then I heard another voice from heaven say:
>
> " 'Come out of her, my people,' so that you will not share in her sins, so that you will not receive any of her plagues; "
>
> Revelation 18:4 (TNIV)

This warning has three lessons of great importance. One: immediately move out of New York City if you have missed the rapture and now realize your mistake. Move inland without delay away from all coastal cities. If you are a Christian this is good advice to follow right now in the slim chance my timing might be off. As I discussed in chapter five, there is limited evidence that everything could possibly be a year earlier than forecasted. Besides, if you are a Christian who keeps watch for the signs, you should have already moved to a safe place away from the unbelieving masses (as Jesus instructed) when you saw the abomination that causes desolation. My thought is why wait until the last minute? There will be clear indications long before this key sign occurs.

Second, a more spiritual meaning is required. Stop praying to idols. Now you may say you are not, but what is really meant by this is to stop thinking and yearning for the material things of this world. Let your heart be content for things will never be the same again. Repent, wakeup, and turn to God for forgiveness.

And lastly, do not take the "mark of the beast" to participate in the economy! Move "off the grid" toward independence from this system. If you must survive God's wrath for three and a half years because you did not heed the warning signs, you must do so without ever accepting this mark if you want eternal salvation![t]

for her sins are piled up to heaven, and God has remembered her crimes. Give back to her as she has given; pay her back double for what she has done. Pour her a double portion from her own cup. Give her as much torture and grief as the glory and luxury she gave herself. In her heart she boasts, 'I sit as queen; I am not a widow, and I will never mourn'.

Revelation 18:5-7 (NIV)

Those who are addicted to this system are so focused on making money, swindling people out of money, acquiring wealth, and living the good life (sex, drugs, and rock and roll; every pleasure they can dream of when money is no object) that they believe nothing will happen to them. ("I am not a widow; I will never mourn"). That today will be just like tomorrow, but they are dead wrong... literally.

Just as it was in the days of Noah, so it will be at the coming of Christ. People are so obsessed with idol worshiping that they simply cannot see the signs of Jesus' return and when they receive this news they reject the truth and those who spread this truth, just as they did to Noah when they watched him build the ark. What a great witness, to the coming devastation of the flood, this huge ship rising on the plains must have been to the people around him.

so shall her plagues come in a single day, pestilence and mourning and famine, and she shall be burned with fire; for mighty is the Lord God who judges her.

Revelation 18:8 (RSV)

270

Again, this is just one more reference to fire consuming Babylon the Great... the woman... the prostitute... the two horned beast of the earth. We also see God uses the word "plague" to describe human emotions (mourning) and not exclusively for diseases as most would envision when they hear this word. Her demise will occur in one day (actually less). How long does it take to feel the effects of a nuclear explosion?

When the kings of the earth who committed adultery with her and shared her luxury see the smoke of her burning, they will weep and mourn over her.

Revelation 18:9 (TNIV)

That is to say, all the kings of the earth except for those kings in Europe! The people of Europe will mourn too, but the antichrist will use this disaster to bring glory to him and deceive the masses.

Terrified at her torment, they will stand far off and cry: " 'Woe! Woe, O great city, O Babylon, city of power! In one hour your doom has come [one hour... a nuclear blast]!' "The merchants of the earth will weep and mourn over her because no one buys their cargoes anymore— cargoes of gold, silver, precious stones and pearls; fine linen, purple, silk and scarlet cloth; every sort of citron wood, and articles of every kind made of ivory, costly wood, bronze, iron and marble; cargoes of cinnamon and spice, of incense, myrrh and frankincense, of wine and olive oil, of fine flour and wheat; cattle and sheep; horses and carriages; and bodies and souls of men.

Revelation 18:10-13 (NIV)

Alas, it is the merchants of the earth, more than anyone else, who will grieve over the harlot's death. They will lose their riches and their ability to make money. It matters little for many unbelievers will not have much money to buy things long before this judgment. When Christians stopped buying goods and services from these businessmen years earlier, the economy will begin a slide that will affect everyone. A serious problem, which unbelievers who have taken the mark will blame

Christians for, that escalates into harassment, mistreatment, and even death for some!

> *"They will say, 'The fruit you longed for is gone from you. All your riches and splendor have vanished, never to be recovered'. The merchants who sold these things and gained their wealth from her will stand far off, terrified at her torment. They will weep and mourn and cry out: " 'Woe! Woe, O great city, dressed in fine linen, purple and scarlet, and glittering with gold, precious stones and pearls! In one hour such great wealth has been brought to ruin!'*
>
> Revelation 18:14-17 (NIV)

These verses are just reiterating things already discussed providing more details, supporting evidence, and clarity. This text continues to point the way to the destruction of New York City like a beacon in the darkness.

> *Every sea captain, and all who travel by ship, the sailors, and all who earn their living from the sea, will stand far off. When they see the smoke of her burning, they will exclaim, 'Was there ever a city like this great city?'*

From these verses, we see this is a coastal city. Nothing more needs to be said that hasn't already been said many times over.

> *They will throw dust on their heads, and with weeping and mourning cry out: " 'Woe! Woe, O great city, where all who had ships on the sea became rich through her wealth! In one hour she has been brought to ruin! Rejoice over her, O heaven! Rejoice, saints and apostles and prophets! God has judged her for the way she treated you.' "*
>
> Revelation 18:18-20 (NIV)

At this juncture, we see why the judgment is imposed. What was the judgment imposed on God's people? They could not buy or sell unless they accepted the mark of the beast. They lost everything. Their money was tied up, frozen, or even taken away when they refused the mark. Effectively the wealth of Christians will decline. They will be hard pressed to keep their homes even if they are paid for because of the inability to pay property taxes. I wonder if there are any states in the U.S. or other parts of the world that don't have property taxes? This will be a time when it will be best to be self-sufficient, much like a farmer is.

The people who take the mark will receive the same punishments when the economy is ruined. They will lose their money and they will be hard pressed to buy food; the same conditions that raptured Christians will have had to live under. This will cause indescribable civil unrest.

Babylon's Fate

> Then a mighty angel took up a stone like a great millstone and threw it into the sea, saying, "So shall Babylon the great city be thrown down with violence, and shall be found no more;"
>
> Revelation 18:21 (RSV)

This is verification of the second Trumpet and Bowl judgment and is a description of a meteor hitting the Atlantic Ocean and the ensuing tidal wave washing away the remains of the NYC disaster—new evidence that identifies it will strike the Atlantic Ocean and not some other ocean.

> and the sound of harpers and minstrels, of flute players and trumpeters, shall be heard in thee no more; and a craftsman of any craft shall be found in thee no more; and the sound of the millstone shall be heard in thee no more; and the light of a lamp shall shine in thee no more; and the voice of bridegroom and bride shall be heard in thee no more; for thy merchants were the great men of the earth, and all nations were deceived by thy sorcery.
>
> Revelation 18:22-23 (RSV)

Here we see this city's merchants were the "world's important people." What city has claim to this fame? Not Jerusalem, Babylon, or Vatican City, but New York City i.e. the financial capitol of the world as we now know. Also, no people will ever live or work there again! Confirmation that there will be nothing left to rebuild or go back to. The desolations of God's first two judgments (nuclear attack and a meteor strike) will be total and complete destruction.

> *And in her* [greed of capitalism] *was found the blood of prophets and of saints, and of all who have been slain on earth.*
> *Revelation 18:24* (RSV)

This verse is probably saying that in the last days Christians who resisted the "mark of the beast" and didn't participate in the economy, will suffer and many will even die because of their choice and faith in Christ.

> *After this I heard what sounded like the roar of a great multitude in heaven shouting: ...*
> *Revelation 19:1* (TNIV)

We have witnessed this phrase before in other Revelation prophecies. This is further evidence of the raptured church—a great multitude (which no one could count) that is now residing in heaven watching Christ meter-out God's punishments. How did they get there? You know![u]

> *Hallelujah! Salvation and glory and power belong to our God, for true and just are his judgments. He has condemned the great prostitute who corrupted the earth by her adulteries. He has avenged on her the blood of his servants.*
> *Revelation 19:1-2* (TNIV)

Notice it's the saints who are praising God for the judgments that have and are taking place and not the angels. He "has" condemned the great prostitute and "has" avenged their killings (for not accepting the mark) by destroying the economy so that unbelievers will suffer the same fate as Christians had in the early years of the tribulation under the antichrist's schemes. Many of the wicked will not easily be able to participate in the buying and selling of goods and services as a result of this disaster, even if they have the required "mark" to do so.

And again they shouted: "Hallelujah! The smoke from her goes up for ever and ever." The twenty-four elders and the four living creatures fell down and worshiped God, who was seated on the throne. And they cried:

"Amen, Hallelujah!"

*Then a voice came from the throne, saying: "Praise our God, all you his servants, you who fear him, both small and great!" Then I heard what sounded like a great multitude, like the roar of rushing waters and **like loud peals of thunder**, shouting: "Hallelujah! For our Lord God Almighty reigns.*

Revelation 19:3-6 (NIV)

God's people are in heaven and the first thing they are instructed to do is praise the Lord, both great and small. Christians have been delivered from the wrath that Christ is administering and it doesn't matter what your economic status in life was before you were raptured. We are all his servants now.

This is just past the midpoint where Christ begins to reign when the antichrist's authority that was granted for twelve hundred and sixty P-days has been taken away, but he has been granted limited authority for another forty-two months to blaspheme God and deceive those who love not the truth.[v]

Also notice our "closing" expression that signals the coming of Christ and the end of this prophecy is near. There are more verses to investigate only because there are so many added details given of Jesus' second coming that were not provided in other prophecies.

Let us rejoice and exult and give him the glory, for the marriage of the Lamb has come, and his Bride has made herself ready;

Revelation 19:7 (RSV)

The bride passed the test of the "mark of the beast" by standing up to the antichrist and rejecting the "new" economic system (a cashless image of the old system) put in prior to the destruction of NYC and the U.S. economy. The bride (the church) was made ready by the righteous acts of the saints during the three and a half years before the abomination of desolation. Testifying boldly to the world of the Gospel of Jesus Christ.

"Fine linen, bright and clean, was given her to wear."

(Fine linen stands for the righteous acts of the saints.)

Then the angel said to me, "Write: 'Blessed are those who are invited to the wedding supper of the Lamb!' " And he added, "These are the true words of God." At this I fell at his feet to worship him. But he said to me, "Do not do it! I am a fellow servant with you and with your brothers who hold to Jesus' testimony. Worship God! For the testimony of Jesus is the spirit of prophecy."

Revelation 19:8-10 (NIV)

For the "testimony of Jesus is the Spirit of prophecy." Interesting. This is important because it tells us we are all prophets in the eyes of God when we testify in the spirit of the truth of Christ. One of the two witnesses before the Lord of the earth and the other—preaching the Word of the Lord! Both Christian responsibilities during the last days... and for everyday!

We also learn that we are not supposed to worship angels and put them on high pedestals. We can infer we should not pray to them as well for our needs, but only to the triune God. Lastly, a blessing is given to those Christians who were invited to the wedding feast. Why, because these Christians had kept watch and understood what was going on around them. They were not asleep at the wheel and were prepared. Their lamps had enough oil.[w] "Well done my good and faithful servant!"

As for the Christians who were caught off guard and mesmerized by the things of this world, there is still hope. Unfortunately they must suffer the wrath of God right along with the unbelievers and do so without taking the mark of the beast. Those who survive these trials and tribulations will inherit the right to enter the Millennial Kingdom. They will be the first generation in the new age. Truly a special group of people.

There is a parable that hints that half of the Christians of the world, which is one third of the planet's population, will be raptured while the other half will have to suffer God's wrath to gain salvation and enter the Millennial Kingdom. This is their punishment for not being ready and keeping watch.(x) Only after witnessing the rapture did they finally understand the truth of God's Holy Word and what their fellow brothers and sisters in Christ were trying to tell them. This is the parable of the ten virgins.(w)

Christ Defeats the Beast

I saw heaven standing open and there before me was a white horse, whose rider is called Faithful and True. With justice he judges and makes war. His eyes are like blazing fire, and on his head are many crowns. He has a name written on him that no one knows but he himself. He is dressed in a robe dipped in blood, and his name is the Word of God. The armies of heaven were following him, riding on white horses and dressed in fine linen, white and clean."

Revelation 19:11-14 (NIV)

The doorway to heaven is open and we get an up close look of Christ's readiness for the last judgments and the final war of Armageddon. The appointed time is at hand. The Lord of lords and the King of kings has already been ruling all the nations of the earth from heaven and now we see the armies of heaven are ready, dressed in fine linen (rewarded for their righteous acts), for Christ to begin ruling the nations of the earth firsthand! These are the returning saints and the angels of heaven under his command.

From his mouth issues a sharp sword [truth and justice(y)] with which to smite the nations, and he will rule them with a rod of iron; he [Jesus] will tread the wine press of the fury of the wrath of God the Almighty [administer God's punishments].

Revelation 19:15 (RSV)

The truth of his words will hurt. Many will already know the error of their ways after witnessing the rapture and being left behind. He will assume direct control of all the nations upon his return. Nations that will still be in existence when he rules from Jerusalem[z]... the center of the earth... the center of the map. Realize that if you add the latitude and longitude of Jerusalem together you get 66.6—man's number in one of its intermediate forms.[62]

He will control sinning and not allow sinful practices to exist. Justice will finally be measured out in purity, not by what he sees or hears, but by the truth of what he finds when he searches into a man's soul.

Jesus "treads God's winepress" or in other words, Christ is in charge of metering out God's wrath (punishment) during the last three and a half years. This is the verse that proves previous claims that Christ is in control from heaven. He is the "man in charge" of executing God's perfect plan on those who have closed their hearts and would not accept him. Just as the unbelieving Gentiles will trample on Jerusalem during the last days of the tribulation, so too will Christ trample on them as he walks along God's winepress of fury!

On his robe and on his thigh he has this name written:

KING OF KINGS AND LORD OF LORDS.

And I saw an angel standing in the sun [will Christ return during the daytime?], who cried in a loud voice to all the birds flying in midair, "Come, gather together for the great supper of God, so that you may eat the flesh of kings, generals, and the mighty, of horses and their riders, and the flesh of all people, free and slave, small and great."

Revelation 19:16-18 (NIV)

It is the time for War. Time for the great War of Armageddon—War on the Megiddo plains of Israel. The killing fields of wars long forgotten are the killing fields of the earth once more. A place where the armies of the earth gather for the Great Supper of God!

And I saw the beast and the kings of the earth with their armies gathered to make war against him who sits upon the horse and against his army.

Revelation 19:19 (RSV)

At this point, the armies of the world that have been fighting each other, stop what they have been doing, and turn to fight Christ and the returning saints from heaven. This is Christ's "official" return in the year 2028 A.D. and not a description of when he returned three and a half years earlier to gather the firstfruits of the harvest at the rapture. His second advent is finally here.

And the beast was captured, and with it the false prophet who in its presence had worked the signs by which he deceived those who had received the mark of the beast and those who worshiped its image. These two were thrown alive into the lake of fire that burns with sulphur.

Revelation 19:20 (RSV)

The antichrist and the false prophet were "captured" alive, not killed. They could not escape from Christ and his army. We see they are real live individuals and not some symbolic reference to something intangible. The two of them were thrown living and breathing into the fiery lake of burning sulfur for all eternity. This suggests that the final destiny for those who reject Christ will be eternal punishment in the physical body as opposed to the spiritual punishment one receives in hell when a person dies now without accepting the Lord as their personal savior.

And the rest were slain by the sword of him who sits upon the horse, the sword that issues from his mouth; and all the birds were gorged with their flesh.

Revelation 19:21 (RSV)

Of the armies that were present during the final battle, Christ's justice will kill them all. Why? Because we know that to enter the military, soldiers will all have to accept the mark of the beast and in so doing they proved by their actions they had not chosen the side of truth. This text is specifically talking about the death of the soldiers who gather to fight and not others. Why? Because we see the birds are gorging themselves on the flesh of the dead and they were gathered only in this area and not everywhere on earth.

As for the rest of mankind who were not at the battle and survived God's judgments, they will be gathered by God's angels for judgment over a forty-day window.[aa] Christ will sentence those individuals found with the "mark of the beast" to death. With just a word, just a command, they will be weeded from the pastures for the new seeds to take root in the Millennial Kingdom. Just as when Christ killed the fig tree because it would not bare fruit.[ab] These wicked people are the fig trees that will not bare good fruit.

Those Christians who survived the last three and a half years of God's wrath (like many converted Jews) and did not take the "mark of the beast" will be allowed to enter the Millennial Kingdom and repopulate the earth. Those Christians who came back with Christ in glorified bodies are now like angels and will serve the Lord for the rest of eternity on earth. They have earned the right to be exempt from the second death.[ac]

Having examined all the prophecies of Revelation that are relevant to the last days, we have learned much about the devastating events that will litter the landscape of the tribulation period. Let's move along our studies and peer into the secrets of a few other end-time prophecies to see what they reveal about the future. Although we will not investigate every end-time prophecy within the Bible, we will look at just a few more which provide additional insights and add strength to the foundation we have already laid.

Chapter 13

Other End-Time Prophecies

HAVING EXPLORED THE TIMING OF the tribulation and determined from St. John's prophecy what prophetic events we might actually expect to unfold, we will next inspect two other prophecies that provide additional details. Let's begin our study of these prophecies from the Bible to see what else can be uncovered that will build upon our mounting knowledge of the last days.

Daniel Unsealed

There are several prophecies in the book of Daniel that are important to eschatology. I will not examine them all here because it would take as much study as we have given the book of Revelation. However, understand that the book of Daniel is the twin to the book of Revelation and with the two together we have learned all the timing necessary to unlock God's timetable. We have referenced many things from this important work, such as the abomination that causes desolation, through the course of our studies as if the reader is knowledgeable of what is written in these prophecies. I apologize for this.

It is impractical to start learning about prophecy from the beginning as this book would become too large and also because many of the prophecies are intertwined, just as God's numbers are. I find most Christians are unaware of most biblical prophecies for the same reason I was. If you grew up in a mainstream denomination, which accounts for the vast majority of Christians worldwide, then you were taught a doctrine that Christ could return at any time. For this doctrine to be true, then all prophecy requirements dealing with the second advent of Christ must have already been fulfilled in the past—a preterism position. Subsequently, these denominations rarely preach or teach on these books of the Bible and hence Christians are unaware of what is written in them unless they do a lot of self-study.

It matters little if these preterism Christian shepherds preached on these subjects because this wrong doctrine clouds their ability to understand what is written and thus they are unable to pass on to their flocks the truth of what is written. Remember, a person cannot find something (i.e. what is to happen in the future) if they are not looking for it. And if they believe they already know it all, they lock the door that keeps the Holy Spirit from reaching them. They are no different than the teachers of the law that Christ chastised for this very problem during his days.

Since the book of Daniel is so important and I do not wish to devote the time for a verse-by-verse analysis of the book, I have provided a study tool in Appendix I in the back of the book. This tool includes the important prophecies of Daniel that deal with the end-times. Understand this: the visions and dreams that Daniel and King Nebuchadnezzar of Babylon received were road maps of future history. These visions not only included important end-time facts, but details of human history from the Babylonian Empire in the fifth and sixth century B.C. through today and beyond.

This study tool is laid out so that you can view side-by-side each prophecy's verses which pertain to the same events in history. For example, if you want to see everything that was foretold about the Greek Empire in each of Daniel's prophecies, you can locate the place in the table that is labeled Greek Empire and scan across the page and find the scriptures that pertain to the Greeks included in each prophecy. Not only are the actual prophecies provided, but the interpretations for these key prophecies are included as well. These are not my interpretations, but rather God's, for in every prophecy of Daniel's an interpretation is also given. Our job is not to formulate our own understanding, but to try and comprehend what is written from a literal standpoint.

There is one prophecy I wish to spend the time to make a detailed analysis of, but only the portion that deals with the last days. This prophecy actually encompasses details of human history from the beginnings of the Persian Empire to the beginning of the Millennial Kingdom. This prophecy is covered in the last three chapters of the book of Daniel and is known as the Kings of the South and the North. We will ignore much of this prophecy and pick up at the point where the antichrist begins his rise to power—ten years before the return of Christ. What we are going to review is not the actual prophecy, but the interpretation given to Daniel of his vision and therefore is a literal explanation. I chose this prophecy for that very reason and because it provides many details of the last ten years of this age that are not given in the book of Revelation.

Many Bible experts assign some of these verses we will investigate to the history of the Greeks, something we discussed in chapter five. In so doing they make several grievous errors. First, they allow for no verses within the whole prophecy to be associated with the Roman Empire. This is akin to a historian writing a history book that omitted all references to the Romans—one of the most powerful empires of all time and the empire that existed during Christ's life. Would any historian do this? Is this how God works? To give important details of man's history, yet skip these chapters of human history? I think not. Secondly, they assign the abomination of desolation to second century Greek history further compounding the mistake by not understanding that future events have happened in the past and that Jesus said this specific event would take place at the end of the age—that point in history which is at our very doorstep.

In my previous verse-by-verse analyses I have provided explanations, thoughts, and insights to the best of my abilities using my understanding of God's overall plan, His methods, and all of the prophecies considered as a whole. I will do the same for this prophecy of Daniel's except that after each verse I will reiterate the verse using modern language modified with my thoughts from details learned from all the other prophetic studies instead of providing a commentary for each verse. These selected verses are all from the King James Version of the Bible. Let's begin.

And in his estate shall stand up a vile person, to whom they shall not give the honor of the kingdom: but he shall come in peaceably, and obtain the kingdom by flatteries.

Daniel 11:21 (KJV)

And he will be succeeded by a vile person who rises to power not through royal bloodlines, but through politics. The antichrist will rise quietly during a time when people are concerned about security, safety, and peace. This leader of the European Union will move up the political ladder through flatteries, politics, and take control using underhanded techniques.

*And with the arms of a flood shall they be overflown from before
him, and shall be broken; yea, also the prince of the covenant.*
Daniel 11:22 (KJV)

During this time a large army (the war of Gog; Russian and Arab
forces attacking Israel) will succumb to his diplomacy and cease hostili-
ties toward Israel. In addition, the prince of the covenant (Christianity)
will be attacked and broken by the antichrist's efforts.

*And after the league made with him he shall work deceitfully: for
he shall come up, and shall become strong with a small people.*
Daniel 11:23 (KJV)

After brokering a peace agreement, the antichrist will rise in
popularity all the while deceitfully working to grab the full power of the
European Union. With just ten leaders from this alliance of countries,
he will seize control of the presidency of the European Council and
become its dictator.

*He shall enter peaceably even upon the fattest places of the prov-
ince; and he shall do that which his fathers have not done, nor his
fathers' fathers; he shall scatter among them the prey, and spoil,
and riches: yea, and he shall forecast his devices against the strong
holds, even for a time.*
Daniel 11:24 (KJV)

The E.U. president will visit many countries under the pretense of
peace and influence them to join his cause with lavish promises and
gifts. He will succeed in expanding his control through politics, something powerful leaders throughout history before him could not do.
Those countries that resist his offers, he will use military pressure and
economic methods for a time to overthrow them and bring them under
his control.

*And he shall stir up his power and his courage against the king of
the south with a great army; and the king of the south shall be*

stirred up to battle with a very great and mighty army; but he shall not stand: for they shall forecast devices against him.

Daniel 11:25 (KJV)

The antichrist will grow in confidence and send his armies against the rebellious Islamic countries (King of the South) who are not following the agreement. The Muslims will be overpowered and their own leader will succumb to the plots devised against him.

Yea, they that feed of the portion of his meat shall destroy him, and his army shall overflow: and many shall fall down slain.

Daniel 11:26 (KJV)

Plots within his own allies—from his own people perpetrated by the E.U. president—will cause him to surrender after many die from armed conflict by the invading European led armies.

And both of these kings' hearts shall be to do mischief, and they shall speak lies at one table; but it shall not prosper: for yet the end shall be at the time appointed.

Daniel 11:27 (KJV)

The antichrist and the Muslim leader will sit across from each other during peace negotiations, both making promises that neither intends to keep. Regardless of both of their treacheries and motives, the appointed time (the end of the age) for their demise will happen just as God has scheduled it.

Then shall he return into his land with great riches; and his heart shall be against the holy covenant; and he shall do exploits, and return to his own land.

Daniel 11:28 (KJV)

Afterward, the antichrist returns back home having secured many concessions from the Arab Federation including control of the oil fields. He will make preparations against the holy covenant by empowering the

Jews to rebuild the temple. He will focus all his energy against Christians by passing laws to make their lives miserable.

> At the time appointed he shall return, and come toward the
> south; but it shall not be as the former, or as the latter.
>
> Daniel 11:29 (KJV)

At the appropriate time on God's plan, he will once again engage the Arabs in battle to squash another uprising against his Israeli ally. But this time the results will be different than his first victory and his later military successes.

> For the ships of Chittim shall come against him: therefore he shall
> be grieved, and return, and have indignation against the holy
> covenant: so shall he do; he shall even return, and have intelli-
> gence with them that forsake the holy covenant.
>
> Daniel 11:30 (KJV)

For the powerful U.S. navy stationed in the Mediterranean Sea off the coast of Cyprus will oppose the European Union and come to the aid of the Arabs protecting their own oil interests. He will be depressed and angered and will return to Europe. He will step up his efforts against Christians and the United States. He will then return again to the Middle East and side with those who also oppose Christianity and Judaism.

> And arms shall stand on his part, and they shall pollute the sanc-
> tuary of strength, and shall take away the daily sacrifice, and they
> shall place the abomination that maketh desolate.
>
> Daniel 11:31 (KJV)

His army will invade and take control of the city and the new temple in Jerusalem. After a few days, he will enter the temple and set up the abomination that causes desolation by abolishing the daily sacrifice to God and pronouncing he is God.

And such as do wickedly against the covenant shall he corrupt by flatteries: but the people that do know their God shall be strong, and do exploits.

Daniel 11:32 (KJV)

With sweet talk and false promises he will defile those who already oppose the truth of God and seal their fate: but the Christians who follow the Lord's Word are strong and will oppose him through actions and deeds.

And they that understand among the people shall instruct many: yet they shall fall by the sword, and by flame, by captivity, and by spoil, many days.

Daniel 11:33 (KJV)

And those Christians who know the truth will spread that truth to many peoples and nations. Yet, the antichrist will hunt these people for a short time: imprisoning them and even killing them—rewarding those who assist in these efforts.

Now when they shall fall, they shall be holpen with a little help: but many shall cleave to them with flatteries.

Daniel 11:34 (KJV)

When these Christians and Jews are being massacred, some people who are neither will help them for fear their freedoms will be taken away. They believe in no God and will not submit to any ruler who claims to be God. Still others will help that are not sincere and cannot be trusted: spies for the antichrist.

And some of them of understanding shall fall, to try them, and to purge, and to make them white, even to the time of the end: because it is yet for a time appointed.

Daniel 11:35 (KJV)

Some Christians and Jews will stumble and be deceived for a time and will miss the rapture. They will come to the truth afterwards so that they may be refined, purified and lead others during the wrath years who knew not the truth; for the appointed time of Christ's return and the end of the age will happen just as God planned it in the fall of 2028 A.D.

> *And the king shall do according to his will; and he shall exalt himself, and magnify himself above every god, and shall speak marvelous things against the God of gods, and shall prosper till the indignation be accomplished: for that that is determined shall be done.*
>
> *Daniel 11:36 (KJV)*

And the President of the E.U. will do as he pleases, and exalt and magnify himself above every god, and against the Creator of heaven and earth. He will speak unheard of things against the Lord. He will greatly succeed in every endeavor until the wrath of God has been poured out on the unbelievers of the world, for those judgments which have been determined must take place as scheduled.

> *Neither shall he regard the God of his fathers, nor the desire of women, nor regard any god: for he shall magnify himself above all.*
>
> *Daniel 11:37 (KJV)*

He will not bow down to the true God or to any god worshiped in the past for that matter: for he will consider himself above them all.

> *But in his estate shall he honor the God of forces: and a god whom his fathers knew not shall he honor with gold, and silver, and with precious stones, and pleasant things.*
>
> *Daniel 11:38 (KJV)*

He will however place his faith in the God of military technology and nuclear weapons and honor it by spending large sums of money on these weapon systems: a new god unknown to those before this time in history.

Thus shall he do in the most strong holds with a strange god, whom he shall acknowledge and increase with glory: and he shall cause them to rule over many, and shall divide the land for gain.

Daniel 11:39 (KJV)

Thus he will attack the mightiest cities (New York City being one) with the aid of this strange new god (nuclear bombs). The antichrist will reward all who acknowledge his authority and make them leaders over many. He will divide the lands he conquers for a price: eternal damnation for those who bow down and follow him.

And at the time of the end shall the king of the south push at him: and the king of the north shall come against him like a whirlwind, with chariots, and with horsemen, and with many ships; and he shall enter into the countries, and shall overflow and pass over.

Daniel 11:40 (KJV)

After a while the Arab nations from the South will resist his control and the antichrist in anger will storm out against them with great military numbers. His armies will spread out into other Arab nations and countries that oppose him in a massive military offensive.

He shall enter also into the glorious land, and many countries shall be overthrown: but these shall escape out of his hand, even Edom, and Moab, and the chief of the children of Ammon.

Daniel 11:41 (KJV)

Many countries will fall under his military onslaught. He will personally enter Israel and Jerusalem and set up command operations. But some areas will not feel his wrath and will remain safe: the mountainous areas of western Jordan.

He shall stretch forth his hand also upon the countries: and the land of Egypt shall not escape.

Daniel 11:42 (KJV)

During his invasions of other countries, Egypt will fall under his control.

But he shall have power over the treasures of gold and of silver, and over all the precious things of Egypt: and the Libyans and the Ethiopians shall be at his steps.
                                        Daniel 11:43 (KJV)

He will take control of all of Egypt including the Suez Canal shipping: the northern African countries will not escape his wrath and will also succumb to his military might.

But tidings out of the east and out of the north shall trouble him: therefore he shall go forth with great fury to destroy, and utterly to make away many.
                                        Daniel 11:44 (KJV)

But news from China and Russia of military troop movements will trouble him: He will set out in great fury to destroy everything he can before they arrive and many will die.

And he shall plant the tabernacles of his palace between the seas in the glorious holy mountain; yet he shall come to his end, and none shall help him.
                                        Daniel 11:45 (KJV)

He will make his last stand in Jerusalem where he has set up his one world government: yet the man of perdition will meet his end at the coming of Christ and none will be able to help him.

And at that time shall Michael stand up, the great prince which standeth for the children of thy people: and there shall be a time of trouble, such as never was since there was a nation even to that same time: and at that time thy people shall be delivered, every one that shall be found written in the book.

Daniel 12:1 (KJV)

And at that time, Michael will stand up: the archangel who protects the Jews. And during the time of tribulation there shall be a time of trouble that has never been equaled since the creation of nations. At that time the Jewish people who have come to accept Christ will be raptured along with the Christians: only those who's names were found written in the book of life.

And many of them that sleep in the dust of the earth shall awake, me to everlasting life, and some to shame and everlasting contempt.

Daniel 12:2 (KJV)

Listen, I tell you a mystery: many will arise from the dead to everlasting life, including myself, but many others who refused to accept the truth will be sentenced to everlasting damnation.

And they that be wise shall shine as the brightness of the firmament; and they that turn many to righteousness as the stars for ever and ever.

Daniel 12:3 (KJV)

Those who were wise and didn't accept the mark of the beast and believed in Christ will be rewarded for their efforts and lead many to righteousness; they will be allowed entrance into the Millennial Kingdom.

But thou, O Daniel, shut up the words, and seal the book, even to the time of the end: many shall run to and fro, and knowledge shall be increased.

Daniel 12:4 (KJV)

But you Daniel, close up the words and seal their understanding of this prophecy until the appointed time at the end of the Piscean Age. A

291

time when knowledge will be increased significantly and people will live in a fast paced world like never before.

> *Then I Daniel looked, and, behold, there stood other two, the one on this side of the bank of the river, and the other on that side of the bank of the river.*
>
> *Daniel 12:5* (KJV)

Then I Daniel looked, and there before me stood two angels: one on this side of the river bank and another on the opposite side of the river.

> *And one said to the man clothed in linen, which was upon the waters of the river, How long shall it be to the end of these wonders?*
>
> *Daniel 12:6* (KJV)

And one of them asked Christ who had authority over all of mankind (the waters of the river), "How long will it take for all these things, from the history of the Persians to your glorious appearing, to occur which were explained to your servant Daniel?"

> *And I heard the man clothed in linen, which was upon the waters of the river, when he held up his right hand and his left hand unto heaven, and swore by him that lives for ever that it shall be for a time, times, and an half; and when he shall have accomplished to scatter the power of the holy people, all these things shall be finished.*
>
> *Daniel 12:7* (KJV)

And I heard Christ, who was clothed in fine linen and suspended over the waters of the river, when he held up his right hand and his left hand toward heaven and swore an oath by the Creator of the Universe who always was and who always is, that the total time for these things to take place will be for a "Time, Times, and half a Time [2,450 P-years]." When the antichrist has succeeded in breaking the Christian church, this will be the sign that all these things have been fulfilled.

And I heard, but I understood not: then said I, O my Lord, what shall be the end of these things?

Daniel 12:8 (KJV)

And I heard, but did not understand so I asked, " O my Lord, what will happen at the end of all these things you have told me?"

And he said, Go thy way, Daniel: for the words are closed up and sealed till the time of the end.

Daniel 12:9 (KJV)

And he said, "Trouble yourself not Daniel for the Father's plans will not be revealed until the appropriate time at the end of the age so that His purposes may be achieved."

Many shall be purified, and made white, and tried [tested]; but the wicked shall do wickedly: and none of the wicked shall under-stand; but the wise shall understand.

Daniel 12:10 (KJV)

Many will come to the truth during the tribulation, but those whose names are not written in the Lamb's book of life will continue to refuse to believe. They will not understand what is going on around them; but the true believers who follow God's laws and believe in Jesus Christ will understand.

And from the time that the daily sacrifice shall be taken away, and the abomination that maketh desolate set up, there shall be a thousand two hundred and ninety days.

Daniel 12:11 (KJV)

From the time the abomination of desolation is completed there will be one thousand two hundred and ninety prophetic days until Christ returns.

Blessed is he that waiteth, and cometh to the thousand three hundred and five and thirty days.

Daniel 12:12 (KJV)

Blessed is the person that comes to the truth late and survives God and the Lamb's wrath during the thirteen hundred and fifty-four days (1,335 P-days) that follow the start of the abomination of desolation. They will be allowed to enter the Millennial Kingdom.

But go thou thy way till the end be: for thou shall rest, and stand in thy lot at the end of the days.

Daniel 12:13 (KJV)

But Daniel, do not trouble yourself any further, for you will die a peaceful death and rest until the judgment when you will arise and be rewarded for your faithfulness and good works as will others like you.

The Olivet Discourse

The Olivet Discourse is really a prophecy from Christ that answers what things will take place at the end of the age just before his return. It provides instructions, signs, and the sequence of events for the last days typically employing literal language as the general rule. We will take a verse-by-verse look at these insights using Matthew's account from the King James Bible.

And Jesus went out, and departed from the temple: and his disciples came to him for to shew him the buildings of the temple. And Jesus said unto them, See ye not all these things? verily I say unto you, There shall not be left here one stone upon another, that shall not be thrown down.

Matthew 24:1-2 (KJV)

These verses predict the destruction of God's temple in 70 A.D. at the hands of the Roman general Titus.

And as he sat upon the Mount of Olives, the disciples came unto him privately, saying, Tell us, when shall these things be? and what shall be the sign of thy coming, and of the end of the world?

Matthew 24:3 (KJV)

Here is where we learn exactly what Jesus will be talking about in the following texts. He will be answering the questions, what will be the signs to watch for of his return and for the end of the age (both occurring at the same time)? The KJV records this as the "end of the world," but a better translation of this wording is the "end of the age." Another way to view this is: the last sign of the end of this age is Christ's return and the disciples want to know when will this happen. Therefore, Jesus' responses to these important questions are answers that pertain only to the end of the age and not descriptions of the past two thousand years (Times of the Gentiles) of history as many Bible teachers incorrectly interpret!

And Jesus answered and said unto them, Take heed that no man deceive you. For many shall come in my name, saying, I am Christ; and shall deceive many.

Matthew 24:4-5 (KJV)

And I saw when the Lamb opened one of the seals, and I heard, as it were the noise of thunder, one of the four beasts saying, Come and see. And I saw, and behold a white horse: and he that sat on him had a bow; and a crown was given unto him: and he went forth conquering, and to conquer.

Revelation 6:1-2 (KJV)

The Olivet Discourse starts with the first signs to watch for: false messiahs and Christ instructs his followers to not be deceived by anyone who makes such a claim. Many false messiahs have come and gone throughout the centuries that followed Jesus' death. Recognizing this fact is nothing more than recognizing the sins of men are always at work. That history repeats itself and God's plans work on many levels; that the last years of this age will be nothing more than a microcosm of the last two thousand years of history. I say it again: they are not evidence that Jesus is giving signs for the last two millennia just because these signs keep reoccurring!

What should be more interesting to eschatology students is that Jesus will effectively list these signs of the end of the age in the same sequential order as those given in the prophecy of the Seven Seals. The first seal reveals the appointed time has arrived for the last false messiah foretold by God's prophets. Symbolically the antichrist rides a white horse exactly as the genuine messiah will[a] and comes in the name of peace (bow with no arrows): the same as when Christ first came.

However, when the true messiah returns, he will not come in peace, but in the name of justice. He will bring death to all non-believers and worshipers of worldly things with his double-edged sword of truth.[b] God's days of turning the other cheek and waiting for man's repentance are over. Seventy times seven we are to excuse our brother's transgressions against us[c] and mankind has used up all seventy of those God has allotted to us for our transgressions against Him. Humanity has reached the end of the Creator's chosen time for patience and forgiveness.

We further notice this rider is "given" authority (a crown) and with that power, his true purpose will be revealed. He will be bent on conquest… conquest of the world that will bring no peace. He will do his killing from afar using the false bow of peace (nuclear weapons), while the true king will do his righteous killing with the sword of truth up close and personal.[d]

> *And ye shall hear of wars and rumors of wars: see that ye be not troubled: for all these things must come to pass, but the end is not yet.*
>
> Matthew 24:6 (KJV)

> *And when he had opened the second seal, I heard the second beast say, Come and see. And there went out another horse that was red: and power was given to him that sat thereon to take peace from the earth, and that they should kill one another: and there was given unto him a great sword.*
>
> Revelation 6:3-4 (KJV)

The antichrist will come initially in the name of peace, but with his actions and ambitions, war will soon follow. Wide scale conflict is the second sign that Jesus lists just as it is when Christ breaks the second seal. However, we are instructed to hang in there (persevere) for this is not the end of the age, but just another sign that we are nearing the end. This is a tumultuous time of wars, anarchy, and riotous behavior in many nations throughout the world.

For nation shall rise against nation, and kingdom against kingdom: and there shall be famines, and pestilences, and earthquakes, in divers [diverse] places.

Matthew 24:7 (KJV)

And when he had opened the third seal, I heard the third beast say, Come and see. And I beheld, and lo a black horse; and he that sat on him had a pair of balances in his hand. And I heard a voice in the midst of the four beasts say, A measure of wheat for a penny, and three measures of barley for a penny; and see thou hurt not the oil and the wine.

Revelation 6:5-6 (KJV)

Following war and anarchy, Jesus tells his disciples there will be all sorts of other signs for his servants to keep watch for. There will be famines, droughts, natural disasters, widespread diseases, starvation, and other hardships in all parts of the world. These will be further evidence the end of the age is upon us.

And when he had opened the fourth seal, I heard the voice of the fourth beast say, Come and see. And I looked, and behold a pale horse: and his name that sat on him was Death, and Hell followed with him. And power was given unto them over the fourth part of the earth, to kill with sword, and with hunger, and with death, and with the beasts of the earth.

Revelation 6:7-8 (KJV)

We see from the opening of the fourth seal, the accumulative effects of all the previous signs. Twenty-five percent of all the people on the earth will die from these terrible calamities. There are presently over 6.6 billion people on the planet and this number will be higher by the time Christ begins unsealing the tribulation scroll.

Using today's population estimates, this means roughly 1.666 billion people will loose their lives. They will die because of humanity's refusal to believe in God and accept Christ, and the antichrist's efforts to remove all who oppose him from becoming the dictator of the world. Continuing on with this important prophecy given by Christ just a few days before his crucifixion,

> *All these are the beginning of sorrows. Then [after these signs] shall they [the unrighteous] deliver you up to be afflicted, and shall kill you: and ye shall be hated of all nations for my name's sake.*
>
> *Matthew 24:8-9* (KJV)

> *And when he had opened the fifth seal, I saw under the altar the souls of them that were slain for the word of God, and for the testimony which they held: And they cried with a loud voice, saying, How long, O Lord, holy and true, dost thou not judge and avenge our blood on them that dwell on the earth? And white robes were given unto every one of them; and it was said unto them, that they should rest yet for a little season, until their fellowservants also and their brethren, that should be killed as they were, should be fulfilled.*
>
> *Revelation 6:9-11* (KJV)

Now we learn of a time when Christians will be threatened by those in the world who will not accept Christ. God's people will be persecuted because of their faith, beliefs and many will suffer; other believers will be put to death. These injustices will occur while the "mark of the beast" is instituted as a means to fix the economic ills of the world caused by the greed and corruption that has deluded so many.

And then shall many be offended, and shall betray one another, and shall hate one another.

Matthew 24:10 (KJV)

Many believers, weak in faith, will be offended by the spiritually strong and give in to the lies of the world. They will be deceived and turn against their brothers and sisters in the faith. They will be left behind at the rapture destined to suffer along with the unbelievers until the time of wrath has been completed.

And many false prophets shall rise, and shall deceive many.

Matthew 24:11 (KJV)

Next, we find Christ talking about the signs of false prophets. Just as the Beast of the Sea (the false messiah: antichrist) comes before the Beast of the Earth (the false prophet) on the tribulation calendar, so does Jesus relate the order of these signs in their proper sequence.

And because iniquity shall abound, the love of many shall wax cold.

Matthew 24:12 (KJV)

Jesus further adds the reasons for why many things will happen. It is because the masses of the world have finally tired of the injustices forced upon them by the rich and powerful. Caring not for others as God had intended when man was created, but only concerned for themselves.

But he that shall endure unto the end, the same shall be saved.

Matthew 24:13 (KJV)

Who is Christ talking about in this text? First, they must be Christians because only Christians will be saved. If we search backwards through the passages from this text we find the last Christians Jesus was talking about are those in Matthew 24:10 who have turned against their fellow brothers and sisters. This suggests those Christians, if they see the error of their ways and endure the wrath of God, are the ones who will be saved and enter the Millennial Kingdom.

A second interpretation of this verse is that it applies to all Christians and there will be no rapture. All believer's and unbelievers alike will suffer under God's wrath. A position I find hard to support given all the evidence of the other prophecies we have investigated.

A third interpretation is that it applies to all Christians again, but the "end" spoken of is the end of the trials the E.U. president has been putting them through. They will be spared further tribulations when Christ rescues them at the midpoint of the tribulation. What we can deduce for sure from this scripture is there will be believers who will be saved at the end and who have endured hardship. The only outstanding question is whether the "end" is the last part of the antichrist's persecutions or is it after both Satan and God's wrath is concluded.

And this gospel of the kingdom shall be preached in all the world for a witness unto all nations; and then [at the point it is finished] shall the end come.

Matthew 24:14 (KJV)

And they sung as it were a new song before the throne, and before the four beasts, and the elders: and no man could learn that song but the hundred and forty and four thousand, which were redeemed from the earth. These are they which were not defiled with women; for they are virgins. These are they which follow the Lamb whithersoever he goeth. These were redeemed from among men, being the firstfruits unto God and to the Lamb. And in their mouth was found no guile: for they are without fault before the throne of God. And I saw another angel fly in the midst of heaven, having the everlasting gospel to preach unto them that dwell on the earth, and to every nation, and kindred, and tongue, and people, Saying with a loud voice, Fear God, and give glory to him; for the hour of his judgment is come: and worship him that made heaven, and earth, and the sea, and the fountains of waters.

Revelation 14:3-7 (KJV)

This text is not the scripture that follows the opening of the fifth seal, but it provides additional details about the 144,000 who are first described after the opening of the sixth seal. There is a lot of activity taking place in the period surrounding the tribulation midpoint. These scriptures are shown together because both discuss preaching the gospel to the whole world. This sign given by our Savior is in the same relative place (near the abomination of desolation) as the 144,000 in Revelation 7 is listed. Jesus' words in Matthew place it just before the five-day abomination of desolation while Revelation prophecies list it immediately afterwards.

I believe this sign will be more visible than most of the other "generic" signs: war, disease, and famine that have already been discussed. Why? Because Jesus says, it will be "a witness unto all the nations" of the truth of his coming kingdom and that the end of the age is at hand. Will the gospel come from the 144,000 sealed by God so they can do Christ's bidding without being effected by the antichrist's actions? I think so, but time will be our witness to the truth of these matters.

When ye therefore shall see the abomination of desolation, spoken of by Daniel the prophet, stand in the holy place, (whoso readeth, let him understand:) Then let them which be in Judaea flee into the mountains: Let him which is on the housetop not come down to take any thing out of his house: Neither let him which is in the field return back to take his clothes.
Matthew 24:15-18 (KJV)

We have now reached the midpoint of the classical seven-year tribulation. Realize you know more about this time in history than Jesus' disciples who he is talking with. We know during the first part of the tribulation Christians are being targeted by the evils within the world and we are to stand firm (persevere) and fight with God's two witnesses—the Word of the Lord and our testimony of Christ's saving blood. However, at this juncture Christians in Jerusalem and Judea are instructed to flee. The time for resistance is over and the time for hiding has come.

This is good advice, regardless of where you live in the world. Why? From our earlier work, there will be thirty-five P-days of intense spiritual warfare at this time. Christians will have to endure until the rapture occurs—although the persecution of Christians will have been escalating for many years before reaching this pinnacle of injustice.

If there is no rapture as many maintain, then there will be three and a half more years until Christ returns. If there is a rapture, I believe those servants who do not keep watch and are napping will be the people who will be left behind. Remember, half of the virgins (Christians) were not ready in the parable of the ten virgins and they were left to suffer with the rebellious unbelievers.[e]

And woe unto them that are with child, and to them that give suck in those days! But pray ye that your flight be not in the winter, neither on the Sabbath day: For then shall be great tribulation, such as was not since the beginning of the world to this time, no, nor ever shall be.

Matthew 24:19-21 (KJV)

We see from this text, Christ warns that at the time of the abomination of desolation (the thirty–five P-days), those days will be the worst in history for a believer (the elect). Obviously, Jesus cannot be claiming that the time when the antichrist's true intentions are unveiled, which is three and a half years prior to his official return, will be the severest time of trouble the world has ever seen or will see. God and His messiah have not even started pouring out Their wrath yet!

This suggests Jesus is advising only the elect (Christians), about the troubles they will encounter. Plainly, this cannot be a reference to an unparalleled time of devastation in human history. What about the flood which killed everyone but Noah and his family? What about the years of God's wrath which are about to start? These were and will be times worse than the midpoint of the tribulation. Therefore, we can conclude by Christ's words that these great tribulations will be the worst of all time as they relate to Christians, but not for everyone else in the entire world.

And except those days should be shortened, there should no flesh be saved: but for the elect's sake those days shall be shortened.

Matthew 24:22 (KJV)

Here is another verse many misunderstand, including me in the past. Maybe it supports the duality of more than one level of understanding for I have used it to calculate the timing of Christ's return. I had always believed it was a reference to cutting the very last days short before Christ returned so that all men would not perish due to their destructive nature.

However, does God care if unbelievers kill each other? Yes and no. He yearns that no man should lose his life and yet He knows as you and I do, that the majority will rebuff Him. When Christ returns he will put to death all who reject him just as the antichrist will try to kill all who refuse to worship him. This idea is hard for many Christians (and even unbelievers) to accept because they are taught there is a loving God and Christ will forgive them for their sins.

However, what is God to do to those who like to sin and who reject Him? Evil cannot be allowed by an eternity of patience can it? What about justice for those who are wronged by those who ignore God's laws? Despite your beliefs, the fact of the matter is when Christ comes he will remove (kill) all those who have rejected his authority during the forty days of cleansing. It logically follows he is not talking about shortening the time to save all of mankind. What's more, since he is discussing the time around the abomination and can only be talking about saving Christians, then he must simply be talking about shortening the time the man of perdition has power to punish his followers. These new understandings flow from the knowledge of knowing where things will take place at this point in the future. Maybe our Lord is stating the appointed time for his return will be decreased, but more than likely he is saying the antichrist will be stopped short of reaching the seventy 'sevens' he needs to finish his work. If he were allowed to reach his goal, then all Christians might be killed and only unbelievers would inherit the kingdom!

Following this logic to its final conclusion means, that when Christ returns and cleanses the earth of all unbelievers and the things that cause sin, there would be no one left worthy enough to enter the Millennial Kingdom. If there is none left deemed acceptable to enter God's Kingdom on earth, then the biblical prophecies would be untrue and God's Word unreliable. We now can appreciate from this logic why it is the time around the abomination that causes desolation that will be shortened. This is the time of persecution and the only way to keep all God's people from being killed is to remove them from harms way and take away the antichrist's authority over those idle servants who were left behind.

Then if any man shall say unto you, Lo, here is Christ, or there; believe it not. For there shall arise false Christs, and false prophets, and shall shew great signs and wonders; insomuch that, if it were possible, they shall deceive the very elect.

Matthew 24:23-24 (KJV)

These verses appear to speak of a period of time that will occur in the latter part of the tribulation under the storm clouds of God's wrath. This is because of the order in which Jesus relates this sign. Christians are not to be enamored with those who will be able to perform miracles. Why? Because supernatural evidence is not part of God's plans, but verification of how Satan works.[f]

Revelation 16:13 says there will come a point when evil spirits will do the bidding of Satan, the antichrist and the false prophet. These demonic spirits will perform unbelievable signs so that the foolish, who know not the truth, will be deceived.

If all Christians are still around to witness these wonders, then this can only mean there was no rapture before Christ's final return. It's possible this is just a warning for Christians who will be left behind because they did not keep God's commands and needed to be polished in the firestorms of His wrath. Notwithstanding what happens in the last days, never believe anyone who says they are the Christ or knows where he is. Christians will recognize the true Master when he comes with his angels.

Behold, I have told you before. Wherefore if they shall say unto you, Behold, he is in the desert; go not forth: behold, he is in the secret chambers; believe it not.

Matthew 24:25-26 (KJV)

Again, do not be easily swayed. Learn what the signs are and study the Word so that you will be ready for all of Satan's tricks. Jesus warns us ahead of time not to be fooled by anyone who comes in his name for when he comes, it will be only at the planned time and not years before and he will come in majesty, not as a man assumes power (like the antichrist).

For as the lightning cometh out of the east, and shineth even unto the west; so shall also the coming of the Son of man be.
Matthew 24:27 (KJV)

Here is the description of how Christ will return to earth and any other method of coming to power only confirms that messiah is a false one. It will be a magnificent arrival the likes of which the world has never witnessed.

For wheresoever the carcase [carcass] is, there will the eagles be gathered together.
Matthew 24:28 (KJV)

At the time Christ returns, billions will have already died and even more will find eternal death during the forty days of personal judgment. This verse looks eerily similar to the scripture from Revelation that calls the birds of the air to gather[g] for the final conflict between good and evil—the battle between Christ and his heavenly subjects and the armies of the world collected on the plains of Megiddo.

Immediately after the tribulation of those days shall the sun be darkened, and the moon shall not give her light, and the stars shall fall from heaven, and the powers of the heavens shall be shaken: And then shall appear the sign of the Son of man in heaven: and then shall all the tribes of the earth mourn, and they shall see the Son of man coming in the clouds of heaven with power and great glory.
Matthew 24:29-30 (KJV)

And I beheld when he had opened the sixth seal, and, lo, there was a great earthquake; and the sun became black as sackcloth of hair, and the moon became as blood; And the stars of heaven fell unto the earth, even as a fig tree casteth her untimely figs, when she is shaken of a mighty wind. And the heaven departed as a scroll when it is rolled together; and every mountain and island were moved out of their places.

Revelation 6:12-14 (KJV)

If one believes the previous explanations put forth in this book, then it must be understood that with these verses Jesus has started supplying additional information to further clarify previous signs already discussed. This is because this verse from Matthew matches very closely with the opening of the sixth seal and these heavenly signs occur when the abomination of desolation is set up midway through Daniel's last 'week.'

What this also means is the words *"they shall see the Son of man coming in the clouds of heaven with power and great glory"* is not the same event as *"for as the lightning cometh out of the east, and shineth even unto the west; so shall also the coming of the Son of man be."* Most Christians who read these verses would disagree with this conclusion, just as I once did; until I began studying the relationship of end-time events. Things we will take a serious look at in chapter fifteen. By viewing both of these scriptures as descriptions of the same event, one should realize that Christ would need to come at the breaking of the sixth seal… before the opening of the last seal. How can Christ be in heaven opening the seventh seal if he is here on earth gathering his people to fight the War of Armageddon? This war occurs long after the sixth seal is removed.

Moreover, when Christ returns for the final battle, the sun will not turn black nor the moon blood red, but it will be a day that is unique. A day when there will be twenty-four hours of light.[h] More evidence that these two descriptions of his coming are not the same, but one a description of the rapture (the saving of the faithful) and the other a description of his final return in 2028 A.D.

And he shall send his angels with a great sound of a trumpet, and they shall gather together his elect from the four winds, from one end of heaven to the other.

Matthew 24:31 (KJV)

This is another description of the rapture. The only question for some is: When will this event occur? I believe the Revelation prophecies' evidence supports a mid-tribulation rapture and we see its position mentioned is after the sixth seal. Now there is some evidence that says this will occur at the last trumpet,[1] which does not jibe with the removal of the sixth seal. If this really does happen at the last trumpet... the seventh trumpet, then Christians should prepare to suffer God's wrath along with the unbelievers. I believe the overwhelming evidence from the prophecies you have read about, suggests this will not be the case. Only time will tell... but be ready always.

Now learn a parable of the fig tree; When his branch is yet tender, and putteth forth leaves, ye know that summer is nigh: So likewise ye, when ye shall see all these things, know that it is near, even at the doors.

Matthew 24:32-33 (KJV)

Jesus is now through giving sequential details of the end-time events and moves on with providing other signs to watch out for. This text is not literal language like many of the verses before. Why? Because Jesus informs us this is a parable and parables always have a deeper meaning for Christians. A meaning that the Holy Spirit reveals to those who belong to Christ. Many sentries who watch for Christ's return recognize this is a reference to Israel (fig tree) becoming a nation again. Moreover, just as no one knows exactly when the leaves will open on the trees in any year, we all know this law of nature occurs in the spring—we just never know the exact day or hour.

The lesson to be learned is this: when history witnessed Israel become a nation again, then the time to Christ's second advent began its countdown. Just as springtime is the season when trees begin to put forth leaves, so shall summer follow... never delayed or in no way late.

the seasons always happen on schedule. So also will the coming of the son of man occur on schedule after witnessing this sign take place!

> *Verily I say unto you, This generation shall not pass, till all these things be fulfilled. Heaven and earth shall pass away, but my words shall not pass away.*
>
> Matthew 24:34-35 (KJV)

This generation is a reference to the generation that sees "all" the signs Jesus has been telling us to watch out for and not just some of the signs (i.e. wars) that Christians throughout the centuries have witnessed before us.

> *But of that day and hour knoweth no man, no, not the angels of heaven, but my Father only. But as the days of Noe [Noah] were, so shall also the coming of the Son of man be. For as in the days that were before the flood they were eating and drinking, marrying and giving in marriage, until the day that Noe entered into the ark, and knew not until the flood came, and took them all away; so shall also the coming of the Son of man be.*
>
> Matthew 24:36-39 (KJV)

If believers are keeping watch, they should be capable of discerning the "season" of our Lord's return, but unable to pinpoint the exact day or hour… just like the parable of the fig tree. At the time of Christ's return billions of people will be oblivious to the signs around them because they are unfamiliar with all that is written in the Bible about these matters. Worse yet, most do not believe in the true God and are unaware of anything the Word says. They will be blind to the signs of Jesus' coming just as the millions who met their death ignored the sign of Noah building a huge ship.

> *Then shall two be in the field; the one shall be taken, and the other left. Two women shall be grinding at the mill; the one shall be taken, and the other left. Watch therefore: for ye know not what hour your Lord doth come.*
>
> Matthew 24:40-42 (KJV)

This wording is a most troubling riddle to solve. Some theologians claim the people who will be taken are Christians at the rapture. I have heard other theology experts claim it is the unbelievers who will be weeded from the fields so that only the saints carry on into God's kingdom on earth in the next age. Jesus said at the time of Noah it was the wicked who were "taken away" by the flood... the first judgment. When these verses are taken in that context, and if the time he is speaking of is the second judgment (forty days of cleansing), then we can surmise it is the unbelievers who will be taken.

However, know this: The percentage (fifty percent) mentioned is not the correct proportion of unbelievers to believers who will be remaining when Christ returns. At the end of the tribulation, the vast majority of people who will still be living are unbelievers.

I have yet to be convinced that either of these positions is the correct understanding. I favor a different explanation that is based on the placement of this block of scripture in relation to Jesus' other words. I believe this passage parallels the parable of the ten virgins in Matthew 25 and their messages are the same. Half of those who profess to be Christian are partaking in things they shouldn't be and will not be ready when the time comes. They are Christians who profess only with their mouths, but not with their actions. Can you truthfully believe Christ is your king if it is more important to sleep-in than to go and worship the Master? Is any excuse acceptable to avoid attending church services for worship at least once a week? Just as half the virgins were ready at the midnight hour, so too we will find half the servants taken unexpectedly from their daily routine will be deemed ready, while the other half will be left behind.

I believe that if we think of the people Jesus is warning not as believers versus unbelievers (as most experts do), but all as Christians, then one can grasp that half will be spared God's wrath (taken) while the other half (left) will have to prove themselves worthy. Looking at all the text that surrounds this information, we find Christ has been talking only to and about his servants. A warning to hide when they see the abomination of desolation, blessings for those servants who put into practice his words, and chastisements for those wicked servants (weak Christians) who mingle with unbelievers, pick-up their sinful behaviors, and do not keep watch. When Christ is telling us, that one will be taken and one will be left, he also adds the warning to keep watch. Do unbelievers ever keep watch? No. Why must a Christian keep watch at this important time? If they are to be the ones to be left, there is no need to keep watch. If the unrepentant sinners are the ones taken away, does it matter if this event surprises a believer? No. Clearly then, it must be the one who is being taken that must watch so they will not be caught off guard and in doing so, we deduce they must be faithful Christians.

Do not deceive yourself. Just because many say they are a Christian, does not mean they will automatically be forgiven for their sins. The Jews were not forgiven when they prostituted themselves with foreign gods, worshipped idols, and lusted after the pleasures of this world and neither will unrepentant Christians who unknowingly do the same things!

A Christian neighbor told me, after I mentioned Christ would return in 2028 A.D., that God would never reveal the time of Christ's return because then Christians would do every evil thing they wanted until the time was near. Then they would straighten up and expect God to forgive them for the sins they had committed. To this I replied, "if this is a Christian's plan then they really are not a Christian!" They might as well keep on sinning to the very end because they may get to swim in the Lake of Fire along with everyone else who refuses Christ. God knows what their plans are and He has made His own plans for them. These plans call for leaving them behind when the faithful are taken.

> But know this, that if the good man of the house had known in what watch the thief would come, he would have watched, and would not have suffered his house to be broken up. Therefore be ye also ready: for in such an hour as ye think not the Son of man cometh.
>
> Matthew 24:43-44 (KJV)

Always keep the faith and observe God's commands for Christ will return at a time even good Christians may not learn of. The knowledge of his return will be out of reach to those who do not study and keep watching, but you are not unaware of these things. This text is really telling us to be a good witness for Christ at all times regardless of when he might return. Why? So that we can maximize the harvest and save as many souls from damnation as possible. For those who see the light of Christ in you will wonder after these things and seek out the light that shines in you. Lift it high at all times and know in doing so you will never need to worry when that day and the hour will come.

> *Who then is a faithful and wise servant, whom his lord hath*
> *made ruler over his household, to give them meat in due season?*
> *Blessed is that servant, whom his lord when he cometh shall find*
> *so doing. Verily I say unto you, That he shall make him ruler over*
> *all his goods.*
>
> Matthew 24:45-47 (KJV)

Christ's servants are Christians plain and simple. Everyone else is considered a dissenter to the truth. God's measure of a man's goodness is not the same measure we use in this world. Hell is full of people who many considered good, for all fall short of the standard the Lord uses.

That standard is simply that you must accept Christ as your personal savior and put into practice the things he taught. "Good" Christians keep watch and are prepared at all times for all circumstances so that no matter when the Lord comes, whether today or twenty years from now in 2028 A.D., they are always ready for his return.

> *But and if that evil servant shall say in his heart, My lord de-*
> *layeth his coming; And shall begin to smite his fellow servants,*
> *and to eat and drink with the drunken; The lord of that servant*
> *shall come in a day when he looketh not for him, and in an hour*
> *that he is not aware of, and shall cut him asunder, and appoint*
> *him his portion with the hypocrites: there shall be weeping and*
> *gnashing of teeth.*
>
> Matthew 24:48-51 (KJV)

This is exactly what I was telling my neighbor who was concerned that fellow Christians, who knew the time, would not live a Godly life until the last moment. Do not fall into the trappings of this world. Do not be tricked into thinking, it has been almost two thousand years since

311

God sacrificed His son to save the world from sin, that the time would never come for Christ to return. Do not join the unbelievers in the world and practice the evil things they do, for the Lord's warning is not without consequences. Those servants who claim they are good Christians, but are without deeds will be treated just like the people who reject Christ at the hour of His wrath.

There are many prophecies that foretell of the coming time of God's wrath. We have investigated a couple that provide timing details and key signs to watch for during the troubled times ahead. The prophets Isaiah, Jeremiah, Ezekiel, Zechariah, and the others of the Bible will not be examined in detail. For students of eschatology these books are necessary reading, to supplement the things we have uncovered from the book of Revelation, so that no stone goes unturned. They are recommended for study so that a Christian will have the full protection of God's Word when needed to withstand Satan's attacks.

Let's turn our attention to Satan's methods and see how the antichrist will use prophecies of the messiah to mislead the world at the end of Piscean Age... the last years of this age.

Chapter 14

Christic and the Antichrist

MANY PROPHECIES OF CHRIST ARE IN reality pseudo-prophecies of the antichrist, while all prophecies of the antichrist are never prophecies of Christ. This statement is probably something you have never heard before. If we are to recognize that when the antichrist comes he will deceive many Christians and Jews by claiming he is the messiah, then he must be able to base his claims on some level of scriptural foundation. The evidence he must offer will logically flow from the twisting of the truth of the biblical prophecies that pertain to the coming of the messiah. It is impossible to convince a knowledgeable Christian or Jew you are the Christ foretold, unless you can demonstrate using scripture, that you meet the requirements of the messiah.

Do you remember when Jesus was fasting forty days in the desert prior to beginning his ministry? What method did Satan use to try and deceive him? The Word of God! We have seen Lucifer's tricks and methods before: the ever so slight twisting of the meaning of scripture, taking verses out of context, or changing a few words so the intended message is altered. Will Christians, in the last days, be able to fight back as Christ did with the Father's weapon of choice—His Word? Many Christians will not be able to since they are unaware of much of what is written in the Bible. They will easily fall prey to these masterful deceptions for even those who are well versed in the scriptures are not without weaknesses.

However, if Christians are prepared for Satan's deceptions by constantly studying God's Word, they will be able to resist and lead others away from an eternity of damnation.[a] One way to prepare ourselves is to view the prophecies of the coming messiah through the eyes of Satan. By doing this, we can see how he might bend these prophecies so that when these deceptions occur, we will be trained to recognize those untruths, and in so doing recognize the antichrist.

Many of Daniel's prophecies provide details of the future end-time leader of the world. These prophecies give detailed information for both

Christ and the antichrist, but usually are viewed as prophecies of Christ with some information about the antichrist included.

Let's go back to the workhorse of Daniel's prophecies. The prophecy of "Seventy Weeks." As we have seen, this prophecy provides the backbone of God's plan by providing the timing for important events: Christ's death, Christ's second advent, the length of the classical tribulation period, and the date for the remaking of heaven and earth. We also know that these 'weeks' can be viewed on multiple levels: as prophetic years, Sabbath years, and Jubilee years and are true for all three different understandings. This prophecy is used to find key times associated with Christ's life and is an incredible example of God's work and planning.

However, the versatility of the prophecy allows us to unlock information about the antichrist as well. I showed you in chapter two, on pages nine and ten, that after sixty-nine 'weeks' exactly Jesus died. That understanding of Daniel's prophecy used P-years as the basis. But if we were to use another elucidation, the parallels of the antichrist could be seen in this prophecy. Although when he comes, he may never relate this version of the prophecy as evidence of his authority because of its negative aspects.

I only point this out so that you may see the duality of using prophecies of Christ as prophecies of the antichrist. If we were to go back and start counting Jubilee years as mentioned in Leviticus from when God told the Hebrews to begin observing them, we would see the duality of this prophecy. The year was 1415 B.C. when the Israelites entered the promise land.[b] If we count sixty-nine 'weeks' (Jubilees), we arrive in the Jubilee year 2028/2029 A.D.[c]—the year of Christ's return and the year of the antichrist's death. Just as Christ was "cut off" after sixty-nine 'weeks' using P-years as the basis, so also will the antichrist die after sixty-nine 'weeks' using Jubilee years as the basis!

Hence, how might Satan twist this prophecy? By not using the starting time mentioned by Daniel (Ezra's decree), but using a different initial time like the one mentioned in Leviticus[d] and applying another level of understanding. Let's consider another illustration. In Daniel's prophecy, we know that Daniel separated out seven of the seventy 'weeks' (set aside ten percent), but never addresses why this was done. Writers would not account for time in the odd manner Daniel does and so we can infer this intentional separation of seven 'weeks' must have some greater purpose. There are reasons to believe it was to signal another starting date for other calculations as we have discussed—but Satan and the antichrist can use this time as well for their purposes.

If you count seven 'weeks' using Jubilee years as the basis; forty-nine Jubilees starting from and including the first Jubilee after Ezra (423/422 B.C.), you reach the Jubilee year 1978/1979 A.D. When counting in Jubilee years, we cannot start at any year of our own choosing (unless otherwise instructed by God). They must begin from a previously recognized Jubilee year. This counting method will always end up at a subsequent Jubilee year. Using this method and including the year 423 B.C. in the count, gets us to the year 1978/1979 A.D.

What is important about this year? It is the time I determined the antichrist was born based on analyses from my first book—the time of his second coming[e] and Satan is always trying to implement forgeries of God's plan. This was a Jubilee year and many Jews believe their messiah will come in a year of Jubilee. However, starting our count from the same 423 B.C. date and moving forward in time the same forty-nine Jubilees, we get to the year of Christ's second coming in 2028/2029 A.D. if we do not include the beginning year. So by either including the first year of Jubilee after Ezra or omitting it in the count, we can get to either the antichrist's second coming or our King's second advent!

Many prophecies are strictly prophecies about the antichrist. Paul prophesies in second Thessalonians that the antichrist will not be "revealed" before the power that restrains him is removed.[f] This power is the Holy Spirit and since the Holy Spirit resides in all Christians, this can only mean that Christians have been "removed" before the antichrist's true ambitions and character are revealed.

This conclusion presents a conflict with other scriptures, but not for those supporters of a pre-tribulation rapture. Their doctrine is fully consistent with the text in second Thessalonians. However, Daniel was told the saints (holy ones) would be handed over to the antichrist for a time, times, and half a time[g] (three and a half years). Pre-tribulation rapture supporters handle this contradiction by claiming these are "new" believers who have converted after the rapture. I do not subscribe to their arguments, as you know, because I do not believe scripture supports their case when examined as a whole. This is just wishful thinking. All of scripture needs to support the correct interpretation; otherwise, that understanding is a false one.

Numerous scriptures must be pieced together to see the whole picture, many of which the pre-tribulation eschatology supporters are unaware of. In our "revealing" dilemma, I believe the problem lies in "whom" the antichrist will be revealed to. I have always viewed this situation from the perspective of a Christian. Clearly, Christians who are knowledgeable of biblical prophecies, and keep watch, understand any European leader who confirms a covenant with the Jews acknowledges by his actions he may be the antichrist. This key sign could reveal the antichrist and signal the beginning of the tribulation countdown. Thus, the only way for Christians to be handed over to him for three and a half years, and not recognize him, is for the three and a half years to take place before this covenant is confirmed. Continuing on this path of logic will only lead to more confusion before we can discover the truth. This difficulty is like a problem in one of those logic puzzle books you buy in the store. They always provide enough information to solve each one; it's just that most people's brains are not conditioned in the ways vital to finding the solution.

This requirement means the antichrist would have to be in charge at least three and a half years before he establishes the covenant so that he has the authority to make life miserable for Christians. It would also mean the total time the antichrist will lead the E.U. has to be at least ten and a half years and that the rapture would need to occur after the first three and a half years; just before he signs the covenant so he wouldn't be revealed. This timing is not consistent with what we know about these events.

How can Christians be persecuted for three and a half years by the antichrist and he remain anonymous? He can't! He cannot hide from a knowledgeable Christian, but possibly he can from those Christians sleeping alongside the unbelievers who know nothing about Bible prophecy. For the unbelieving world, his real identity and intentions will remain invisible. Hence, what Paul must be saying is the revealing of the antichrist does not include believers who keep watch for they will know who he is at least three and a half years prior to when Paul says it will occur. Therefore, this revealing can only be applied to those who are blind to the truth. What will open their eyes at this very moment after ignoring Christian pleas for three and a half years that he is the antichrist? Perhaps witnessing the rapture of the faithful at the midpoint of the classical tribulation? This will be the revealing spoken of by Paul.

Everyone in the world will see this and they will know at that point Christians were telling the truth.

When the power of the holy people has been broken, then God's wrath will happen as prophesied. This power will be broken when Christ comes for the church. This prophecy also provides further predictions about the antichrist, as does the prophecy in Daniel 8 and Daniel 10-12. When you analyze them from the perspective that the antichrist will try to mimic (forge) what has been foretold about Christ, then you can see even more things that have been hidden. The antichrist will be the most evil person to ever walk the earth and yet loved by millions. How?—by deceit of the grandest magnitude. Let's see how this deception might work overall since various pieces have been discussed in other chapters.

You know that Christ will return in the autumn of 2028 A.D. and the reason of his coming is to stop the antichrist from destroying the world, to ruin Satan's plans, to judge humanity for their transgressions, and take the throne of his kingdom. We also know Satan's objectives are to implement a counterfeit plan "ahead" of the real plan—three and a half years earlier as discussed when the timing of the tribulation was analyzed. Moreover, just as Christ will set up his earthly throne in Jerusalem,[h] so shall the antichrist try to lead the world from Jerusalem.[i]

This transfer of power will occur in the spring of 2025 A.D. around the time of the Passover. Why this time? Well there is a lot of symbolism involved here. First, the rapture will occur around this time. This means the first resurrection for those who have died in Christ will mirror the timing of Christ's own resurrection! Secondly, Christ entered Jerusalem on Palm Sunday as the coming King, but was rejected by the Jews. The antichrist will take his turn at trying to achieve what Jesus did not and enter as king as well. Ultimately, the Jews will reject him too after he declares himself the messiah and sets up the abomination that causes desolation. However, many gentiles will not reject him but rather embrace him.

What is the abomination foretold by the prophet Daniel? The antichrist will abolish sacrifices to God and pass laws to worship only him. Just as there will be no need for animal sacrifices when Christ returns, there will be none required for the antichrist except the sacrifice to bow down to him! The antichrist's army will set it up and thus a foreign army will be in Israel. This is another sign to watch for.

We know from our timeline that the fake tribulation period begins seven years earlier than this in 2018 A.D. This is so it can be shown the antichrist met the timing of Daniel's last 'week'. It also might mean there is some agreement signed in 2018 A.D. as well (the real one being signed in 2021 A.D.). Since we know from Daniel the antichrist cannot take control of the E.U. until after he signs the real agreement (covenant), it can be deduced that these events will happen very close together.

As a result of signing such a historic agreement to finally bring peace to the Middle East, achieving what no one has been able to do,[j] everyone will be so enamored with him that they will gladly cede power over to him.[k] So what might have been his position in the EU that would give him the authority to broker these peace deals prior to taking full control of the empire? The most logical answer is he will be a foreign minister in the European Union. That is where we need to begin looking... looking for a successful foreign minister that is moving up through the ranks between 2018 and 2021 A.D. who is about forty years old in 2019 and most likely a politician from Greece of Jewish decent![l]

Let's go back to 2025 A.D. when the leader of the European Union declares he is God... the abomination that causes desolation for the world and the start of God's wrath. Clearly, anyone making such boasts in reality will be saying that if they are God, then they are in charge of the whole world. This claim will not go unchallenged by many countries of the world as we know, with China and Russia being the best examples of this resistance.

Just as Christ first began his ministry in 29 A.D. and died in 33 A.D., those who unwittingly do Satan's work by moving back the fifteenth year of Tiberius to 25/26 A.D. do this intentionally to support their position that the crucifixion occurred in 30 A.D. This allows them to say Christ began his ministry three years earlier in 26 A.D. Given that God's plan calls for adding roughly two thousand years from the beginning of Christ's ministry (29 A.D.) and shortening that time so everything will not be destroyed, he can be expected to return in late 2028 A.D. Applying identical logic to the false times that many scholars now support, we can conclude that adding two thousand years to 26 A.D. and shortening the time would get us to the year 2025 A.D. The year the antichrist will claim he is the messiah. Using wrong information i.e. the false times of Christ's first coming and applying the correct understanding from God's plan, gets us to an incorrect date for Jesus' return. However, by using this twisted logic the antichrist now has the support he needs to convince many that he is the messiah.

Did you notice what was done? This was not a complete lie because total fabrications are easy for people to spot. The best lies are the ones weaved from the fabric of truth, but have infrequent threads of lies stitched within so they are hard to make out. The half-truths politicians and advertisers bombard the world with daily. What Satan did here is use a copy of God's plan to validate his claim as the Messiah in 2025 A.D. by starting with a wrong date (the lie)—a date that is already being established and accepted as truth by many unsuspecting Christians at this time.

We know that Psalm 90 gives us some keys for unlocking the Lord's hidden timing. By adding the eighty years mentioned in this text for a generation to the time Israel became a nation in 1948 A.D., gets us to 2028 A.D. as Christ's coming. How might the antichrist deal with this? Well, Psalm 90 also says the time could be seventy years and by adding that time we get to 2018… the start of the false tribulation and the beginning of the antichrist's ministry (political ascension). So he is covered again. Christ's ministry was three and a half years long when he came the first time and will be an additional three and a half years long ruling from heaven for a total of one 'week', one 'seven' P-year period.

The antichrist ministry will be the same. Watch this lie… this forgery of Christ's. The antichrist will begin his ministry of peace in 2018 A.D. and, three and a half years later (2021 A.D.), he will come to Jerusalem to sign the covenant (peace plan) for the Middle East. He will complete his remaining three and a half in authority (from 2021 to 2025 A.D.) on earth. Thus, his first three and a half years will be predicated on peace (just as Christ's was) and his next three and a half years will be ruling with power… conquering power for a total of seven years or one 'week.'

However, you may ask, what about the time from 2025 to 2028 A.D. when the antichrist is ruling? Although he appears to be ruling, he is not. Remember Christ has taken control in heaven and the antichrist has no authority any longer. How can you rule dead people? You see with the removal of the Holy Spirit (Christians at the rapture) the majority of people will be spiritually "dead"... unbelievers. In addition, many of those unbelievers will now realize they have been deceived and turn on him. The revealing will open many eyes. Many Jews and new Christians will flee to the mountains (countryside away from civilization), and the other nations of the world: the Kings of the North (Russia), South (Arabs) and East (China) will converge on the Middle East for the War of Armageddon... World War III... the last war of this age where the last battle will be fought at the return of Christ.

How many or which prophecies will the antichrist use to convince Israel he is the promised messiah? No one knows for sure until the time comes. I offer these ideas to you as an exercise for you to prepare yourself so you will not be uninformed like those in the world who have no hope. To train you for all possibilities in what to watch for so that you will not be deceived from obtaining the salvation, through Christ, the Creator has planned for you.

Chapter 15

The Tribulation Events

HAVING CONCLUDED AN EXHAUSTIVE analysis of the particulars of the tribulation timing and the prophecies of Revelation (and others), it might be best to see how all these things interrelate so that Christians can have a detailed map to follow of the end-time events. By combining the timeline work with the studies of the end-time prophecies, we will be able to visualize the future in terms most should be able to understand.

How best then to accomplish this objective? Well, I determined some engineering software from my previous career would work nicely in carrying out this goal. There are computer programs that allow project managers to plan for large and complex projects. This software allows anyone to logically break down intricate projects into smaller tasks so they can be more effectively managed. It also allows one to see, ahead of time, where conflicts of interest might cause trouble so those difficulties can be reorganized to prevent future problems.

For instance, if you were to build a house, everyone knows the roof has to go on after the walls are up. It is obvious the interior wiring needs to be installed after the house shell is finished and the basement has to be dug before the main structure is built. If you were to list all the things that have to be done, you would quickly find out there are many tasks that must be done in order to successfully build a home and having a detailed plan reduces the chances of unwelcome surprises. You sure wouldn't want the roofers showing up when you were still digging the hole for the basement would you? This type of planning software is designed to efficiently manage the various aspects of projects so that things can be organized properly and conflicts avoided.

This software has other benefits besides the ability to arrange tasks on a timeline in their logical order. It has the capability to assign dates and use "conditional" tasks. Conditional tasks are events that are predicated on other events occurring in advance. Just as our house-building example required the basement to be installed before the roof, so must some prophetic events occur before others can take place.

Using this computer-planning program, I began the process of breaking down individual end-time prophecies into sequential tasks and inputting them into the software. I did this for all the prophecies contained in the book of Revelation, Daniel's prophecy of the "Kings of the North and the South," and the "Olivet Discourse." All the prophecies we examined in verse-by-verse detail.

There are many other prophecies that deal with the time before the end of the Piscean Age we are currently living in, but these prophecies were enough to get a very good rendition of what can be expected. Other prophecies in the Old Testament provide redundant information and supplementary details to the event timeline. They were inserted afterwards to bring as much clarity to this future period as possible.

After doing this, times and dates were added based on the timelines developed earlier, along with educated guesses as necessary. For example, it is impossible to know exactly how long some of the "seal" openings in Revelation will last, so they were assigned durations based on deductive reasoning. Realizing the first seal signifies the beginning of the anti-christ's rise, this seal can last either ten years, as has been shown, or seven years depending on whether we are to believe what other eschatology scholars claim—that the starting point is when the antichrist confirms the covenant mentioned by Daniel. If we begin counting at the seven-year mark prior to Christ's return, then all the things associated with opening the scroll's remaining seals would have to take place during this shorter timeframe. On the other hand, beginning at the ten-year mark, allows the latter seals to be spread out over a longer time period. Making the wrong choice ultimately leads to the possibility of events being off by as much as three years!

By applying the knowledge we have learned, we can adjust and align all the prophetic signs so that it can be determined where the best place is to position the start of each seal opening on the timeline and further-more, guestimate the length of time it takes for each to be accomplished. This approach applies to the other prophecies as well.

After completing this exercise, some things of interest were observed. First, the order of things (signs) Christ lists in Matthew 24 to watch for is nearly identical to the order given in the Seal prophecy! They are not just some random, general list of signs to keep looking for, but a specific, sequential list of events for the end of the age. The problem most theologians have is in recognizing them as such. I hope our look at these two prophecies in previous chapters has convinced you of this fact.

I once heard a sermon based on the scripture from Matthew twenty-four. The crux of the sermon was that the signs given (wars, famine, disease, false messiahs) have been occurring regularly since the first century, so Christians should be ready at all times for Jesus' return because every sign has been fulfilled. What we have here is an example of Christian training and a preterism doctrine affecting the interpretation of prophecy—something that should be guarded against.

Remember, the question put to Christ by his apostles in the Olivet Discourse was, "when will it be the end of the age and how will a person know it?" The answer cannot be: mankind has been living at the end of the age since the time of Christ so be ready! This is sheer nonsense. There can only be "one" end of the age and not one end of the age that has lasted almost two thousand years—which is the only conclusion that can be drawn from this sort of reasoning. We know the end of the age has not come because at the actual end, Christ will return. Thus, the real issue is not that these signs are always in place so be faithful at all times (which we should for many other reasons, just not for this reason), but in recognizing these wars, famines, etc. that Christ mentions are "uniquely different" from those during the past two thousand years. I'm unsure what the differences are, but I am confident that Christians who keep watch will recognize the "real" signs because of the power of the Holy Spirit.

Let us take a side-by-side look at the prophecies of the Seven Seals and the Olivet Discourse using a chart generated by the project managing software. This graphic representation, known as a "Pert Chart," will allow us to see the relative position of end-time events and the timing associated with each event. It also allows us to compare the various events of each prophecy with one another.

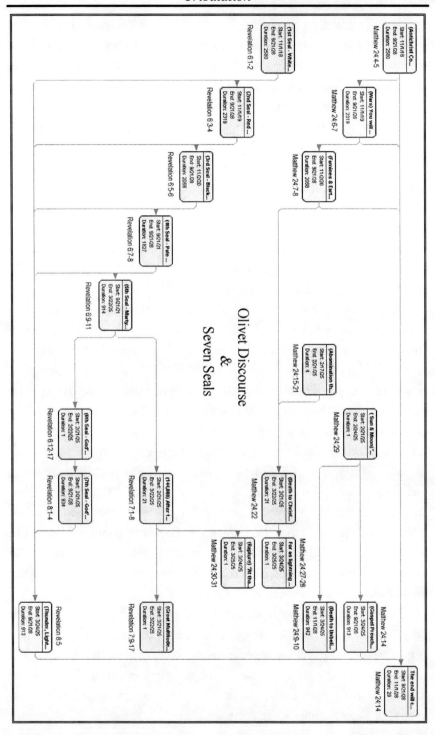

Olivet Discourse
&
Seven Seals

This is the beginnings of the final Pert Chart that incorporates all the prophecies we discussed. That chart is too large to present in its entirety in this book. However, we will add a few more prophecies to this initial version of the chart, as space permits, so the reader can get an idea of how different prophecies depict the same events. The final version of the chart, with all the studied prophecies shown together, is provided on the website www.christsreturn2028.com for the serious eschatology student.

The next iteration of this chart incorporates the prophecy of the Seven Trumpets. This prophecy was inserted into the first chart at its proper location. This revised chart includes a few event markers from the beginning chart so that the reader can see how this added prophecy "fits" within the overall timeline. When the Trumpet prophecy is added to the timeline of the Seven Seals and the Olivet Discourse, one can observe that the Seal openings occur during the first two thirds of the ten-year tribulation, while the Trumpet judgments occur during the last third.

This is not to say the Seal decrees will be completed before the Trumpet judgments commence, as many scholars interpret these events. Let me explain: the opening of the first seal begins at the start of the tribulation years and does not cease when the second seal is opened, but continues for the duration of the tribulation period while other judgments are added. Think of it this way: if a person were making a cake, they would first add the flour to a bowl, followed by some eggs. By adding the eggs, they in no way remove the effects of the flour. The flour is always present and interacts with the eggs—so too with many of God's punishments.

The opening of the first seal signifies the coming of the false messiah: the antichrist. His efforts will continue throughout the entire tribulation period until Christ returns and are not stopped by sending forth the second rider on the red steed who takes away peace from the earth. The second rider only adds to the problems... builds upon the terror of the last days. Not all of God's judgments work this way. Other punishments have finite durations that need to be completed before the next one starts. There are many events in both categories. Keep this in mind as we study these end-time events. Moving ahead, let us examine our first chart that has been modified with the addition of the Seven Trumpets.

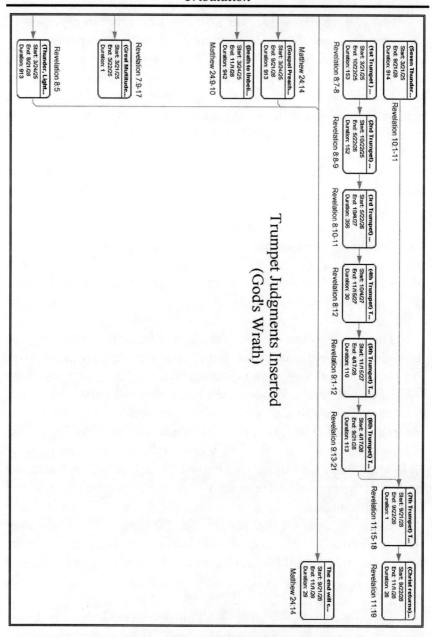

Only a portion of this expanded chart is shown (from the preaching of the gospel to the end of the tribulation) so that the reader can focus on the location of the Trumpet judgments in relation to the other events shown on the first chart. Furthermore, if we were to include the prophecy of the Seven Bowls to this chart, it would confirm visually that these punishments run parallel to the Seven Trumpets prophecy. That is to say, they will occur at the same time because they are really describing the same events with different symbolism. Let us zoom in on that area of the timeline.

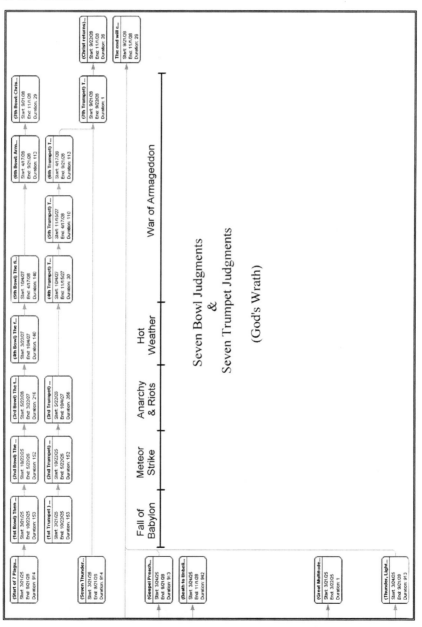

When studying these Pert charts, those tasks (events), which are in the same column, denote either the same event or events that occur at the same time or very close to the same time. For instance, on our last chart the first Bowl judgment is the same thing as the first Trumpet judgment and they occur as the ministry of the 144,000 begins spreading the gospel to the entire world along with the gentiles trampling Jerusalem.[a] Adding the rest of the prophecies we studied to the chart and providing them within this book will not be possible as I said before. Therefore, we will use a written format to convey the timeline, rather than a visual chart. Remember, a completed chart has been provided on the Internet.

Sequence of End-Time Events

Let us continue by viewing a sequential list of the tribulation events in plain, everyday language instead of the symbolic language used in the prophecies. The planning software was used as the foundation for determining the order of these events and dates.

----

1) The antichrist begins his ascent during troubled times. He is thirty-nine years old and the foreign minister from Greece.[b] [Autumn of the year 2018] *Daniel 8:23-24, Daniel 8:9, Daniel 7:8&24, Zechariah 11:16-17, Revelation 6:1-3, Revelation 13:1-10, 2nd Thessalonians 2:3*

2) A major war in the Middle East (War of Gog) breaks out between Israel and an Islamic coalition led by Russia. A major earthquake strikes Israel causing large military losses. [2018 –2021] *Ezekiel 38-39, Daniel 11:22*

3) The antichrist confirms a treaty (covenant) with Israel and her enemies because of this war and takes power with the aid of ten European leaders. [Autumn of the year 2021] *Daniel 9:27, Daniel 11:23, Revelation 17:12-13*

4) The "Two Witnesses" begin their tribulation ministry. [Sept. 2021] *Revelation 11*

5) Israel builds a temple to God after the Dome of the Rock Mosque is

destroyed in the War of Gog (timing and circumstances are an educated guess). [Sept. 2021 - Jan. 2025]
2nd Thessalonians 2:4

6) Strange weather conditions will continue to affect people's lives. Earthquakes, severe droughts, and heat waves will cause food shortages and famines. [Nov. 2018 - Jan. 2025]
Matthew 24:7-8, Mark 13:8, Luke 21:25 Revelation 6:8, Revelation 18:8, Revelation 16:8-9

7) The antichrist attacks various countries and cities around the world. [2022 - 2023] *Daniel 11:24*

8) Another minor war in the Middle East breaks out. [Summer of 2024] *Daniel 11:25-26*

9) The antichrist begins phasing in laws to control the economy and further revoking people's freedoms by requiring strict identification. [November 2018 - February 2025]
Daniel 11:28, Revelation 13:14-18

10) Twenty-five percent of the world's population dies because of Satan, the antichrist's actions, and God's punishments. *Revelation 6:7* [July 2018 – April 2025]

11) The abomination of desolation is set up (3.5 P-days) restricting religious freedoms. Riots breakout in Israel and other parts of the world due to Jews and Christians revolting to the things the antichrist is doing. [February 2025] *Matthew 24:15-25, Daniel 11:31-36, Daniel 9:27, 8:10-12, Daniel 8:24-25, Daniel 7:8&11, 2nd Thessalonians 2:4, Revelation 6:12-17, Revelation 11:7-14, Revelation 13:5-10 & 15-18, Daniel 7:25*

12) War in Heaven (1.666 P-days). [February 2025]
Revelation 12:3-4, Revelation 12:7-9

13) A tremendous earthquake rocks Jerusalem and the Middle East while the antichrist is desecrating the third temple of God. [February 2025] *Revelation 11:13, Daniel 11:30*

14) "Mark of the Beast" laws go into full affect and become mandatory. Marshal law enacted in many parts of the world. Time to flee for the mountains and countryside. [February 2025]
Rev. 13:16-18, Matt. 24:16-21, Mark 13:14-17, Luke 21:20-21

15) Silence in heaven for half an hour (thirty minutes... thirty days). [February - March 2025]
Revelation 8:1

16) Christians (and others) slaughtered worldwide for thirty P-days. [February - March 2025]
Luke 21:17-19, Revelation 13:7-10, Daniel 11:33-35

17) One hundred forty-four thousand men are prepared during the same thirty P-days to preach the "Gospel of Christ" to the world. Weather problems are put on hold (eye of the storm). [February - March 2025] *Revelation 7:1-8, Revelation 14:1-7, Matthew 24:14, Mark 13:10, Luke 21:21:13*

18) The rapture occurs at the end of the thirty P-days. The sanctuary (Christ) is reconsecrated (Christ begins ruling from heaven). [late March 2025]
1st Thessalonians 4:13-17, Revelation 6:9, Revelation 19-1, 1st Corinthians 15:51:54

19) God and Christ's (Their) wrath begins. [late March 2025]
Revelation 6:16-17, Revelation 11:18, Revelation 15:7-8, Revelation 19:15

20) New York City is destroyed by nuclear blast as well as other major cities. Babylon the Great is destroyed. [April/May 2025]
Revelation 17:12-18, Revelation 18, Daniel 11:39

21) Radiation poisoning carried around the world causes sores. [April - October 2025] *Revelation 16:2*

22) Meteorite fragments begin showering the earth as a large meteorite approaches the earth and breaks up. [October 2025 -May 2026]
Revelation 8:7

23) Meteorite hits the Atlantic Ocean further wiping out evidence that New York City ever existed. Other coastal cities suffer severe destruction from tidal waves. [October 2025 - May 2026] *Revelation 8:8-9, Revelation 18:21*

24) Creatures of the sea die from the meteorite and/or initial nuclear strikes. [October 2025 - May 2026] *Revelation 8:8, Revelation 16:3*

25) Demons are let loose to deceive the world into gathering for one final battle and incite riots worldwide. [2026] *Revelation 16:13-14, Mark 13:22*

26) The antichrist's time is short. Anarchy and riots are commonplace. He begins attacking other countries: Northern African countries are first... Egypt, Libya, etc. Jordan is spared (probably due to a treaty). [2026 - 2027] *Daniel 11:40-42*

27) The War of Armageddon begins between nations. [2027 - 2028] *Zechariah 12:2-6*

28) Electrical power plants are damaged from military strikes and heavy demand from hotter than normal temperatures. Power is disrupted to about a third of the world. [2027] *Revelation 8:12, Revelation 16:10*

29) Euphrates river tributaries dry-up and that allows a two hundred million man army led by China to advance on Israel. [2028] *Revelation 9:13-16, Revelation 16:12, Daniel 11:44*

30) The armies of various nations (China, Russia, Arab federation) gather in the Middle East to oppose the E.U. [2028] *Daniel 11:44, Rev. 14:17-20, Rev. 15:14-16, Zechariah 14:2*

31) An earthquake rocks Jerusalem and the world. [Sept. 2028] *Revelation 16:17-20, Zechariah 14:3-5*

32) Christ returns with angels and redeemed Christians. [September 2028] *Revelation 19:11-16*

33) Christ captures the antichrist and the false prophet and throws them alive into the Lake of Fire... they are the first to enter this new prison of no escape. [September 2028] *Revelation 19:20*

34) The remaining combatants of Armageddon (those who have the mark of the beast) are killed by Christ and fed to the birds. [September/October 2028] *Revelation 14:17-20, Revelation 19:21*

35) The angels gather the rest of mankind who are deemed unworthy to enter the Millennial Kingdom (those who have the mark) and they are sentenced to death. [September/October 2028] *Matthew 13:36-43, Revelation 14:9-11*

36) All those who survived Armageddon and did not take the mark of the beast are allowed to enter the new kingdom on earth. [November 2028] *Revelation 11:15*

It is highly unlikely that anyone can correctly determine with precision what will take place in the last days, let alone put those things in their proper order with dates and without mistakes. Yet, I provide this information to the best of my abilities in the hope that it will provide you with a good enough idea of what is to take place and when things will occur. By doing this, a Christian can prepare for the times ahead and find the urgency to face up to their friends and relatives who have not accepted Christ as their savior.

You have now learned what events to expect during the last days and about when to expect them. Very few mysteries remain that I can help you with. True understanding comes from studying the Word and submitting to God's will. In doing so, the Holy Spirit provides the confidence you'll need to accomplish the good works that were planned for you to do at the beginning of the world.[c]

However, I would like to devote the next few chapters to cover some "technical" issues that pertain to God's methods and thoroughness, by searching deeper for reasons as to "why" things are designed the way they are. Many of these ideas are speculative in nature, but are supported with the knowledge uncovered throughout the book. These are advanced ideas, which if true, have major ramifications.

Chapter 16

The Library of Knowledge

CHRIST HAS RETURNED AND THE year is 2028 A.D. What a glorious time for all—well maybe not. Isn't this the visual picture you get when you, as a Christian, think of our Savior's return? What a great time it will be when Jesus returns. This is the reaction of almost all of the Christians I tell that Jesus is coming back in 2028. They have no clue that before this happens their lives will be changed beyond recognition.

Their thoughts are the same ones I always had until I began writing for the Lord. But the fact of the matter is when Jesus returns, this earth-shattering event will follow untold devastations and two-thirds of our neighbors, friends, and relatives will perish in everlasting damnation after all that was said is done. Billions of people have been deceived by Satan and, because of their love of the things of this world instead of the truth, they will burn forever in eternity.

Of those who suffer through these holocausts, some will finally find the peace that man has searched for his entire existence... those new Christians who have come to know Jesus as their personal Savior. They will be blessed with the knowledge they were accepted by Christ to be the first generation to enter God's earthly kingdom and repopulate it. As for the others who survived who accepted the mark of the beast and followed the man of perdition, their pain and suffering will only deepen into an endless sea of suffering. They will have no rest forever. Their spiritual death warrant was signed the day they accepted the mark and their physical death is now at hand.[a]

Forty Days

As many picture end-time events, they bear the same burden I constantly fight. That is viewing a single prophetic event as a snapshot in time on God's calendar. For instance, when Daniel speaks of the "abomination of desolation" it always conjures up visions of something that will not take a long time. In my case no more than a day's time.

But after all the stuff that has been uncovered, I now know this key event will take place over a five day period and if we include all the preliminary things necessary for this to occur, one could reasonably argue the time is much longer.

So too with Jesus' return. Although his return will take a very short period of time, if we include all the time required to straighten things out before moving into the peaceful time of the Millennial Kingdom, the total time will be forty days. A time you are now quite familiar with. But these forty days should be the most terrifying for a non-Christian. They encompass the opening and closing of God's slaughterhouse. The final piece of God's righteous judgment on this generation's wicked who discarded the truth of His authority and the requirement that atonement for our transgressions must be made. Restitution made with the blood of Christ for believers, but made for the unbeliever with their own blood.

> *Then he left the crowd and went into the house. His disciples came to him and said, "Explain to us the parable of the weeds in the field." He answered, "The one who sowed the good seed is the Son of Man. The field is the world, and the good seed stands for the people of the kingdom. The weeds are the people of the evil one, and the enemy who sows them is the devil. The harvest is the end of the age, and the harvesters are angels. "As the weeds are pulled up and burned in the fire, so it will be at the end of the age. The Son of Man will send out his angels, and they will weed out of his kingdom everything that causes sin and all who do evil. They will throw them into the blazing furnace, where there will be weeping and gnashing of teeth. Then the righteous will shine like the sun in the kingdom of their Father. Whoever has ears, let them hear.*
>
> Matthew 13:36-43 (TNIV)

So why does it take forty days for the angels to weed the fields… to remove all people and all things that cause sin in the world? Why doesn't Jesus just snap his fingers and make it so? I'm unsure as to all the reasons why God's plan is the way it is. Maybe if this chore were done too quickly, those remaining would not appreciate the gravity of the moment and would be unable to relate this historic period with enough conviction to their children in future generations.

Maybe the killing fields are vast and the task is too great in the physical world that God has created. Regardless of the reasons why it will take forty days, the facts remain the same. Jesus will return and during the following forty days the goats (unbelievers) and sheep (saints) will be separated so that the goats will enter God's slaughterhouse and the sheep will graze under the watchful care of Christ.[b]

Not knowing the reason why this period of time will take forty days is in all probability like not knowing why God chose the number forty for so many other things. Why did Jesus remain forty days on earth after his resurrection? My best guess is he needed to remain some period of time so that enough people saw him and could personally attest to the fact he had indeed risen. If he just ascended into heaven the same day of his resurrection, after a few of his closest friends saw him, that may not have been enough time to install his church on this earth. He had to reveal himself to enough disciples (more than five hundred) to reach "critical mass" so the explosion of Christianity, as we know it, would occur. But again, why forty days and not some other number?

There is one thing for certain; God is consistent and redundant with His planning. This knowledge does make it easier to unlock the secrets of His plan. I have mentioned on many occasions that I have been unable to find the common denominator as to why God chose the number forty for various things that upon close inspection have nothing in common. Some claim the number forty signals a "time of testing," but they base this assertion on a few instances where this number is used. Clearly, these last forty days of this age are anything but a time of testing. The testing is over. This is the time for grading and, be aware, the system used is pass/fail!

However, if we were to reexamine God's use of numbers we might find the answer buried in a larger stack of knowledge that satisfies the question of why? I know it is the best answer I have been able to uncover to date and I have been searching for this truth for many, many years. Longer than any other unanswered question in the Bible. Yet, in the end, it will lead to another difficulty that I have no idea how to address. This mystery into the number "forty" comes from another side journey. Let me relate this story to you for it starts in a familiar place—with God's number as many of my travels do.

The Library is now Open

As I work my way through the last days I continue to observe things I find fascinating. I am resigned to the fact I will always see them this way and that many who read my books will not find them as interesting as I do. Many of these oddities on their own would not be recognized as God's handy work without understanding all of God's methods beforehand.

If I told you God's number was 333.333 in its completed form, most would say, "huh?" But by showing the many, many uses of this number in God's plan that are backed up in Revelation, Zechariah, and other scriptures, I would hope by now that the importance of this number and how the Creator of the Universe uses numbers has opened your eyes.

We have a good idea of why God uses this number because it symbolically signifies the eternal trinity of the Lord. But how is it that God decided to use other numbers that are as significant as well? Most of you reading this book, or who have read my first book, had no idea this was such an essential number beforehand. But many of you are very familiar with other numbers that are important to God like the number "forty" or the number "seven."

So the real question is, "Why are these numbers so vital to God's plan?" Well I can't be one hundred percent sure, but I can speculate with extreme confidence that they are all based on God's number, man's number, and their relationships… the relationship of God with mankind as well as the connection between these two numbers.

In my first book, I wrote the reason why God uses the number forty is that it is another form of man's number. Forty really is six times 6.666. But I have come to recognize the actual reason may be much more complex than this earlier explanation.

We have discussed the number of completion (ten) at great length and the higher forms of this number—one hundred and one thousand. I showed this on page 95 that when you add man's number and God's number together you get the various forms of this number. This is because man is destined to be with God for all eternity and without Him we would always remain in our sinful, uncompleted form. This is one reason why the antichrist, who intentionally chooses to separate himself from God, is recognized by the number six hundred and sixty-six. He will forever remain apart from the love of God, as will all those who reject Jesus.

In addition, I stumbled upon the notion that the number seven could be viewed as the number of completion (page 132) when viewed as time on a clock. What I failed to realize, until yesterday (October 23, 2008), was the number of completion is not necessarily ten, but can be other numbers. The number ten, one hundred, and one thousand were just the standard cases of a more general rule that I had uncovered, but was unable to fully understand. A rule I should have seen when I found the number seven "substituting" for the number of completion on a clock.

We have seen that when you add God's number and man's together we get the number of completion, but in the past this was done using only the same "scale" of each number. Different scales (levels) of each number were not intermixed until I realized for the first time that the number seven could be derived by adding different scales of each number. By doing this, the result could be viewed symbolically as a variant of the number of completion.

What do I mean by scale or level of a number? Not having a clear term to use, I am only trying to differentiate for the reader between numbers scaled by ten. For example, the number 6.666 is a lower scale or level than the number 66.666 just as this number is one level lower from 666.666. What all this really leads to is the understanding that different scales of these two numbers can be joined together and we are not limited to just adding numbers of the same scale. We used this method when we were examining God's plan and different levels within God's plan, just not when it came to the pure mathematics of number of completion. By applying this broader understanding we can find the general rule of the number of completion versus specific cases that we have been using.

So how is this done? By investigating every possible combination of God's number and man's number. There are twenty-four possible combinations of interest when carrying out this analysis. Half of the possible combinations have to do with no difference between levels (standard use of the number of completion) or a one level difference (special cases). The remaining half deals with level differences that are two or greater. Let's go ahead and look at all the promising combinations and their results.

Table of the Numbers of Completion (Table 16-1)

	Level	Man's Number	God's Number	Result
Section 1	Number of Completion (Standard case)			
Identical Level	10^3	666.666 +	333.333 =	1,000
Identical Level	10^2	66.666 +	33.333 =	100
Identical Level	10^1	6.666 +	3.333 =	10
Identical Level	10^0	.666 +	.333 =	1
Section 2	Number of Completion (Special cases)			
1 Level Difference	10^3	6,666.666 +	333.333 =	7,000
1 Level Difference	10^2	666.666 +	33.333 =	700
1 Level Difference	10^1	66.666 +	3.333 =	70
1 Level Difference	10^0	6.666 +	.333 =	7
1 Level Difference	10^3	666.666 +	3,333.333 =	4,000
1 Level Difference	10^2	66.666 +	333.333 =	400
1 Level Difference	10^1	6.666 +	33.333 =	40
1 Level Difference	10^0	.666 +	3.333 =	4
Section 3	Number of Completion (Unknown uses)			
2 Level Difference	10^2	6,666.666 +	33.333 =	6,700
2 Level Difference	10^1	666.666 +	3.333 =	670
2 Level Difference	10^0	66.666 +	.333 =	67
2 Level Difference	10^2	66.666 +	3,333.333 =	3,400
2 Level Difference	10^1	6.666 +	333.333 =	340
2 Level Difference	10^0	.666 +	33.333 =	34
Section 4	Number of Completion (Unknown uses)			
3 Level Difference	10^1	6,666.666 +	3.333 =	6,670
3 Level Difference	10^0	666.666 +	.333 =	667
3 Level Difference	10^1	6.666 +	3,333.333 =	3,340
3 Level Difference	10^0	.666 +	333.333 =	334
4 Level Difference	10^0	6,666.666 +	.333 =	6,667
4 Level Difference	10^0	.666 +	3,333.333 =	3,334

Note: The smallest number in each pair determines the level. No level higher than three or lower than zero is included in the table.

From this table you should notice some interesting things. The first part, where there is no difference between the scale of man's number and God's number, yields the "standard" forms of the number of completion. The next section deals with a one level difference between these important numbers. When man's number is the higher of the two, we find all the variations for the number of perfection—seven. When the distinction is the other way around and God's number is higher, various forms of the number forty are revealed. A number I have yet to properly label. Regardless of the reasons why God chooses this number for His purposes as opposed to another special number, the various forms of the number forty do appear in our table. And we can therefore deduce from the use of other numbers derived in the table that "forty" symbolizes a "special" act of completion of some sort.

These combinations of man's number and God's number we have already discussed cover exactly half of all the possibilities. As for the other half of possibilities which address differences greater than one level, they yield numbers not easily recognized as being used by God. This is the new quandary that at present goes unanswered. It appears on first inspection that the numbers from half the table generate many of the important numbers used in the Bible, but the other half does not.

If this concept is the reason why God uses the numbers forty, seven, or even seventy for many things (because they symbolize the completion of some designed aspect of His plan), then why are the other half of the numbers in the table not used? As I stated before, I am convinced this is the best answer as to why God uses these numbers based on knowing His methods, but I am troubled with not knowing why these other unused numbers of completion sit idle within God's plan. With slight modifications to these leftover numbers some can be observed within God's plan—or Satan's forgeries of God's plans. However, I will not belabor the point any further because the speculation is far too great.

So why will it take forty days after Christ returns to rid the earth of evil? The same reason it took Jesus forty days in the desert and the flood took forty days and forty nights. It is because this is the time designated for these events; for these tasks to be completed to achieve God's purposes.

When the first judgment came in the great flood, it took forty days to complete the first cleansing of evil men from the earth. We now know this task will again take forty days (and nights) at the second judgment to remove all the wicked men and women from the planet. This act will complete the second judgment of mankind, while the final "white throne" judgment will take place after the one thousand P-year reign of Christ on the earth. Maybe this too will take forty days, but this information is not given in the Bible.

What else can we learn about the number forty? I have talked previously about some of these things like forty hours is 144,000 seconds and forty prophetic hours is 146,100 seconds (forty hours plus an additional thirty-five minutes). But here are some additional things to ponder when we include in our examination the other levels of this number (four, four hundred, and four thousand).

> 4 years = 1,461 days
> 40 years = 14,610 days
> 400 years = 146,100 days
> 4,000 years = 1,461,000 days

As you can see, all these numbers are multiples (levels, scales) of the converted prophetic number 144,000 [63] when changed into days. This understanding presents some contradictions with other work that I want to highlight so that the reader can make up their own mind. Many authors ignore the problems of their work, but my only objective is to find the truth of God's words and I pray that the Lord will forgive me if I write untruths which lead others astray. Many times I am writing as I am learning, so things I have written in the past turn out to be untrue in the future. Where I know of problems or find out later, I try to let the reader know so they can make their own decisions.

As time moves forward, some things I have written in the past I no longer believe, such as my position on the rapture which has changed over the years. Also, I wrote that Christ was not dead forty hours, which I calculated in my first book. I now believed it to be forty hours and thirty-five minutes because when the number 144,000 is used for time, the real number is 146,100.

We now see that when the number "forty" is used as years, the real time shows up without making any prophetic conversions. When God punished the Israelites in the desert for forty years, this time was real time and did not need to be converted as the forty hours needed to be. What are the ramifications of finding this now? Just this: if there are three equal trimesters in God's overall plan of twelve thousand P-years just as there are in His plan for mankind, then these must be three periods of four thousand P-years. However, from our work so far, we notice that if the number four thousand is used to account for years, then the time is already in real time and not prophetic time. This understanding would lead to the idea that the total time for God's plan is twelve thousand years long and not twelve thousand P-years (12,175 years) long!

But we also know that if the plan is twelve thousand years long, this translates into 144,000 months and they would need to be converted. One calendar measurement (years) needs no conversion while the other (months) does require a time correction. Around and around the circle we go. Chasing our tail with no idea how to catch it. I do not believe this quandary will ever be solved, but I bring it to your attention for you to contemplate and move on.

We have seen how the number forty is viewed using years as the time basis. Let's look at these identical numbers again using "hours" as the time scale. We know that forty hours is 144,000 seconds, but I think seeing this information all in one place helps one to see the majesty of God and opens a person's eyes to the truth of these concepts.

> 4 hours = 14,400 seconds
> 40 hours = 144,000 seconds
> 400 hours = 1,444,000 seconds
> 4,000 hours = 14,400,000 seconds

We see from this work that when the assorted forms of forty are viewed as hours, that the various scales of the number 144,000 are uncovered. One timescale (years) reveals the number 144,000 in its converted form, while this timing (hours) yields the number in its symbolic form. This is one of the reasons why I believe, when the time is in hours, it needs to be converted.

If we were to look at the remaining periods of calendar time: "days," "weeks," and "months" we would not see the number 144,000 emerge. I am unaware of the Bible using the various scales of the number forty with these other units of calendar timing, except the day unit. Let us look at the "month" unit nevertheless, before focusing on the day unit.

> 4 months = .333 years
> 40 months = 3.333 years
> 400 months = 33.333 years
> 4,000 months = 333.333 years

Interesting? We now see a familiar number show up i.e. God's number in its various forms. Moreover, if we were to transform these times further we would uncover yet another number that we are quite familiar with …12,175!

> .333 years = 121.75 days
> 3.333 years = 1,217.5 days
> 33.333 years = 12,175 days
> 333.333 years = 121,750 days

What about viewing these numbers as "weeks"? As of this moment I have been unable to see how transforming the various scales of the number forty into "weeks" reveals any similar supporting evidence. Maybe this is why God uses the term "weeks" when using the various forms of the number seven in Daniel's prophecies. We know those numbers are just as important and showed up in our "completion" table along with the number forty.

What about analyzing these numbers as "days"? This is a time period God uses many times in the Bible and again we have a problem when we associate it with the number forty. These conversion methods do not work for forty days and its different forms. So we are left with a mystery still not completely solved.

We know God uses many methods and views different numbers as representatively the same, but the knowledge of God's view of forty days is still incomplete. If we were to apply the rule a "year for a day" we could grasp that if forty years is equal to the number 14,610, then so are forty days from God's perspective. It may be just this simple. In fact, we know God uses other time analogies besides a "year for a day." Consequently, one might conclude that all times which deal with the number forty and its various aspects are viewed symbolically as the sealing number 144,000, and also the number of completion ten, if we were to bundle all this knowledge together.

Let's examine a few more instances of the number forty, how God uses it, and how other numbers can be viewed from God's perspective in many of the same ways.

The Master Clock of God

If we were to look back to the Master Clock of God, we would find there are 12,175 years that equate to seven hundred and twenty minutes. This means 16.9097222 years pass by for every minute that ticks off on this abstract clock.[64] This is another very interesting number because it is God's number and the number forty in disguise. Let's review. How many days are in forty prophetic hours? There are 1.666 prophetic days.[65] Converting P-days into real time by means of multiplying by the conversion factor 1.01458333 [35] yields 1.69097222 days.[66] This quantity is exactly ten times smaller than the amount we just calculated for the number of years associated with one minute on this conceptual clock of the Master's.

We know from the Bible and confirmed from our earlier work, that a "year is like a day" and a " day is like a year." We also are familiar with a "thousand years are like a day" and a "day is like a thousand years." Furthermore, we now know that there are other levels between one (10^0) and one thousand (10^3). They are ten (10^1) and one hundred (10^2). Combining these two understandings of God's methods allows eschatology students to grasp that not only are "one year and a thousand years like a day" but also "ten years and a hundred years are like a day!" So 1.69097222 days can be like:

> 1.69097222 years (10^0), or
> 16.9097222 years (10^1), or
> 169.097222 years (10^2), or even
> 1,690.97222 years (10^3)!

Since 1.69097222 days is equal to forty P-hours we can deduce from God's perspective that these higher levels are also symbolically identified with the number forty.

Let's keep shuffling through the stacks of paper in the "Library of Knowledge." We know that God's Master Clock has twelve hours on it, but we need only to concern ourselves with the time between the hours of six and twelve because the first six hours deal with creation and are of no importance for this analysis. At six o'clock Adam was created and at twelve o'clock Christ will return. We calculated that the great flood occurred on this clock at 7:38 and this was between 7:33 and 7:40, which are the times designated for the thirty-five days of conflict on the Tribulation Clock. So let's begin counting from Adam's creation (the start of humanity), which is six o'clock, and move to the time 7:40. How much time has elapsed? That's easy. There are sixty minutes from 6:00 to 7:00 and forty minutes from 7:00 to 7:40 for a total of one hundred minutes. We know that one hundred is an intermediate form of the number of completion and that each of these minutes is equal to 16.9097222 years. Therefore, the total number of years from Adam to 7:40 on God's Master Clock is one hundred minutes times 16.9097222 years per minute or 1,690.97222 years! One can now see that the time 7:40 on our hypothetical clock is also symbolically linked with the number forty as well.

I first realized the flood occurred on the forty-ninth S-day counting from Adam based on how God administers punishments and from studying the ages of Noah's ancestors in the Book of Genesis. And since each S-day is also based on God's number (33.333 P-years), it turns out they each equal 33.819444 real years (after multiplying by our conversion factor).

Hence, forty-nine S-days is forty-nine times 33.819444 years, which equals 1,657.152777 years. How much time does fifty S-years equal—a Jubilee of S-years? You guessed it; it equals 1,690.97222 years exactly![67] So now we see that a Jubilee of S-years also is symbolically linked to the number forty and the time 7:40. A time that, if we have done our analyses correctly and understood the Word of the Lord, is associated with the timing of the rapture of the saints.

Summarizing these additional ideas discussed, we have determined that a Jubilee of S-days (fifty) is equivalent to: 7:40 on God's Master Clock, equal to 1,690.97222 years (half of God's number), which is also like 1.666 P-days or forty P-hours. And what's more, forty P- hours are 144,000 P-seconds! In other words:

144,000 P-seconds = 40 P-hours = 1.666 P-days =
1.690.9722 days and applying

"a day is like a thousand years" ≈

1,690.9722 years = 50 S-days ≈ 7:40 on God's
Master Clock

All these times are equivalent to God as viewed from His perspective.

One important matter left pertains to the number forty and God's uses of that number. It is a complex matter I have been struggling with that involves higher mathematics. It is much like the exercise at the end of chapter six in that it requires understanding the things already shown to you, how God works, His methods, the Creator's use of numbers, some things we have brushed around the edges of, and other methods that God employs, but we have never discussed. It requires applying many of these things at one time and yet knowing these things, which are going on simultaneously, are hard to portray in a systematic way.

The Off Limit Sections

I woke up the morning of Saturday, January 24, 2009 after reading and editing this chapter the night before. I do a lot of this work just before I sleep. I had been distressing about how I was going to present these last realizations clearly. I also knew that I had not found the room of full understanding, but was still learning as I sought out the mysteries of God. Hidden knowledge that has been off-limits and locked-up until now, as it has been for those servants of God who came before us, was now wide open. As has been the case many times, I got an inspiration while sleeping just as I awoke. I had not been dreaming of these things at all during the night, but was startled out of sleep with a direction to follow.

We have investigated these ideas many times in my first book and throughout this book, but even so, many of these things if not thoroughly studied and committed to memory are easily forgotten. Even though they may have been analyzed, they were analyzed separately and not together as a whole. You should know by now that God does have a plan. A plan that is unchanging and that plan was designed much like the physics of the universe was designed—it cannot be changed by us. Water boils at 212°F and a day is twenty-four hours long. Physical truths like these are just the way things are. This is also the same for God. He has methods and is working His plan and that is just the way things are. Sure, God gave us free will to choose between Jesus (the truth) or to refuse the truth as many do, but as for God's purposes, He is consistent, fair, and unwavering when it comes to His overall plan for salvation.

There are a few things that must be reviewed so they are fresh in your mind. The first is when God does something, He tells us in advance of His intentions. I have discussed several examples of what was foretold to the prophets in terms most Christians should be able to understand. Clearly God told Noah ahead of time there was going to be a flood many years in advance so that he had time to finish building the ark at the designated spot on the Lord's plan.

Now the earth was corrupt in God's sight and was full of violence. God saw how corrupt the earth had become, for all the people on earth had corrupted their ways. So God said to Noah,

"I am going to put an end to all people, for the earth is filled with violence because of them. I am surely going to destroy both them and the earth."

Genesis 6:11-13 (TNIV)

The Lord then said to Noah, *"Go into the ark, you and your whole family, because I have found you righteous in this generation. Take with you seven pairs of every kind of clean animal, a male and its mate, and one pair of every kind of unclean animal, a male and its mate, and also seven pairs of every kind of bird, male and female, to keep their various kinds alive throughout the earth.* **Seven days from now** *I will send rain on the earth for forty days and forty nights, and I will wipe from the face of the earth every living creature I have made."*

Genesis 7:1-4 (TNIV)

So God not only tells us many years in advance of His plans as He did for Noah so the ark could be built in time, He also informs us just before He acts of the exact time; as He did for Noah seven days before the flood. This is precisely like the prophets words of warning for civilization many, many years before the coming of the "Day of the Lord's Wrath." And now, as the end of the age nears, we are being forewarned again. Could Noah calculate when the flood was? Not at first, but surely he could after the Lord spoke a second time. Why? Because I'm confident Noah could count to seven!

If you are a Christian watching for the signs and yearning for our Savior's return, can you not count twenty-five hundred and twenty P-days from the antichrist's confirming a covenant with Israel or twelve hundred and ninety prophetic days from the abomination that causes desolation to determine when Christ will return? Of course you can and so those people who say you can "never" know the time of Christ's second advent either do not study or understand the Word, do not keep watch—or cannot count!

You should also know that in addition to God telling us in advance, sending signs beforehand of important milestones on His calendar, the

Lord always tacks on some extra time to complete the event… to seal the deal. Let's look back to the flood account because it is the closest event we have in past history that compares with what is to come. Remember, the flood was the first judgment for humanity and we are fast approaching the second judgment for the world. The last and third will occur after the one thousand P-year reign of Jesus.

> *By the first day of the first month of Noah's six hundred and first year, the water had dried up from the earth. Noah then removed the covering from the ark and saw that the surface of the ground was dry. By the twenty-seventh day of the second month the earth was completely dry.*
>
> Genesis 7:13-14 (TNIV)

We see here that the floodwaters were dried-up by the first of Nisan and yet God did not allow Noah to leave the ark for fifty-six more days. Why fifty-six? If we were to study the flood account more closely we would find that the Great Flood started on the seventeenth day of the second Jewish month (Iyar: April/May). So the total time for the flood was one year (actually one 360 P-day year) and ten days. Subsequently, we notice that although it rained only forty days and forty nights, the flood took one year and an extra ten days to be finished in God's eyes.

We could state this another way; the flood took exactly one year and the Creator added another ten days to "complete" the punishment. Obviously ten is the number of completion and we would expect nothing else from our Lord if we understood His methods. Since the number forty is also a special form of the number of completion, we can see other parallels that support this idea. Daniel's last week will take seven P-years to complete before Christ returns and then another forty days will be tacked on to finish the second judgment before the elect can inherit a life of peace in the Millennial Kingdom.

Those who are observant might also notice that the duration of the flood was based on calendar mathematics and not some other number. It was one P-year long plus ten days and the tribulation will be seven P-years long plus another forty days. These are things we have discussed before and should not be new. The point of this lesson is they are examples of God's methods. Methods we will be using again and again through the rest of this chapter. But a new lesson has been slipped in. That message is: **important times for events are based on the calendar and God's number!** This is the new revelation I grasped this morning—a new key I needed to enter other rooms in the library.

Now this next endeavor is an advanced exercise in God's math. It is more complex than the simple linear math we have been using. But we need to discuss it nevertheless. It will expose God's majesty in a new way that is absolutely incredible. So where to start? As usual, I am unsure of how to present these ideas, but let's begin with a simple multiplication of God's number and one year on the calendar. What are the answers we get when we multiply an intermediate form of God's number with one year and one prophetic year?

$$365.25 \text{ days} \times 33.333 = 12,175 \text{ days}$$
$$360 \text{ P-days} \times 33.333 = 12,000 \text{ P-days}$$

Applying God's method of a "year for a day," we now get twelve thousand prophetic years or 12,175 years. To simplify our work we will add a superscript to any units of time when this technique is being used. The superscript $^{(yfd)}$ will be used to denote a "year for a day" and the converse $^{(dfy)}$ will be used to signify a "day for a year." Hence:

$$365.25 \text{ days} \times 33.333 = 12,175 \text{ years}^{(yfd)}$$
$$360 \text{ P-days} \times 33.333 = 12,000 \text{ P-years}^{(yfd)}$$

These are things we already know. That God's overall plan is based on His number and calendar mathematics. Let's continue searching God's Library of Knowledge floor-by-floor. If we were to look at all the forms of God's number and multiply them by one year or one prophetic year we would get all the levels of the number 12,175 as shown on page 95.

What happens though when we multiply a year by God's number squared? Mathematically squaring a number is nothing more than multiplying a number by itself. We have been doing this with the number of completion throughout the book and should know this is one of the many methods God uses. So let's do this using a lower scale of God's number from the one used in our first example. Otherwise the numbers would get rather large.

$$365.25 \text{ days x } 3.333 \text{ x } 3.333 = 4{,}058.333 \text{ years}^{\text{(yfd)}}$$
$$360 \text{ P-days x } 3.333 \text{ x } 3.333 = 4{,}000 \text{ P-years}^{\text{(yfd)}}$$

Interesting. You should easily recognize these numbers for they are the time for one third of God's overall plan for mankind. Restating:

$$365.25 \text{ days x } (3.333)^2 = 4{,}058.333 \text{ years}^{\text{(yfd)}}$$

If we were to substitute God's number for its most basic form (a third or .333), then we can calculate even more familiar numbers.

$$365.25 \text{ days x } (.333)^2 = 40.58333 \text{ days or years}^{\text{(yfd)}}$$
$$360 \text{ P-days x } (.333)^2 = 40 \text{ P-days or P-years}^{\text{(yfd)}}$$

Is this the reason why God uses the number forty or just further confirmation of His planning? This math suggests that God uses this number because it is quite simply "built into" His plan. It is nothing more than a building block of His genius. We now realize it is not only a special case of the number of completion, but a number based on the trinity and the calendar. What number is $(.333)^2$ or $(3.333)^2$? They are God's number in its other form: .11111 and 11.11111! God is one and God is three and now we see mathematical proof of this theological truism.

There is more to this story, but before going on I want to take a side trip down another aisle of the library and show you something else. We need to go back to the table of the Numbers of Completion (pg. 338).

If we were to move God's number one level higher, expanding on this idea, we would find that $(33.333)^2$ is exactly one hundred times larger than $(3.333)^2$ which is one hundred times larger than $(.333)^2$. The result of $(33.333)^2$ is $1,111.111$ and is equivalent to the sum of all the various multiples of the normal number of completion! That's if we include all the lower levels into infinity below level one (10^0). God does use the number of completion for the level 10^{-1}. This is one tenth and the amount used for tithing. A most important number and idea we will discuss in detail shortly. As for summing all the numbers of completion for the standard case, let's do this quickly:

1,000	10^3
100	10^2
10	10^1
1	10^0
.1	10^{-1}
.01	10^{-2}
.001	10^{-3}
+ etc.	
1,111.111	$= (33.333)^2$

Therefore, we find that God's number (33.333) squared is the sum of all the numbers of completion. What happens if we sum all the numbers of the special case "forty" from our table? Without going into the details, this number comes out to $4,444.444$ [68]. The other special case is the number of perfection "seven" and using the same mathematical technique yields the number $7,777.777$ [69]. Many claim this is the Lord's number and yet we cannot find evidence in the Bible of its use as we can with the number three. If we were to continue calculating the sum of the other numbers from the table that God doesn't use, we might find reasons why these numbers go unused. They do not generate this single digit pattern of numbers.

Since $(33.333)^2$ equals $1,111.111$, then we also know that the square root of $1,111.111$ is 33.333. The square root is a mathematical technique that allows us to examine this equation from the opposite direction… in the reverse order. Just as addition and subtraction are opposites, so is squaring a number and taking the square root opposite functions. So, what is the square root of the sum of all the multiples of the number "forty" (4,444.444)? Get out your calculator before reading on. The answer is a number that you are quite familiar with and yet we probably could not guess what it is just by looking at this number. It is the number of man![70]

So God's number squared confirms that "God is one" and "God is three" and that it is the sum of all the numbers of completion. The number forty, which is derived from God's number and calendar mathematics, the relationship of God to man, and the sum of all variants of the number forty is associated to man's number 66.666![70] If you read my first book you learned that there are many numbers associated with Jesus that are based on God's number, man's number, and the number of completion.[c] Jesus is both God and man! Is this why Jesus uses the number forty for many things? You have to choose for yourself. As for me, I have long ago decided.

If we were to examine the number 7,777.777 and the rest of the other numbers from the "Number of Completion" table, we would not find numbers that are familiar to us by applying these understandings. For instance, the square root of the number 7,777.777 is 88.19171 and is not a number that is recognizable at this time. It has never come up before just as the number 777 does not reveal itself in the Bible. We know God uses the number seven for special cases and it is a form of the number of completion, but it is not readily associated with Jesus and this may be the reason why. Making these same calculations on the remaining unused numbers from our table yields unfamiliar numbers as well.

Let's get out of this section of the library and return to the main study area. We noticed that when we squared the two lowest multiples of God's number (.3333 and 3.333) and multiplied them by the number of days in a year, we got forty and four thousand; numbers we see plenty of occurrences within God's plans. However, the other two forms of this number, four and four hundred, cannot be calculated in this manner.

The number four is not used very often nor is the number four hundred. That is not to say God doesn't use them or cannot use them, for surely they are variants of the number forty and are not off limits. It is only highlighting an observation. How often have you read in the Bible where four days or four hundred days are mentioned? How about four years? We have not used theses time-frames in our work nor does God use them regularly in His plans. I am aware of one instance where our Maker uses four hundred years[d] to achieve His purposes, but there are no others. There is only one way to generate these two forms of the number forty using calendar math, and that is to multiply a year with two different scales of God's number as opposed to the same scale used when a number is squared. As for the case of the number four, we have to use a scale (10^{-1}) of God's number that is ten times smaller than the most basic level of God's number... something we have never done before. The method works mathematically, it's just not a recognized method God has used before. This may be why we see few instances of these two variations of the number forty being used in the Bible for timing pertaining to important events. Here is what that math looks like for the curious student.

$$360 \text{ P-days} \times 3.333 \times .333 = 400 \text{ P-days}$$
$$360 \text{ P-days} \times .333 \times .0333 = 4 \text{ P-days}$$

Moving on to the Revelation section of the library, we find that using this higher math uncovers more knowledge of God's mysteries. We have squared God's number and found some fascinating things that deal with the number forty. What happens when you square both numbers: both God's and a year from God's perspective?

$$(360 \text{ P-days})^2 \times (3.333)^2 = 1,440,000 \text{ (P-days)}^2$$
$$(360 \text{ P-days})^2 \times (.333)^2 = 14,400 \text{ (P-days)}^2$$

Wow! Did you expect to see scaled numbers of 144,000? Just as some of the other things we looked at generated numbers that were entirely unexpected, so too with this analysis. To actually get 144,000, and other levels of this number, requires mixing different scales of God's number. For example:

$$(360 \text{ P-days})^2 \times (3.333) \times (.333) = 144,000 \text{ (P-days)}^2$$

Examining these calculations we can learn a few more things. First up, this technique does not work using real years. If we square 365.25 days and multiplied it by the square of God's number, we do not get the expected number 146,100. This is additional evidence that the number 144,000 is truly a prophetic number i.e. a symbolic number and not a real number. Secondly, what are (P-days)^2? Normally when we find this number from Revelation within God's designs, it is not associated with a unit of measure that is squared. I am unsure what this really portends, but this may be an indication that multiplying numbers that deal with time by God's number negates some aspect of the prophetic effect. I would have to study this thought further to see if the Holy Spirit would illuminate this dark hallway for me, but at this time I only bring to light this observation and move along.

Let's climb the stairs in God's Library of Knowledge and find out what is at the top of the staircase in the off limit sections of the Master's library. What is the result we get if we cube God's number—just like we cubed the number of completion ten to get to the highest form God uses? This means multiplying God's number by itself three times.

$$(3.333)^3 = 37.037037037$$

This is a most curious number sequence and we touched on it at the end of chapter six. This number is the backbone of God's plans for His Messiah… His son… Jesus our Savior! But the full truth of the matter is the plans surrounding this number are very complex just as the higher math is. To understand the full ramifications of this knowledge requires us to understand all of God's methods and something else we have not talked about… tithing. I will begin with some preliminary math and then finish our studies by exploring the area dedicated to tithing.

Let's begin just as we have the previous calculations by taking this number and multiplying it by one year.

$$365.25 \text{ days} \times (3.333)^3 = 13{,}527.777 \text{ days}$$
$$360 \text{ P-days} \times (3.333)^3 = 13{,}333.333 \text{ P-days} \ \&$$

$$365.25 \text{ days} \times (.333)^3 = 13.52777 \text{ days}$$
$$360 \text{ P-days} \times (.333)^3 = 13.333 \text{ P-days}$$

Do you recognize any of these numbers? We speculated about them at the end of chapter six. Some would say speculation, but seeing how using God's number with this higher math generates these numbers should make one realize it is not just conjecture. To search out more familiar variations of these numbers requires us to mix scales of God's number—something that has been done before. Let's look at the most familiar case of these numbers and the math associated with them.

$$365.25 \text{ days} \times (3.333)^2 \times (.333) = 1{,}352.777 \text{ days}$$
$$360 \text{ P-days} \times (3.333)^2 \times (.333) = 1{,}333.333 \text{ P-days}$$

If we were to add forty prophetic hours, these would be the exact times mentioned in the last chapter of Daniel! Just as God tacks on forty P-days to the end of Daniel's last 'week', which is a higher level within His plans (the seven P-years of the tribulation), so also an additional forty P-hours is added for this lower level of the plan. God is consistent and so are His methods. When we add the additional forty P-hours we get 1,354.469 days[71] or thirteen hundred and thirty-five P-days[72] exactly (see page 64).

We know this is the time Christ needs to complete Daniel's seventieth 'week' by ruling in heaven during the wrath years of the tribulation. But what else is there to learn in these off limit areas locked up for thousands of years until now? That this number or multiples of this number are used in every important plan related to Christ!

We know that Christ was physically dead forty P-hours or forty hours and thirty-five minutes of real time. We speculated on his time spent in hell applying God's methods and numbers briefly at the end of chapter six. Seeing this new information provides stronger support for all the calculations I have presented. Are you surprised to learn that one third of the number forty also comes to the exact same numbers—just one hundred times smaller?

$$40.5833 \text{ hours x } (.333) = 13.52777 \text{ hours} \quad \& $$
$$40 \text{ P-hours x } (.333) = 13.333 \text{ P-hours}$$

This is the same as:

$$365.25 \text{ days x } (.333)^3 = 13.52777 \text{ hours}^{(dfh)} \quad \& $$
$$360 \text{ P-days x } (.333)^3 = 13.333 \text{ P-hours}^{(dfh)}$$

if a "day is like an hour."[dfh] We know God uses these types of time substitutions and His blueprints mathematically support similar analogies. I am unsure if this can be proven mathematically from the Lord's plan because I have not tried, but I think you can agree it almost certainly is true from God's perspective knowing how the Master works.

Now I just realized another potential problem as I wrote this, for as you know, many times I learn as I write. This new hitch is that Christ may not have been dead forty hours as I calculated in my first book, and may not have been dead even forty prophetic hours, as I have come to believe up to this point. For if we were to apply God's rule of requiring additional time, this number might really need to be 13.54469 hours instead of 13.52777 hours! This discrepancy is extremely small, but strange because it generates a number we are familiar with—just not on such a minute scale: .0169097222. [73] This difference is one hundred times smaller than forty prophetic hours. Just as these times are one hundred times smaller than Daniel's thirteen hundred and thirty-five P-days (which includes forty P-days), this amount of possible missing time is also numerically identical to forty P-hours.

In view of the fact that we are examining "hours," then how many extra, or potentially missing, minutes are we talking about? This converts into 1.01458333 extra minutes![74] This is the exact number as our prophetic conversion factor—only expressed in minutes! This is crazy. These numbers keep twisting around and around each other. We started with cubing God's number and ended up with God's conversion factor expressed in minutes.

We also know that 13.52777 hours is one third of forty prophetic hours as well as the result of cubing God's number. If we were to grasp that this missing time may be really only a third of the real missing time because this number is one third of forty, then we would have to multiply this time by three to find the total time missing. Consequently, 1.01458333 minutes multiplied by three equals 3.04375 minutes of extra time that may need to be included to "seal" the full forty P-hours. This new number is exactly ten times smaller than the average number of days in a month and one thousand times smaller than the time spans we looked at in chapter one that deal with God's overall plan. Are these just more coincidences?

Since we have been investigating forty prophetic hours and one third of forty P-hours, what we have been really doing is discussing the time Jesus was physically dead. In chapter six I reasoned the most plausible explanation for the times relating to Christ's death, using Bible accounts and the redundant ways God uses numbers within His plans, was that Jesus died at 2:58 pm on Friday and rose at 7:33 am on Sunday. In the grand scheme of things, it matters little whether or not we can determine these details. Knowing only allows us to understand God's methods more fully and in so doing pinpoint future tribulation events more accurately... something that can be done without this knowledge in the future as prophetic signs present themselves.

This logic also applies with taking into account possibly three additional minutes. I only belabor the point to address another thought. If these extra three minutes must be included with the forty P-hours and we move the time of death of Christ to exactly 3:00 pm, then the time of our King's resurrection would have been 7:38 am, a time you should be familiar with i.e. the time of Noah's flood on God's Master Clock. This may have been the time the rains began to fall, stretching these ideas even further.

During our studies in these areas of the Master's Library of Knowledge we have used only God's number and one year to generate important numbers linked to God's plans and prophecy. We have uncovered more methods that God uses and we have climbed far above even the most thorough understandings of most eschatology students and teachers. The time is short and equipping the saints for what is to come and finding the lost can only be a faithful Christian's mission in these last days. Learning what is possible to know so that we can be prepared for what is to come. There is but one final exercise that reveals God's methods… tithing.

The Tithing Area

*The LORD said to Moses, "Speak to the Levites and say to them: 'When you receive from the Israelites the tithe I give you as your inheritance, you must present **a tenth** of that tithe [a tenth of a tenth…] as the LORD's offering.*

Numbers 18:25-26 (TNIV)

*Be sure to set aside **a tenth** of all that your fields produce each year.*

Deuteronomy 14:22 (TNIV)

'A tithe of everything from the land, whether grain from the soil or fruit from the trees, belongs to the LORD; it is holy to the LORD. Whoever would redeem any of their tithe must add a fifth of the value to it. Every tithe of the herd and flock—every tenth animal that passes under the shepherd's rod—will be holy to the LORD.

Leviticus 27:30-32 (TNIV)

At the end of every three years, bring all the tithes of that year's produce and store it in your towns, so that the Levites (who have no allotment or inheritance of their own) and the foreigners, the fatherless and the widows who live in your towns may come and eat and be satisfied, and so that the LORD your God may bless you in all the work of your hands.

Deuteronomy 14:28-29 (TNIV)

Every knowledgeable Jew and Christian knows that the Lord requires a tithe of ten percent. Why this amount? Well ten percent is just a smaller scale of the number of completion. Understand that God does not need your money or things. He already owns all that you have whether you realize it or not. He wants your commitment to Him... your love for Him just as you want love from those you care about. The Creator knows that those who are willing to part with this amount have their hearts in the right place. They are centered on the things of God and not the things of the world. Those who are addicted to money and the things of this world will not give up even this small portion. Many make excuses for why they cannot give, but very few are honest with themselves for the Maker knows all and cannot be deceived. He knows where your choices lie. Do you cheat God from what is required so that you can enjoy some pleasure of this world?

Regardless of the answer, God follows this requirement Himself and we will see how in the most important piece of His plan. Let us look back to Jesus' life we briefly examined at the end of chapter six. We know Christ was around thirty when he began teaching and he taught for three and a half years. This would have made him about thirty-three and a half years old when he was crucified. We recognize as well he will reign in heaven for another three and a half years to fulfill Daniel's prophecy. Adding these times together gets us to approximately thirty-seven years of total time. The total time is most likely $(3.333)^3$ years or 37.037037 years exactly as I wrote before. If you remember, I subtracted Daniel's 1,335 P-days from this number and there were 12,175 days left. But what I didn't stress at that time was this: Daniel's remaining time for Christ was one tenth of the total amount. What does this realization indicate?

The real question is why did God choose to have Christ "cut-off" from reaching the required seventy 'sevens' at the very moment he did?

Why not choose some other time? Because God required ten percent of Christ's "life" to be given unto Him! Let us suppose that Christ completed the full seventy sevens on earth as prophesied. Knowing how old he was when he began his ministry means he would have been right around thirty-seven years old if he had completed the full seventy 'sevens' all at once—just as the antichrist will attempt. If we were to estimate the time God requires as a tithe (giving sacrifices which Christ was), this would require giving unto God approximately 3.7 years of Christ's life. Remember, what He expects from us is no less than He requires of Himself. Sacrificing ten percent of our time, talents, and wealth.

How many days are there in 3.7 years? There are 1,351.425 days.[74] This is very close to Daniel's time of thirteen hundred and thirty-five P-days (1,354.469 days) and is short by roughly three days. In fact, this crude calculation comes to exactly 3.04375 days[76] difference, which was unexpected since this was only an estimation. This number is equivalent to three prophetic days.

It is unwise to assume these two numbers should not be the same— understanding God's methods. With just a quick calculation, we got to within three days of the time it would take to reach the Millennial Kingdom God revealed to Daniel. Knowing what the thirteen hundred and thirty-five prophetic days represents only leads to this tithe and Daniel's thirteen hundred and thirty five prophetic days being one in the same.

By grasping this hidden knowledge, we can understand that the total time to account for from Christ's birth to Christ's second coming is probably 37.037037 years! Having written this, I have been trying to reconcile this time with everything I know written in the Bible about Christ and apply God's methods to verify this time without success. I know this is the right room of the library to find this information, and yet I am still unable to piece the particulars together without errors. I expect if I were able to rectify these last details, the ramifications would be that Christ's return could be pinned down to an exact day and hour… something I have been unable to do using the other methods we have investigated.

Do you remember when we summed all the standard numbers of completion and got 1,111.111? This represented God's number 33.333 squared. What answer do you suppose we would get if we used this same summing mathematical technique on God's number? Let us go ahead and do that:

33.333333 God's number
 3.333333 10% of the previous number (tithe)
  .333333 10% of the previous number (tithe)
  .033333 10% of the previous number (tithe)
  .003333 10% of the previous number (tithe)
  .000333 10% of the previous number (tithe)
  .000033 10% of the previous number (tithe)
  .000003 10% of the previous number (tithe)
+   etc. 10% of the previous number (tithe) into infinitum
37.037037 = $(3.333)^3$

When we cube the Lord's number, we get the same number we would get if God repeatedly took a tithe of His number into infinity and then summed up all those tithes together! Understand this: there is no other number to which you can apply the "tithing" concept in this way and have it come to ten percent of the same number cubed! The point is, God requires tithing of Himself just as He does of us. This is the most important example of its use... planning for Christ. We have seen how using these advanced ideas of God's number is in harmony with what the Bible says about Christ and the last days. By grasping these ideas, we have revealed things that have been hidden since the creation of the world. Can all these manipulations of God's number be merely chance when studied along side scripture and the other oddities we have unlocked along our journey? Statistics would back up finding this many coincidences are not possible if they are just false ideas.

We have been discussing the various forms of the number of completion and other related matters that deal with God's number. However, I want to reveal one last thing that relates to man's number. The

KJV says the number of man is "six hundred, three score, and six" whereas newer translations simply list the number as six hundred and sixty-six. This subtlety is not important to most scholars, but with our last work the King James text might be thought of more like this:

600.	Man's number (incomplete form)
60.	10% of the previous number (tithe) (3 score)
6.	10% of the previous number (tithe)
666	

We see that this calculation has only three levels that were included in the biblical text (10^2, 10^1, 10^0). If we were to continue with the remaining levels into infinitum, as was done for God's number, we would get 666.666 as the complete number of man we have been using throughout the book.

God does have a plan. A plan since the creation of the world and it is a precise plan. You and I are now seeing finished in our life-times essential parts of the plan, and the most significant part of that plan is the Plan for the Messiah and our salvation. A plan that flows from the knowledge that it is based on the calendar and God's number in its highest form... 37.037037.

Chapter 17

Searching for the Impossible

WE HAVE STUDIED GOD'S OVERALL PLAN for Mankind as well as smaller pieces within the plan. We have learned a lot about the plans for the end-times and how God uses numbers. However, there is another plan that is very important for your salvation. This plan deals with the coming of the Messiah. If we could see this plan, we might catch a glimpse of why things happened the way they did and why things will happen the way they will.

You now know everything I can teach you about what is to come and the plan for the messiah will not change anything. However, seeing the plan for Christ will shed more light on the majesty of our Lord. Although I have been unable to construct this plan as confidently as other plans, nevertheless I provide what I believe in the hope that the Holy Spirit may use this information to open the eyes of others so they may see with greater clarity the additional things that have been uncovered.

We have investigated the details surrounding Christ's crucifixion on Friday, April 3, 33 A.D. and have come to recognize that our Lord was physically dead for 144,000 prophetic seconds. This was forty hours and thirty-five minutes of real time. We also know from Isaac Newton's writings quoting old historians that Christ was born in 2 B.C. and consequently this knowledge supports that Christ was thirty-three years old at the time of his death. What else can we learn about our savior's life as it relates to God's plan and the way our Lord works? Plenty! By taking all the things we know and have learned, and combining them in different ways, we can search for the impossible knowledge that may allow us to see from a different perspective how everything fits together.

I recently had a sister in Christ contact me who also believed Jesus would return in the year 2028 A.D. Why? Well based on her studies and the Holy Spirit's guidance, she is confident 2028 A.D. is the next year of Jubilee and that Christ will return in a year of Jubilee. This important year will be the fortieth Jubilee from the time when Christ began administering the Gospel to the Jews. However, by contacting me, she unknowingly provided a key piece to the mystery of why Christ's life turned out the way it did. In other words, a piece that may explain why God designed things the way He did for our Messiah's coming because there is purpose to all that He does. The Lord is not haphazard with any aspects of His planning. This quandary was something I have been working off and on at trying to more fully understand.

What was this gem of information that I am so gracious to her for? She wrote in her correspondence that Christ officially began teaching on Yom Kippur. Interesting. For those of you who are unknowledgeable of the Jewish sacred days, this is the Day of Atonement.[a] In my first book, I provided evidence that this may be the day Christ returns. Also in that work, based on my studies of the biblical prophecies, I had found that Christ's birthday may have been on October 13, but it could also have been October 3 since I received an insight on that day and could not figure out why the Holy Spirit chose that very day in 2007 A.D.[b] This was the last time I received a strong inspiration when writing "2028" and it appears now to point toward this being the birthday of Christ.

Newton claimed Christ was born in the month of September based on his understanding of Daniel's prophecy, while I had calculated it to be in early October based on my work in the first book. Both very close… within a month. The dates are unimportant as of yet, but I mention them now as foundational information that will be built upon later for knowing these dates is important.

"About" Thirty

Scripture relates Christ was "about" thirty when he began his ministry.[c] This is a phrase I have been struggling with as I try to figure out all the details of the Messiah as they relate to God's methods and planning. The difficulty I have come to recognize is that it is a calendar problem of my own upbringing. For instance, my view of birthdays is biased by the calendar in use today. Therefore, if I were born on, let's say, September 20, then I would celebrate each successive birthday on that same day of the year. However, Jesus was a Jew and the Jews use another calendar system. So let's speculate about Christ's date of birth to see what I mean about accounting for birthdays. Suppose Christ was born on September 20, 2 B.C. (the autumnal equinox) using the Julian calendar that was in place at that time. This calendar is very close to the Gregorian calendar we use today. The Gregorian calendar has 365.2425 days in it and accounts for leap years more accurately than the Julian calendar, which had 365.25 days in it and had to be manually adjusted every so many years (by decree) to correct for the accumulative drift in the seasons. Thus, for our purposes, they are close enough to make the necessary points without continually specifying which calendar system is being referred to.

The Jewish calendar, as we have learned, is based on the phases of the moon and the number of days in a year fluctuates based on a complex set of rules. What day was the twentieth day of September in 2 B.C. based on the Jewish calendar? It was Saturday, the twenty-second day of Tishrei (the seventh month of the sacred Jewish calendar). This is the first day after the celebration of the fall Feast of Tabernacles that God commands to observe.[d] This is also the feast the prophet Zechariah mentions will be of special significance after Christ's return.[e]

If we were to use our present day understanding of birthdays, we would assume that thirty years from this hypothetical date of Jesus' birth would be September 20, 29 A.D. However, what day was this on the Jewish calendar? It was Elul 23 and not the twenty-second day of Tishrei where we started. Even though it is thirty years exactly as we count birthdays nowadays, it falls short of thirty years as the Jews account for time. In fact, it is twenty-eight days short. In the year 29 A.D., the twenty-second day of Tishrei fell late in the year on October 18. What does all this mean? If we are to believe Christ began teaching on the Day of Atonement in 29 A.D., which is always the tenth day of Tishrei, then the term "about thirty" fits nicely with the data we have been using to illustrate the differences in accounting for birthdays. The start of Jesus' ministry would have been twelve days short of his thirtieth birthday if it were on Tishrei 22—or in other words, "about" thirty.

Why did I pick this day to illustrate the differences on how birthdays are viewed between the Jewish calendar and the modern calendar, which has been slightly altered from the calendar used at the time of Christ? It was because on October 3, 2007 A.D. when I received my last inspiration when writing "2028," it was the twenty-second day of Tishrei (after sundown).

Why the Day of Atonement?

Why is the Day of Atonement significant other than it is the day required to "pay" for man's transgressions and is the day of announcement for Jubilee years (blowing the shofar—the ram's horn)? Let's find out. Yom Kippur in 29 A.D. occurred exactly 1,277 days prior to Christ's resurrection on April 5, 33 A.D. This is effectively 1,260 prophetic days (three and a half years) and a time mentioned in Revelation. If we were to examine similar time-spans for other combinations of years around this period in history, we would learn something else. Let us do this quickly.

Table 17-1

Tishrei 10 (Yom Kippur)	Nisan 16 (Day of Resurrection)	Difference
23 A.D.	27 A.D.	1,306 days
24 A.D.	28 A.D.	1,276 days
25 A.D.	29 A.D.	1,304 days
26 A.D.	30 A.D.	1,304 days
27 A.D.	31 A.D.	1,276 days
28 A.D.	32 A.D.	1,306 days
29 A.D.	**33 A.D.**	**1,277 days**
30 A.D.	34 A.D.	1,276 days

If a person is to believe Christ's ministry encompassed four Passovers (about three and a half years long), then the table above covers all combinations of years that scholars have studied. If one believes the number of Passovers were fewer, or more, as a small minority of theologians do, then the differences would be significantly lower or higher. From our previous studies, we can be confident the time is three and a half years.

Looking closely at the data, we see one time-span that is unique compared to all the others. It is the time period we know to be true: from the Fall of the year 29 A.D. to the Passover in 33 A.D. Not only does this time stand apart like a beacon, it is also the only time-span that can be considered 1,260 prophetic days.[77] This is an important distinction, as we will learn. For those who claim Christ died in another year like 30 A.D., which is popular with many scholars of today, the time differential is 1,304 days and does not support what we are about to learn. We will see that this is further evidence these other years are false teachings. Let us begin developing a timeline by including the information we have already discussed.

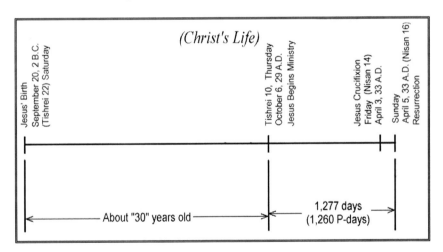

Christ's Baptism

What else do we know about Christ's life? We know before beginning his ministry, he was baptized and immediately went into the desert to fast for forty days and forty nights. At the end of this fasting period, Satan tempted him.[()] After fulfilling the required forty days and rejecting the devil's offers, Christ was nursed back to health by angels. How long did this take? This recovery time is not mentioned in the Bible and is thought to be unknowable.

However, understanding God's methods and His use of numbers we can make an educated guess. We know that this time falls between the finishing of the forty days in the desert and the beginning of Jesus' ministry on the tenth day of Tishrei. We should also understand recovering from starvation and dehydration (fasting) for this length of time would require more than a couple days of care to get a person's strength back to a point where they could function at a normal level. Taking into account all these common sense ideas, we come to the answer of thirty-five P-days.

The Temples' Destruction

Why this time? If we combine the forty days of fasting with the thirty-five P-days of resting to restore Christ back to health and subtract them from Yom Kippur, we should theoretically arrive at the day Christ was baptized. That day was Friday, July 22 or Tammuz 22 (the fourth month of the biblical Jewish calendar)—a most curious day. In the year 587 B.C., when the first temple was destroyed, according to II Kings 25:8-9, it was the seventh day of Av (the fifth Jewish month). This would have been Thursday, July 21 on the Gregorian calendar. These dates are not the same using the Jewish calendar, but are effectively the same day using today's calendar. What this really means is the position of the sun with respect to the earth was the same because our calendar is more precise in this regard than the Hebrew calendar. However, Jeremiah 52:12-14 says this event occurred three days later on the tenth day of Av. Surprisingly, July 22 is the day when the earth moves from the constellation of Cancer into the constellation of Leo… the Lion… the King, symbolism that cannot be ignored!

As for the second temple, according to Jewish history it was destroyed on the ninth day of Av in 70 A.D. However, Isaac Newton records this event as occurring August 11, 70 A.D., which was the sixteenth day of Av. Jewish tradition supports that both temples were destroyed on the ninth day of Av, but we know the Bible does not agree nor does Isaac Newton. So were both temples actually destroyed (burned) on the same Jewish day separated by six hundred and fifty six years of history or were there in fact a few extra days included that separated these pivotal events? Can temples be destroyed in one day or do they take several days, which might explain the differences in dates? Regardless of the answers to these questions, the Jews in all their wisdom have deemed the ninth of Av as a day for mourning the destruction of both temples and a day of fasting.

Why is all this talk about the destruction of temples relevant? Because Christ is the new temple of God.[g] If we understand that Christ was baptized on the same day that the destruction of the temples took place, we can see additional symbolism that the new temple of God (Christ) picked-up where the old temples left off (ceased to exist). The new temple of God was christened, possibly on the very day of their destruction, as the replacement for the physical temples. Presently, the Jews commemorate the ninth day of Av (Tish'a B'Av) as the time to recognize the tragedies of these events by fasting on this day. Projecting this idea back to the time of Christ, how fitting it would be if Christ began his forty-day fast on this day of fasting to symbolize the substitution of the physical temples with a spiritual temple.

We know that July 22, 29 A.D. was summertime and a good time to perform baptisms. What day was it forty days later? It was September 1, 29 A.D. and theoretically the starting time of Christ's recovery. Let us add this new information to our chart.

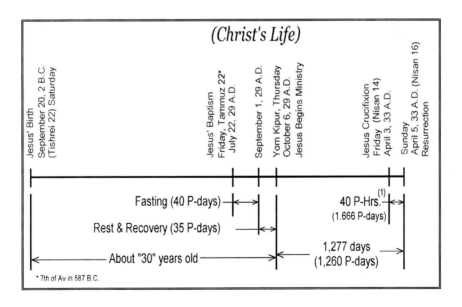

Daniel's Seventy 'Weeks'

What else do we know about Christ's life from the Bible? We know that after his resurrection, he stayed on earth for another forty days and ascended into heaven afterwards. As for more details about Christ's life, there are few left that the Bible speaks of directly, which are of value to these studies. However, there is still much more to learn.

We know that Christ was cut-off from reaching the seventy weeks at the midpoint of Daniel's seventieth 'week' using one understanding of Daniel's prophecy (Sabbath years). This left three and a half years that needed to be fulfilled from heaven before he could return. However, what if Christ had not been cut-off and the seventy 'sevens' were completed without interruption. By speculating on this matter, we might be able to see the complete plan God had in mind for the coming of the Redeemer.

We have been using twelve hundred and seventy-nine days as the converted time for Revelation's twelve hundred and sixty prophetic days, but based on this new work, it appears the number that should probably be used for this conversion is twelve hundred and seventy-seven days. This would mean no leap year days are accounted for within this portion of the timeline and they would have to be accounted for elsewhere on the timeline. In any case, let us add the twelve hundred and seventy-seven days to the day of resurrection (Nisan 16) as if nothing happened and see what date we get. That end-date is Wednesday, Tishrei 24 in the year 36 A.D. or October 3, 36 A.D. i.e. the very day of the year I received my last inspiration and within two days of the Jewish day that was determined for Christ's birthday using this same inspiration.

This would suggest that the plan called for Christ to take control effectively on his thirty-seventh Jewish birthday, but at thirty-seven years and thirteen days of age using the modern calendar. Realize there is something significant going on with these extra thirteen days. When Daniel's sixty-nine 'sevens' were converted from prophetic years to real years, an extra thirteen days were also generated in that conversion.[6] Is this a coincidence? I don't believe so. There have been others. What about converting three and a half P-days and finding out the time was thirteen hours and ten minutes?

Let us go ahead, add this new information to our timeline, and add up the various times to highlight how the Lord works.

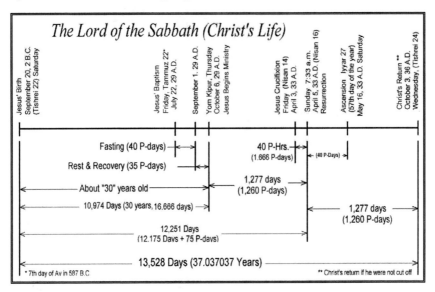

If we were to examine this timeline closely, we would begin to see many wondrous things about the way the Lord works. The first being the total time from the birth of Christ to his resurrection was 12,175 days plus 75 P-days! This was 12,251 days.[78] When we add the remaining 1,277 days needed to complete Daniel's seventy 'weeks', we get a new total of 13,528 days. For most people this number would not ring a bell, but we must be ever guarding against number blindness lest we let important information pass by our eyes. This number is really 37.037037 years[57] or God's number raised to the third power (3.333^3)! Not only that, but it is Daniel's number of 1,335 P-days minus forty P-hours exactly… just ten times greater.

We also know from our previous work, that after Christ returns, there will be a forty-day window where Christ will cleanse the world of all unbelievers. If we were to add this piece of information, then we would have all the details we need to see the complete "Plan for the Messiah" and why things will take place in the manner they will. As for the plans of the tribulation, they are based on these plans for the Christ! They are intertwined and we can see them if we thoroughly search this essential timeline. Let us add this last piece of information, as well as further document the things already on the timeline.

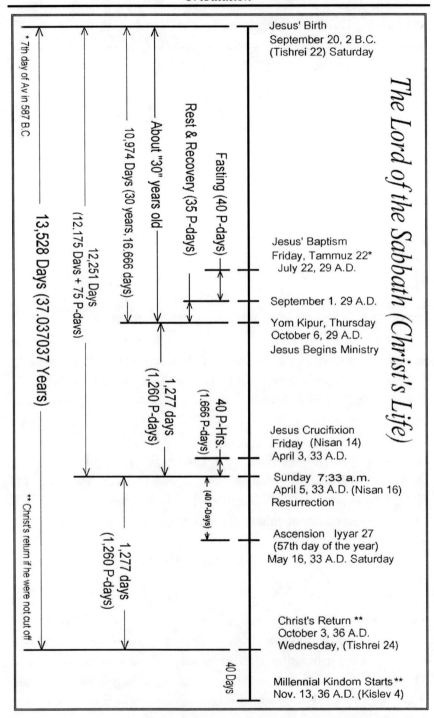

The Lord of the Sabbath (Christ's Life)

Jesus' Birth
September 20, 2 B.C.
(Tishrei 22) Saturday

* 7th day of Av in 587 B.C.

Fasting (40 P-days)

Rest & Recovery (35 P-days)

About "30" years old

10,974 Days (30 years, 16.666 days)

12,251 Days
(12.175 Days + 75 P-days)

13,528 Days (37.037037 Years)

Jesus' Baptism
Friday, Tammuz 22*
July 22, 29 A.D.

September 1. 29 A.D.

Yom Kipur, Thursday
October 6, 29 A.D.
Jesus Begins Ministry

1,277 days
(1,260 P-days)

40 P-Hrs.
(1.666 P-days)

Jesus Crucifixion
Friday (Nisan 14)
April 3, 33 A.D.

Sunday 7:33 a.m.
April 5, 33 A.D. (Nisan 16)
Resurrection

(40 P-Days)

Ascension Iyyar 27
(57th day of the year)
May 16, 33 A.D. Saturday

1,277 days
(1,260 P-days)

** Christ's return if he were not cut off

Christ's Return **
October 3, 36 A.D.
Wednesday, (Tishrei 24)

40 Days

Millennial Kindom Starts **
Nov. 13, 36 A.D. (Kislev 4)

372

The Lord of the Sabbath

Jesus said he is Lord of the Sabbath,[h] and we see from our chart that he was born and ascended to heaven on the Sabbath—more evidence that supports his claim… although the Master needs no such proof. Exploring the timeline further, beginning from the end and working backwards in time, we see an excellent rendition of the tribulation timelines developed in chapters 5 and 6. The only difference is the placement of the thirty-five P-days. On the tribulation timeline, these days fall just before the second occurrence of 1,260 P-days whereas on this timeline, they occur before the first 1,260 P-day period. Otherwise, these timelines are identical!

Notice that if God's planning is based on the highest form of His number (37.037037), then the prophetic conversion times are probably 1,277 days rather than 1,279 days. In addition, I believe there is strong evidence that Christ rose at 7:33 am and if this is true, then it might be possible to figure out the hour of his birth as well as the hour of his return. This is something Jesus said was not possible during his time and so I will keep these calculations to myself for they mean little in the grand scheme of things. Also, this plan hints that Christ will return either on October 3 or his birthday. In 2028 A.D., Tishrei 1 falls on September 21 (autumnal equinox) and this would place the start of the Millennial Kingdom forty P-days later on November 1. Even stranger, in the year 2027 A.D. the second day of Tishrei is October 3 on the Gregorian calendar and September 20 on the old Julian calendar. The drift between these calendar systems has increased to thirteen days over the years—the same number of extra days that come from converting 37.037037 into years and days. There are other oddities on this timeline, such as Christ ascended into heaven exactly 2,435 years after Noah entered the ark. We have struggled with this number many times before and this analysis is provided in Appendix II.

Although not shown on the timeline, you should also be aware of the fact that a Jewish male child was required to be circumcised on the eighth day (after seven full days) following his birth.[h] After another waiting period of thirty-three days[i] (forty P-days in total), the mother was expected to bring a sin offering to the temple to make atonement in accordance with the purification rite laid out in the Mosaic laws. All the times mentioned were doubled if a woman gave birth to a female. As one should expect, all these requirements were followed in the birth of Christ.[i] What this means is that if Jesus was born on September 20, as I now contend based on all the evidence, then he would have been consecrated to the Lord on November 1! The same day of the year the consecration of the Millennial Kingdom of Christ will take place in 2028 A.D.

We have examined Christ's life in a way never before seen by applying information from the Bible and the methods God uses for other things. Some of this information is reliable, while other data is speculative and based on circumstantial evidence, revelation, and logical deduction. As for proof that these things are true, this can only be confirmed when Christ comes back. Having searched for the impossible knowledge by following where the numbers of the Bible have led, I have reached the end of that journey. However, I wish to revisit a few matters of importance for a Christian in the last days. Things that God's servants need to know to be ready for the troubled times ahead.

Chapter 18

The Remaining Signs

The Days of Noah

As it was in the days of Noah, so it will be at the coming of the Son of Man. For in the days before the flood, people were eating and drinking, marrying and giving in marriage, up to the day Noah entered the ark; and they knew nothing about what would happen until the flood came and took them all away. That is how it will be at the coming of the Son of Man.

Matthew 24:37-39 (TNIV)

AT THE END OF DAYS, JESUS PROPHESIES it will be just like it was when the great flood came. However, what exactly was he saying? We have seen many parallels in the way God uses numbers between the great flood and the last days. There is even more we can learn about these similarities in just these few sentences.

Noah was a prophet whether he wanted to be or not. How do we know this? Well it is impossible to build a boat... a ship... an ark as big as the Bible indicates it was in a short period of time. Noah would have had to start building the ark years in advance. So we know God gave him the order to construct the ark ahead of time. God obviously told him a flood was coming well in advance of this judgment on mankind for their wicked, rebellious hearts.

Clearly, Noah must have been wealthy enough to obtain the necessary materials to build the ark from people who were destined to die in the flood... unbelieving friends, family, acquaintances, and even strangers. What do you suppose those people asked Noah when he was obtaining supplies and building the ark? Can you hear them?

"Noah, what are you doing?"

"I'm building an ark because God is going to send a flood to wash away the sins of the whole world."

"That's ridiculous. You're a fool. There is no God. You're just an idiot wasting your time and money (but we'll be happy to take your money from you)."

You should realize that Noah was prophesying to everyone of the coming wrath of God well in advance by his actions and his testimony of God's plans. It matters not that the Bible doesn't call Noah a prophet. It only matters that by his actions and testimony we know that he was one.

In our last days, there will be many prophets… many witnesses of God who will proclaim the news of the saving grace of Jesus Christ before the time of the coming wrath. Those are the requirements of God's servants who keep watch: to announce to all (by words and deeds), the coming wrath while testifying about the truth of Christ. These are the same requirements for all Christians at the end—as they were for Noah: to be prepared for all things that come their way.[a]

How many are keeping watch for the Master? How many know the Word and are prepared? How many are listening to the Holy Spirit's call? During Noah's time, no one listened but Noah and his family. I am not sure how much of God's warnings Noah's family believed, but I am confident they trusted Noah and assisted him and their reward was to enter the ark with him.

What about Noah's extended family? Obviously they were not on the ark. This will be the same during the remaining years before Christ returns. Many of your friends and family will suffer God and the Lamb's wrath and will spend eternity in the Lake of Fire. When the antichrist comes to power, they will be untrustworthy and will cause trouble for you and other Christian family members.

Many of your friends and family will not heed your pleas and will believe you are foolish just as they did of Noah. However, take heart, for many others will listen and find salvation through you. Most people (including the majority of Christians), know nothing about the events that are coming just as they did not understand in Noah's time. For those who knew Noah, lived near him, or were related to him, clearly they saw the sign (ark) and knew what disaster was forecasted (flood). Nevertheless, their love for the things of this life, and not of God, blinded them from seeing the truth—just as it does now.

How many people died in the great flood? Have you ever thought about this? Maybe thousands, millions, perhaps many more? Did children and babies die in the flood? Or course they did. Many are taught Christ can come at any time… even now, as you read this. Let us suppose that Jesus did come unexpectedly right now. What do you think would happen? Is Jesus going to gather all Christians and take them to heaven where they will live happily ever after?

Some might argue Jesus will begin ruling the nations and straightening everything out—removing the "bad" people from the earth, while never once considering they may be one of the "bad" people in the Lord's eyes. Are you a good person and believe this? You should understand that hell is populated with many people who are considered good from man's perspective. It is not enough to just be a "good" person by following man's laws and expect to find eternal life. One must obey Christ's teachings and worship the King of kings.

Subsequently, good people watched Noah build a great ship right in front of them and yet they could not recognize the ark was a sign of their impending doom. This is how it is in today's world. Unbelievers have no idea the time of their destruction draws near and as for most Christians, they are also trapped in the shadows of unawareness. Some Christians do keep watch, but due to forces I do not understand, they are unable to comprehend what signs to watch for and mislead many with their false ideas.

We have studied the timing for the tribulation years as well as corresponding events that will take place during the last days. However, what specific signs should a Christian watch for? I will provide a list of signs, in no particular order, to keep watch for. Some signs are end-time events and will occur, while others are only possible signs and may never happen. For instance, one sign we will discuss is watching out for two real life prophets. As I wrote previously, I believe these witnesses from Revelation chapter eleven are symbolic. On the other hand, there is a possibility my understanding is in error. Therefore, Christians ought to keep watch for their appearance, even if they never come, on the chance they are actual people.

377

Christians must guard against the blindness of Noah's last days, which prevented people from recognizing the ark as a sign of their destruction. Studying the Word is the only defense against this problem.

Daniel's Covenant

Daniel's covenant is projected to occur in late summer to early autumn of the year 2021 A.D., but could possibly happen a year earlier. This will be a peace treaty mediated by a European leader. No need to look for this pivotal sign at any other time of the year! There will also be another peace treaty to watch for in the year 2018 A.D. that will be similar in scope and will signal the rise of the antichrist. A treaty signed at any other time of the year is not the one foretold by Daniel and is not pertinent to end-time prophecies.

The Two Witnesses

Many believe there will be two prophets who will preach the good news of Christ's return during the last days. I do not share this literal view. Nevertheless, Christians should keep an eye out for this sign on the chance those who claim a plain fulfillment of Revelation 11 are correct. If these two witnesses come, they will be stationed in Jerusalem and will begin preaching to the world on the very day a European Union leader confirms Daniel's covenant with many nations. More than likely, these prophets will be Jews. They will both start their ministry at the same time.

Anyone who claims they are one of these prophets and does not meet these basic requirements is a false prophet and should not be believed! If these witnesses are merely symbolic, watch for an overwhelming flood of information on the Internet from Christians who claim, they know who the antichrist is and assert it is the same leader who brokered the Middle East peace accord in 2021 A.D.

Earthquakes in Jerusalem

Three devastating earthquakes are prophesied for Jerusalem in the future. Pay close attention to any news reports of earthquakes occurring in God's city.

One will occur after the "two witnesses" have finished doing God's work fulfilling the prophecies. If you are unable to identify whom the two witnesses are and refuse to believe other brothers and sisters in Christ who point them out to you, then maybe an earthquake will open your eyes. Urgency is critical at that very moment. If there has not been a rapture by this time, then the next window for the rapture to occur (based on a mid-tribulation doctrine) is about a month later. If the rapture doesn't take place then, there will be no rapture until Christ returns for good. The first earthquake should take place in the year 2018 A.D.

The last quake will take place when Christ returns and is really no advanced sign at all since Christ himself will be the last sign. This warning will come too late to change a person's fate.

Restricting Laws

Watch for laws that are passed which restrict freedoms. These laws will continue to be implemented in phases so the masses will not complain or recognize their freedoms are being gradually taken away. I do not know what the laws will encompass, but most will be in the form of controlling people's lives. For instance, a law is going into affect in the United States that will call for all Americans to have a passport when entering Canada and Mexico (and I assume conversely for Mexicans and Canadians). This regulation has not been needed before. Starting in June 2009, Americans will no longer be able to travel to bordering countries without a passport. The grease for this regulation is already showing up in television commercials, which claim passports will make it easier to travel back and forth between neighboring countries. This is untrue.

I live near the Canadian border and have traveled between our two nations many times by simply answering a few questions. Having each person present a passport, that will require inspection by border agents, will only increase crossing times and cause long delays for travelers. I know this because I once crossed the border with four European passengers and the time to go through the checkpoints increased considerably.

They were treated sternly until the agent reviewed their passports and they had finished answering many more questions than I had to. Is this what the American government means by making travel easier or do they mean it will make tracking you easier for them? These are the types of laws to watch out for because the last one will deal with the mark of the beast.

Another example is forcing people to present a photo or electronic card ID for every financial transaction so their identity can be verified. This is an unnecessary tightening of current practices. I have used the same bank for over twenty years. Have the manager and bank personnel forgotten who I am or who their regular customers are? The existing system in place is perfectly fine. When I use a bank branch office I do not typically frequent, they ask for ID. However, I am not required to present photo identification to bank employees who know me personally.

Why is this requirement being phased into place? Because it will condition people to accept being identified for everything in the future! I know this regulation is coming because I know people who work in the banking industry. There is little to do about these government efforts except to resist implementation to the best of ones ability.

Greasing the Stairs to Hell

Pay careful attention to government sponsored television commercials, articles, and news shows. These established methods of communication have but one goal... to "sell" the citizens in that country to "accept" what is being promoted. These are the same techniques big businesses use to brainwash people into buying their products. Very few of these marketing tools are ever truthful! Do you think I am being overly harsh with my assessment? Then you should recall all the political campaign commercials during election years. How many politicians ever do entirely what they said they were going to do if elected? None! They only say what voters want to hear to get into public office.

Have you watched military recruiting commercials from the American government that entice young men and women to join the armed services by making them sound so wonderful?—That is until our young people are sent to the battlefield. War is never great. These are half-truths and these methods are the grease used to slip in laws that will take away freedoms in the name of "safety." Indications as to how near the end-times are, can be discerned if we pay close attention to these advertisements.

A New Temple in Jerusalem

There is strong evidence that indicates a new Jewish temple will be rebuilt in the last days. Although there is no prophecy as to when this will occur, my best guess as to the timing is based on deductive reasoning. The War of Gog prophesied by the prophet Ezekiel will occur in the Middle East and during that war the Dome of the Rock Mosque will be destroyed. That damage will pave the way for Jews to rebuild the temple in the same location.

Some prophecy experts state a new temple will be built adjacent to the Dome of the Rock because this mosque is not really in the same location as the first two temples. I say, why would God leave this abomination standing only to destroy it after Jesus returns? This chore can be done ahead of time. Do you think temples built to worship Satan will be left intact after Christ returns or have you never thought about this? If Christianity is the "true" religion, then all other religions are false and an abomination unto God. All things which caused sin in the past and tempted man will be destroyed or eliminated when Christ rules.[b]

This effort to build a new temple will be aided by the antichrist and looked upon with favor by unsuspecting Jews. If this scenario occurs as I have portrayed it, then the rebuilding will start sometime after the War of Gog around 2018 A.D. and be completed no later than 2025 A.D.

Mark of the Beast

Keep constantly on guard for the mark of the beast for your eternal salvation rests on recognizing this system and refusing to take part in its implementation. Although I know not what form this "mark" will take, I believe every true Christian will know it when they see it. A physical identification mark that is associated with the number 666 and will be needed for everything: travel, banking, shopping, employment, bill paying, borrowing, identification, personal tracking, etc. Never accept the mark for any reason.

Miracle Workers

During the last days, there will be a few people who have the ability to perform supernatural feats. This sign should start appearing late in the tribulation period. If you witness these individuals, then there was no rapture. Do not follow any of them or believe what they say. Although, much of what they utter may be true, it is the lies that are weaved within the truths that will lead you to the gates of hell. These people are the antichrist, the false prophet, and maybe even a few others who are possessed by unclean spirits and are allowed by God to deceive those who love not the Word of God.[c] If anyone claims to be a Christian, but cannot see the lies behind their miracles, then they are not the faithful servant the Lord will reward at his coming. Believing in these miracle workers is quite simply not knowing and following the Word of God, for how can anyone claim to be a Christian if they rarely study the Bible? Would you allow someone who said they were a doctor, but had never studied medicine, operate on you? Would you permit someone who claimed they were a dentist, but had never learned about dentistry, fix your teeth? Of course not! A person who claims to be a doctor, dentist, lawyer, or anything requiring training does not make them so. The studying, training, and practicing makes them who they are. Why do so many believe Christ will acknowledge they belong to him when they ignore studying and practicing his teachings? Something they themselves would never do.

Wars in the Middle East

Always pay attention to news about military confrontations coming out of the Middle East. The last big wars on earth will be fought in this region of the world. There appear to be two large wars still to be fought before Christ returns: the War of Gog and the War of Armageddon

(World War III). These are not the minor battles between Israeli and Palestinian supporters we see everyday on the evening news. They are wars that involve multiple nations.

Furthermore, a foreign army must occupy Jerusalem before Christ can return.[d] Presently no such army occupies Jerusalem or Israel. When you see a foreign army move in, this will be a sign that Christ will be returning in another forty-two months.

Calculating the Number of his Name

During the time of the tribulation, God will raise up wise servants who will be able to calculate man's number using the name of the antichrist. That is not to say every person whose name is equivalent to this number is the antichrist. It only means when you see those on the Internet (or by other means), who can calculate the number of the beast from the name of a European leader, then we must investigate their evidence carefully. This leader may be the antichrist and we must use scripture to confirm if he fulfills all the requirements of the end-time prophecies. Again, this is only possible by knowing what is written in the prophecies. I have provided many checks to confirm the identity of the man of perdition throughout this book and in my first book.

Spreading false information about anyone who is not the genuine antichrist only desensitizes people for when the real antichrist is revealed. In other words, Christ's servants who cry "antichrist" too many times because they are unfamiliar with the complete Word of God will only hurt their credibility in the future.

The Yellow Flag of Islam

Revelation symbolism suggests an army will fight on the Armageddon battlefield under a yellow flag. This army must be a coalition from the Islamic countries by process of elimination. Pay attention to the politics involved in these Middle Eastern countries for the formation of a military union that adopts a yellow flag. This will be an extra sign that the end is near.

Jewish Ancestry

During the years (and prior year) the European Union holds elections, watch for young European politicians of Jewish ancestry who have charisma, and powerful allies. From this group of nations the antichrist will arise. Pay more attention to Greek politicians because the antichrist should come from this nation if we have understood the prophecies correctly. However, do not neglect the slim possibility this tyrant may be a government leader from any country within this group of nations. Christ's servants who keep watch will be able to recognize this human abomination in God's eyes before most will be able to. Do not believe or dismiss any brothers or sisters in Christ who make claims about the identity of the antichrist without testing everything they say. Let the Holy Spirit and the Bible weigh the truth of the evidence they present just as those Christians who calculate the number of the beast.

Abomination of Desolation

The abomination that causes desolation will definitely occur in the February to April timeframe. As for the year, the evidence supports the year 2025 A.D., but again it could be a year sooner. What is the abomination that causes desolation? It is a sign that is hard to grasp for many. The best interpretation of this key sign is the antichrist will enter Jerusalem and claim he is the messiah i.e. the leader of the world, and may even claim he is God. In other words, he will take control of Jerusalem with his army and enter the temple if one has been built. He will make blasphemous declarations and they will be carried on television worldwide. When you witness this event, the time to escape to a safe haven away from major metropolitan areas has come.

China on the Move

Keep watch of the military movements of Russia and China during the next twenty years. China moving toward the Middle East for any reason will be a signal the end is very near. This signal will bring clarity to the truth of the "end-times" I have written about as well as settle the rapture controversy. You will only be able to witness this sign if there was no rapture, you converted to Christianity during the tribulation, or you were not deemed "ready" to be taken in the rapture. Do not lose heart for any reason, but stiffen your resolve to live for the Lord. Ask for forgiveness and you will find your salvation at his return.

Peter the Roman

This sign is included even though it is not supported by Bible prophecy. It has to do with the prophecy attributed to St. Malachy I discussed briefly on page 224. The current Pope, Benedict XVI, was elected to the head of the Catholic Church in April 2005 A.D. at seventy-eight years of age. It is unrealistic to believe he will hold the papacy for as many years as his predecessor John Paul II did due to his old age. Should Pope Benedict's successor take the name of "Peter" and be of European decent (possibly Italian) this will be another sign the last days are here. Should the next Pope not take this name, then more than likely, this is a false prophecy and no outward sign at all pertaining to the tribulation.

We have explored fifteen signs to watch for during the last days. I have not included general signs like weather disasters, moral decay, riots, wars throughout the world, epidemics, and others for they provide no detailed clues as to how close we are to the end of the age. I have tried to focus on unique signs that you might be able to recognize (given enough study). This is so you will have some knowledge of how close the end is in the remote chance the date calculations are off significantly. Furthermore, I have not rehashed some of the cataclysmic events discussed earlier; like meteors and nuclear bombs striking major population centers. These are signs that are self-evident the end of the age has come.

I have attempted to provide additional signs that you may not have realized before. "Ark" like signs that will aid you in knowing beyond a shadow of a doubt that Christ's second advent is near. By witnessing things that were foretold in the Bible thousands of years ago, you can know with your mind and not just with your heart, there is a Creator who is working His plan. This is the solid evidence needed to convince unbelieving loved ones of the truth. If you can show them that events happening around them were foretold in the Bible eons ago, maybe they will finally grasp that everything foretold in the Bible is true—including the salvation of believing in Christ.

Imagine if one day the snow on top of Mount Ararat in Turkey melted and Noah's ark was discovered under the snowcap thousands of feet up the mountain.[e] Wouldn't all the scoffers who reject the Bible's teachings have the evidence needed to prove what is written in the Bible is true? How many would come to the Lord given this new evidence? I think you know the answer. Showing them proof that prophecies are being fulfilled right before their eyes will also provide similar proof. Some will always find reasons for not believing because they love the material things and pleasures of this world. They belong to Satan, and yes, some will be our relatives and friends. Nothing is more depressing and painful than learning your loved ones will not listen to your pleas and spend an eternity in suffering. However, many will come to the Lord because of you.

There is but one thing left to do—to learn what is expected of a Christian during the trials ahead. Knowing much of what is to come and doing nothing with this information entrusted to you will not be looked upon kindly when the Master returns.[f] The time for training is now and clearly you are working at preparing yourself by studying. We must also put into action what we learn and teach others to do the same in the time remaining.

Chapter 19

What is Expected of Christians

As YOU KNOW, THE MOST LIKELY timing for Christ's return is at the end of the tribulation years in 2028 A.D. But what does Christ demand of his faithful servants between now and the end of the age? There are many things expected of Christians. Some are obvious like living a righteous life, but others are not so apparent.

> It [the beast] was given power to make war against God's people and to conquer them. And it was given authority over every tribe, people, language and nation. All inhabitants of the earth will worship the beast—all whose names have not been written in the Lamb's book of life, the Lamb who was slain from the creation of the world. Whoever has ears, let them hear.
>
> "If anyone is to go into captivity, into captivity they will go. If anyone is to be killed with the sword, with the sword they will be killed."
>
> This calls for patient endurance and faithfulness on the part of God's people.
>
> Revelation 13:7-10 (TNIV)

Patient Endurance and Faithfulness

The first thing that can be expected under the beast's (antichrist) authority is it will not be an easy time for anyone; especially Christians. During those days, Christians can expect to be persecuted for their beliefs to the point some will be imprisoned and some put to death. When you read these verses closely, one should get the sense that God has predestined some believers to be "super" witnesses. Witnesses that prove they are faithful servants through deeds and not just with lip service.

"If anyone is to go into captivity (planned by God ahead of time), then into captivity they will go (what is planned will take place)." If you are a Christian who God plans on using to achieve His goals of witnessing to non-believers or "lukewarm" Christians through your actions, then you can expect to have a harder time than other Christians. Jesus tells St. John in Revelation that we need to be ready for this eventuality. It calls for patient endurance and faithfulness to keep up the good fight against Satan. Knowing the amount of time that has been determined in God's plan can provide hope to those who are called by Christ for this service. Don't give up, don't give in, and don't loose sight of the goal of your salvation and winning unbelievers for Christ.

For we know, brothers and sisters loved by God, that he has chosen you, because our gospel came to you not simply with words but also with power, with the Holy Spirit and deep conviction. You know how we lived among you for your sake. You became imitators of us and of the Lord, for you welcomed the message in the midst of severe suffering with the joy given by the Holy Spirit. And so you became a model to all the believers in Macedonia and Achaia. The Lord's message rang out from you not only in Macedonia and Achaia—your faith in God has become known everywhere. Therefore we do not need to say anything about it, for they themselves report what happened when we visited you. They tell how you turned to God from idols to serve the living and true God, and to wait for his Son from heaven, whom he raised from the dead—Jesus, who rescues us from the coming wrath.

1st Thessalonians 1:4-10 (TNIV)

The Thessalonians shall be a model for Christians during the last days. In both of Paul's letters to the Thessalonians, we can learn many things on how to act and the need to be ready at all times for Christ's return. This is because these Epistle letters specifically address the concerns the Thessalonians had about the second advent of Christ. Paul's messages provide prophecies and instructions for this future period. So, what are the lessons to be gleaned from his words of encouragement? First, that the message of Christ is not to be proclaimed with only words, but with actions that backup those words! The Thessalonians were shown how to act and then began imitating the disciples' actions. In so doing, they became an example for all believers and their faith in God became widely known because of their lifestyle. You can conclude their actions led more unbelievers to Christ than just merely proclaiming the Gospel. We see they welcomed the message of Christ, with joy even in the midst of severe suffering, knowing that they would be rescued from the appointed wrath of God should that time come during their lifetimes. Clearly, they were not being tormented under God's wrath; nevertheless, they were still suffering. This is an important distinction that Christians of today should take notice of. We are not promised to be saved from persecution and suffering as pre-tribulation rapture theology claims, but only from God's appointed wrath that will come upon the world.

You should understand, that there are reasons why things may be going well for you and reasons why life is treating you poorly. Knowing the reasons why gives us clues on where we are in our walk with Christ. For instance, if life is treating you well it may be the Lord is blessing you for your obedience to His will. However, it may also be Satan is easing your way to keep you turned away from the Lord… believing you don't need Him. Christ said it was easier for a camel to go through the eye of a needle than for a wealthy man to enter the kingdom of heaven. Why? Because the rich man can have anything his heart desires and those desires grow from the roots of unrepentant sin… the fruit that Satan cultivates. Many rich men crave only the materialistic things of this world and if they cannot control these urges, then by their actions we understand who their true master is. Jesus said, we can only have one master and where the desires of our heart are, there our true master will be found.

If you believe you are a good Christian who goes to Church once a week (for a few hours), gives of your tithe, but during the rest of the week you rarely acknowledge Christ to friends, family, and even strangers, then you may need to look in the mirror. If you seldom study your Bible, overindulge regularly, neglect family responsibilities, constantly obsess on the pleasures of this world, or the desires of the flesh, then your heart may not be centered on Christ.

For Christ's servants, much has been given (the knowledge of salvation) and much is expected (to be beacons for the unbelievers of this world). We are predestined to be the lamps unto a dark world for unbelievers to find their way out of the darkness of eternal damnation. We cannot hide the light of truth at all times except for a few hours a week. We must do our best to live as examples for Christ in all that we do so that those who do not know Christ will seek the truth that guides us: just as the Thessalonians did. We must persevere when the going gets unpleasant, be subservient to His will, and be content with the cards we are dealt.

Our Duty as Watchmen

Before the fall of Jerusalem to the Babylonians, Ezekiel was instructed in 593 B.C. to warn all the Jews of the impending destruction from God's wrath (advanced warning of His plans six years before the final destruction of the city and the temple). Similar circumstances to those punishments will be metered out in the near future, not only to the Jews as it was back then, but on a larger scale that encompasses all mankind. From the Lord's directives to Ezekiel, we discover other things we are obligated to do as faithful Christians. We must inform people of what is going to take place.

At the end of seven days the word of the LORD came to me: "Son of man, I have made you a watchman for the house of Israel; so hear the word I speak and give them warning from me. When I say to the wicked, 'You will surely die,' and you do not warn them or speak out to dissuade them from their evil ways in order to save their lives, those wicked people will die for their sins, and I will hold you accountable for their blood. But if you do warn the wicked and they do not turn from their wickedness or from their evil ways, they will die for their sins; but you will have saved yourself. "Again, when the righteous turn from their righteousness and do evil, and I put a stumbling block before them, they will die. Since you did not warn them, they will die for their sins. The righteous things they did will not be remembered, and I will hold you accountable for their blood. But if you do warn the righteous not to sin and they do not sin, they will surely live because they took warning, and you will have saved yourself."

Ezekiel 3:16-21 (TNIV)

Ezekiel was warned that he would be answerable for not spreading the news of Israel's destruction to both unbelievers and believers alike who had strayed from the truth and by their actions, blocked the Spirit of God from revealing to them the signs of their ruin. He was in fact his brother's keeper. Although a person's salvation was determined by their own choices and not Ezekiel's actions, Ezekiel was accountable for telling them the things they needed to know to help them make the right choice.

Having read this far, you too are now one of God's watchmen for the knowledge of the prophecies of the last days has been given to you. What you do with this knowledge is up to you but be forewarned, as Ezekiel was, that keeping this knowledge to yourself has unpleasant consequences that far outweigh the troubles you will experience in spreading this news.

Warning for Lazy Christians

The Lord answered, "Who then is the faithful and wise manager [Christian leader or teacher], *whom the master puts in charge of his servants to give them their food allowance at the proper time? It will be good for that servant* [the leader] *whom the master finds doing so when he returns. Truly I tell you, he will put him in charge of all his possessions. But suppose the servant says to himself, 'My master is taking a long time in coming,' and he then begins to beat the other servants, both men and women, and to eat and drink and get drunk* [follows the ways of this world]. *The master of that servant will come on a day when he does not expect him and at an hour he is not aware of* [because he is not paying attention]. *He will cut him to pieces and assign him a place with the unbelievers.* **"The servant who knows the master's will and does not get ready or does not do what the master wants will be beaten with many blows.** *But the one who does not know and does things deserving punishment will be beaten with few blows. From everyone who has been given much, much will be demanded; and from the one who has been entrusted with much, much more will be asked.*

Luke 12:42-48 (TNIV)

The time is soon coming, when all will answer for what they have done… and God's teachers, prophets, religious leaders, and watchmen will be judged more harshly for what we have taught others about the written Word that may have led some astray, as well as those things we failed to instruct them on. Christ expects more from those who know the truth, that the end approaches swiftly, than he does from those servants who do not know. Furthermore, Christ also expects more from those entrusted to lead his flock while he is away—to teach the truth of Christ to the best of their ability and to not ignore the signs of his return. Pastors, priests, and Christian teachers are instructed to educate Jesus' servants in the true Word of the Lord. They are not to get bogged down in the things of the world and find themselves unprepared for the Master's return as many Christian leaders do these days. The message is clear: keep watch at all times and be ready so that when the appointed time arrives, they can hear the Master's call and aid any who will listen to the truth. Realize that Christ first came not to bring peace to the world (that will happen at his second coming), but to bring the knowledge of the truth that would divide the world. This is so that at the judgment, it will be clear the right decision was made for every person's salvation.

We are not to be lazy servants (Christians) spending our days chasing after the same things the unbelievers of the world are deluded into doing. For if we do, we can expect to be punished more harshly than our brethren and assigned to the raging fires of hell along with everyone who rejected Christ's authority.

> *I have come to bring fire on the earth, and how I wish it were already kindled! But I have a baptism to undergo, and what constraint I am under until it is completed! Do you think I came to bring peace on earth? No, I tell you, but division. From now on there will be five in one family divided against each other, three against two and two against three. They will be divided, father against son and son against father, mother against daughter and daughter against mother, mother-in-law against daughter-in-law and daughter-in-law against mother-in-law.*
>
> Luke 12:49-53 (TNIV)

Throughout this book, I have written many times that the most important thing a Christian can do during the last days is to witness to their loved ones that the end is near, and that Christ is the only way to save themselves from an eternity of suffering. However, in so doing we learn this will be a painful, but necessary responsibility. Not all of our family will believe us. Jesus' words are further evidence of what is expected of his servants.

Timothy's Training

Brothers and sisters in Christ, have you read and studied the words of wisdom from our brother Paul to Timothy given long ago? His lessons concerning how a good Christian is suppose to live are as important today as they were the day he first gave them. On these instructions, I believe that Christians in today's world may need reminding about some of these lessons.

The first is the inability of most people to realize their compulsion with achieving the American dream (when uncontrolled) is nothing more than idol worshiping. Timothy's training on this matter is as follows:

> Command those who are rich in this present world not to be arrogant nor to put their hope in wealth, which is so uncertain, but to put their hope in God, who richly provides us with everything for our enjoyment. Command them to do good, to be rich in good deeds, and to be generous and willing to share. In this way they will lay up treasure for themselves as a firm foundation for the coming age [not this age, but the thousand P-year reign of Christ], so that they may take hold of the life that is truly life.
>
> 1st Timothy 6:17-19 (TNIV)

Paul does not instruct Timothy that Christians need to be poor to enter the kingdom of God, but rather that a good Christian must follow the Word of God and keep a proper perspective in regards to their wealth the Lord has blessed them with. We should never let the material things of the world have power over the decisions we make or let them displace the love in our hearts for Christ.

> *But godliness with contentment is great gain. For we brought nothing into the world, and we can take nothing out of it. But if we have food and clothing, we will be content with that. Those who want to get rich fall into temptation and a trap and into many foolish and harmful desires that plunge people into ruin and destruction. For the love of money is a root of all kinds of evil. Some people, eager for money, have wandered from the faith and pierced themselves with many griefs.*
>
> 1st Timothy 6:6-10 (TNIV)

Be content with what the Lord has given you. That is not to say we should not work toward a better life for our family and ourselves. It is only to say as you walk along the road of life be satisfied with the things God has provided for you when following His path—for without contentment the desires of the heart may spiral out of your control. What is expected of Christians can be summed up with this simple counsel. Look on your wealth and possessions as things that are only borrowed for they are not yours, but God's. When you borrow something from a friend or neighbor, you do so because you have a need for it rather than a desire for it. When that item is returned, your feelings for the thing(s) you borrowed are usually one of gratitude and nothing more. Viewing money and wealth as anymore than this, is not the way of a Christian heart and will not be helpful in the remaining time the world has left. This is because much of a Christian's wealth will be lost or taken away when the mark of the beast becomes law.

Lamenting on the loss of your money and material possessions, when this occurs, will keep you from doing the things the Lord requires of you and will tempt you into accepting the mark of the beast. If you cannot control your urges for the things of this world now, you must start working on your addictions before the time comes or else you will be blind to the truth of the mark of the beast. Paul also gives testimony to how people will be during the last days. Here is what he prophesies on this matter.

But mark this: There will be terrible times in the last days. People will be lovers of themselves, lovers of money, boastful, proud, abusive, disobedient to their parents, ungrateful, unholy, without love, unforgiving, slanderous, without self-control, brutal, not lovers of the good, treacherous, rash, conceited, lovers of pleasure rather than lovers of God–having a form of Godliness but denying its power. Have nothing to do with such people. They are the kind who worm their way into homes and gain control over gullible women, who are loaded down with sins and are swayed by all kinds of evil desires, always learning but never able to come to a knowledge of the truth. Just as Jannes and Jambres opposed Moses, so also these teachers oppose the truth. They are men of depraved minds, who, as far as the faith is concerned, are rejected. But they will not get very far because, as in the case of those men, their folly will be clear to everyone.

2nd Timothy 3:1-9 (TNIV)

Paul hit the nail right on the head! This is an accurate description of the inhabitants of today's world... both unbelievers and many Christians as well. Mankind today is a lover of pleasure and money, not God, and in pursuing such endeavors most will have nothing to do with God. We find not only that idol worshiping is prevalent in the world, but also that Christians must be careful around those who do not accept God. Their power from the evil one will bring you down if you are unprepared for it and dabble in their lifestyle. We are not to spend countless hours with them, nor are we to ignore them or have nothing to do with them, but we are to inform them of the Word of God from arms length. Do not live like them. As you attempt to lead others to the light of Christ, be on guard lest their influence will change you instead. For Paul writes,

The sins of some are obvious, reaching the place of judgment ahead of them; the sins of others trail behind them. In the same way, good deeds are obvious, and even those that are not obvious cannot remain hidden forever.

1st Timothy 5:24-25 (TNIV)

Understand many sins are easy to recognize in others, but some are not so obvious. They are traps for the unknowledgeable Christian... the weak Christian... the servant who doesn't put into practice the lessons of the Bible, but unwittingly allows their unbelieving friends to negatively influence their walk with Christ. Paul tells Timothy to flee from these traps...

> *But you, man of God, flee from all this, and pursue righteousness, godliness, faith, love, endurance and gentleness. Fight the good fight of the faith. Take hold of the eternal life to which you were called when you made your good confession in the presence of many witnesses.*
>
> 1st Timothy 6:11-12 (TNIV)

What is expected of us in the last days? The same things that have always been expected of God's people since the beginning of time. Following this holy path will lead to eternal life—even if you are powerless to recognize the end-time signs or unable to believe anything I have written about these important matters. For those who do understand and believe the things I have revealed, you have the added burden of witnessing to every one of these mysteries.

Study the Bible

> *But as for you, continue in what you have learned and have become convinced of, because you know those from whom you learned it, and how from infancy you have known the Holy Scriptures, which are able to make you wise for salvation through faith in Christ Jesus. All Scripture is God-breathed and is useful for teaching, rebuking, correcting and training in righteousness, so that all God's people may be thoroughly equipped for every good work.*
>
> 2nd Timothy 3:14-17 (TNIV)

> *In the presence of God and of Christ Jesus, who will judge the living and the dead, and in view of his appearing and his king-*

dom, I give you this charge: Preach the word; be prepared in sea-
son and out of season; correct, rebuke and encourage—with great
patience and careful instruction. For the time will come when
people will not put up with sound doctrine. Instead, to suit their
own desires, they will gather around them a great number of
teachers to say what their itching ears want to hear. They will
turn their ears away from the truth and turn aside to myths. But
you, keep your head in all situations, endure hardship, do the
work of an evangelist, discharge all the duties of your ministry.

2nd Timothy 4:1-5 (TNIV)

Preach the word at all times and guard against those who teach false doctrine and do not understand the truth as you do. We must test everything in the last days and this can only be done by constant study of the Bible. Study scripture and learn the signs of the last days. Speak in an out of season about the blood of Christ. Do not continue to study those things you already know and which are taught repetitively in many churches today. These lessons are for the new servants who come to the Lord. Search for the sustaining knowledge that provides insight into the last days so that you may be a blessing to those who are lost without this knowledge.

Last Words of Wisdom

The time grows short as the second hand approaches the midnight hour on God's master clock. There is much to do before the return of the King. He will be here before you know it. You have learned much about the way God works and His plans for mankind. You have seen what is planned for the last days and when things are likely to occur. Will all these things take place exactly as I have portrayed them? Probably not, for no one knows exactly what is to take place except the Master Planner of all things. I have however given you a good rendition of the last days and with that knowledge you should be able to watch for the signs that will allow you to see, with greater clarity, what is to take place as the tribulation draws near. Does knowing what will take place matter?

Not for an obedient Christian who keeps God's commands for they are constantly striving to do the right thing and always searching for Christ's lost sheep. As for the servants of God who do not believe in their heart Christ will ever return and practice detestable things along side the unbelievers, knowing when Christ will return should bring them back to the things that are important to the Lord with greater commitment and urgency.

If you knew Christ was returning tomorrow to put to death all who refuse to believe in him or disobey his teachings, what would you do today? Would you go to work? Would you continue with your life as if nothing was different? Would you spend all day in prayer begging for forgiveness? Or... would you try to convince your friends and family who do not know Christ that it is never to late to accept the gift of salvation? Do you really want to spend eternity without those you love? What if Christ were not coming back tomorrow, but two days from now? Would your actions be any different? At what point in time do you begin changing what you are doing (or not doing) before Christ returns? Is it days before, weeks or months before, or years before? You now know the time is less than twenty years away and waiting until the end is not what God wants from you. Urgency is required of you for Christians know how hard it is to reap where the Lord does not sow i.e. to find new believers within the fields of the unbelievers. Sowing the knowledge of Christ's return now may bear the fruit you are looking for in the last days—the fruit of salvation for your loved ones who do not yet know Christ.

May the Creator of the Universe, send the light of the Holy Spirit to, empower you to do the good works He has planned for you during the closing days of the Third Age of man.

Appendix I
Daniel's End-Time Prophecies (RSV)

	PROPHECIES	INTRODUCTION
Nebuchadnezzar's Dream	2 In the second year of the reign of Nebuchadnez'zar, Nebuchadnez'zar had dreams; and his spirit was troubled, and his sleep left him. 2Then the king commanded that the magicians, the enchanters, the sorcerers, and the Chaldeans be summoned, to tell the king his dreams. ...	
Daniel's Dream of Four Beasts	7 In the first year of Belshaz'zar king of Babylon,	Daniel had a dream and visions of his head as he lay in his bed. Then he wrote down the dream, and told the sum of the matter. 2Daniel said, "I saw in my vision by night, and behold, the four winds of heaven were stirring up the great sea.
Daniel's Vision of a Ram and a Goat	8 In the third year of the reign of King Belshaz'zar a vision appeared to me,	Daniel, after that which appeared to me at the first. 2And I saw in the vision; and when I saw, I was in Susa the capital, which is in the province of Elam; and I saw in the vision, and I was at the river U'lai.
Daniel's Seventy 'Weeks'	9 In the first year of Darius the son of Ahasu-e'rus, by birth a Mede, who became king over the realm of the Chalde'ans. 2In the first year of his reign,	I, Daniel, perceived in the books the number of years which, according to the word of the LORD to Jeremiah the prophet, must pass before the end of the desolations of Jerusalem, namely, seventy years. 3Then I turned my face to the Lord God, seeking him by prayer and supplications with fasting and sackcloth and ashes. 4I prayed to the LORD my God and made confession, saying, "O Lord, the great and terrible God, who keepest covenant and steadfast love with those who love him and keep his commandments, 5we have sinned and done wrong and acted wickedly and rebelled, turning aside from thy commandments and ordinances; 6we have not listened to thy servants the prophets, who spoke in thy name to our kings, our princes, and our fathers, and to all the people of the land. 7To thee, O Lord, belongs righteousness, but to us confusion of face, as at this day, to the men of Judah, to the inhabitants of Jeru-
The Kings of the South and the North	10 In the third year of Cyrus king of Persia a word was revealed to	Daniel, who was named Belteshazzar. And the word was true, and it was a great conflict. And he understood the word and had understanding of the vision. 2In those days I, Daniel, was mourning for three weeks. 3I ate no delicacies, no meat or wine entered my mouth, nor did I anoint myself at all, for the full three weeks. 4On the twenty-fourth day of the first month, as I was standing on the bank of the great river, that is, the Tigris, 5I lifted up my eyes and looked, and behold, a man clothed in linen, whose loins were girded with gold of Uphaz. 6His body was like beryl, his face like the appearance of lightning, his eyes like flaming torches, his arms and legs like the gleam of burnished bronze, and the sound of his words like the noise of a multitude. 7And I, Daniel, alone saw the vision, for the men who were with me did not see the vision, but a great trembling fell upon them, and they fled to hide

INTRODUCTION (Continued)

Nebuchadnezzar's Dream	Daniel's Dream of Four Beasts	Daniel's Vision of a Ram and a Goat	Daniel's Seventy 'Weeks'	The Kings of the South and the North
			salem, and to all Israel, those that are near and those that are far away, in all the lands to which thou hast driven them, because of the treachery which they have committed against thee. 8To us, O Lord, belongs confusion of face, to our kings, to our princes, and to our fathers, because we have sinned against thee. 9To the Lord our God belong mercy and forgiveness; because we have rebelled against him, 10and have not obeyed the voice of the LORD our God by following his laws, which he set before us by his servants the prophets. 11All Israel has transgressed thy law and turned aside, refusing to obey thy voice. And the curse and oath which are written in the law of Moses the servant of God have been poured out upon us, because we have sinned against him. 12He has confirmed his words, which he spoke against us and against our rulers who ruled us, by bringing upon us a great calamity; for under the whole heaven there has not been done the like of what has been done against Jerusalem. 13As it is written in the law of Moses, all this calamity has come upon us, yet we have not entreated the favor of the LORD our God, turning from our iniquities and giving heed to thy truth. 14Therefore the LORD has kept ready the calamity and has brought it upon us; for the LORD our God is righteous in all the works which he has done, and we have not obeyed his voice. 15And now, O Lord our God,	themselves. 8So I was left alone and saw this great vision, and no strength was left in me; my radiant appearance was fearfully changed, and I retained no strength. 9Then I heard the sound of his words; and when I heard the sound of his words, I fell on my face in a deep sleep with my face to the ground. 10And behold, a hand touched me and set me trembling on my hands and knees. 11And he said to me, "O Daniel, man greatly beloved, give heed to the words that I speak to you, and stand upright, for now I have been sent to you." While he was speaking this word to me, I stood up trembling. 12Then he said to me, "Fear not, Daniel, for from the first day that you set your mind to understand and humbled yourself before your God, your words have been heard, and I have come because of your words. 13The prince of the kingdom of Persia withstood me twenty-one days; but Michael, one of the chief princes, came to help me, so I left him there with the prince of the kingdom of Persia 14and came to make you understand what is to befall your people in the latter days. For the vision is for days yet to come." 15When he had spoken to me according to these words, I turned my face toward the ground and was dumb. 16And behold, one in the likeness of the sons of men touched my lips; then I opened my mouth and spoke. I said to him

Nebuchadnezzar's Dream	Daniel's Dream of Four Beasts	Daniel's Vision of a Ram and a Goat	Daniel's Seventy 'Weeks'	The Kings of the South and the North
			INTRODUCTION (Continued)		BABYLON
			who didst bring thy people out of the land of Egypt with a mighty hand, and hast made thee a name, as at this day, we have sinned, we have done wickedly. 16O Lord, according to all thy righteous acts, let thy anger and thy wrath turn away from thy city Jerusalem, and for the iniquities of our fathers, Jerusalem and thy people have become a byword among all who are round about us. 17Now therefore, O our God, hearken to the prayer of thy servant and to his supplications, and for thy own sake, O Lord, cause thy face to shine upon thy sanctuary, which is desolate. 18O my God, incline thy ear and hear; open thy eyes and behold our desolations, and the city which is called by thy name; for we do not present our supplications before thee on the ground of our righteousness, but on the ground of thy great mercy. 19O LORD, hear; O LORD, forgive; O LORD, give heed and act; delay not, for thy own sake, O my God, because thy city and thy people are called by thy name."	who stood before me, "O my lord, by reason of the vision pains have come upon me, and I retain no strength. 17How can my lord's servant talk with my lord? For now no strength remains in me, and no breath is left in me." 18Again one having the appearance of a man touched me and strengthened me. 19And he said, "O man greatly beloved, fear not, peace be with you; be strong and of good courage." And when he spoke to me, I was strengthened and said, "Let my lord speak, for you have strengthened me." 20Then he said, "Do you know why I have come to you? But now I will return to fight against the prince of Persia; and when I am through with him, lo, the prince of Greece will come. 21But I will tell you what is inscribed in the book of truth: there is none who contends by my side against these except Michael, your prince. 11 And as for me, in the first year of Darius the Mede, I stood up to confirm and strengthen him.
31"You saw, O king, and behold, a great image. This image, mighty and of exceeding brightness, stood before you, and its appearance was frightening. 32The head of this image was of fine gold,	3And four great beasts came up out of the sea, different from one another. 4The first was like a lion and had eagles' wings. Then as I looked its wings were plucked off, and it was lifted up from the ground and made to stand upon two feet like a man; and the mind of a man was given to it.			

	MEDO-PERSIAN EMPIRE	GREEK EMPIRE	ROME
The Kings of the South and the North			
Daniel's Seventy 'Weeks'			
Daniel's Vision of a Ram and a Goat	3I raised my eyes and saw, and behold, a ram standing on the bank of the river. It had two horns; and both horns were high, but one was higher than the other, and the higher one came up last. 4I saw the ram charging westward and northward and southward; no beast could stand before him, and there was no one who could rescue from his power; he did as he pleased and magnified himself.	5As I was considering, behold, a he-goat came from the west across the face of the whole earth, without touching the ground; and the goat had a conspicuous horn between his eyes. 6He came to the ram with the two horns, which I had seen standing on the bank of the river, and he ran at him in his mighty wrath. 7I saw him come close to the ram, and he was enraged against him and struck the ram and broke his two horns; and the ram had no power to stand before him, but he cast him down to the ground and trampled upon him; and there was no one who could rescue the ram from his power. 8Then the he-goat magnified himself exceedingly; but when he was strong, the great horn was broken, and instead of it there came up four conspicuous horns toward the four winds of heaven.	9Out of one of them came forth a little horn, which grew exceedingly great toward the south, toward the east, and toward the glorious land.
Daniel's Dream of Four Beasts	5And behold, another beast, a second one, like a bear. It was raised up on one side; it had three ribs in its mouth between its teeth; and it was told, 'Arise, devour much flesh.'	6After this I looked, and lo, another, like a leopard, with four wings of a bird on its back; and the beast had four heads; and dominion was given to it.	7After this I saw in the night visions, and behold, a fourth beast, terrible and dreadful and exceedingly strong;
Nebuchadnezzar's Dream	its breast and arms of silver,	its belly and thighs of bronze,	33 its legs of iron,

Nebuchadnezzar's Dream	Daniel's Dream of Four Beasts	Daniel's Vision of a Ram and a Goat	Daniel's Seventy 'Weeks'	The Kings of the South and the North
EUROPEAN UNION (REVIVED ROMAN EMPIRE)				
its feet partly of iron and partly of clay. 34As you looked, a stone was cut out by no human hand, and it smote the image on its feet of iron and clay, and broke them in pieces; 35then the iron, the clay, the bronze, the silver, and the gold, all together were broken in pieces, and became like the chaff of the summer threshing floors; and the wind carried them away, so that not a trace of them could be found. But the stone that struck the image became a great mountain and filled the whole earth.	and it had great iron teeth; it devoured and broke in pieces, and stamped the residue with its feet. It was different from all the beasts that were before it; and it had ten horns. 8I considered the horns, and behold, there came up among them another horn, a little one, before which three of the first horns were plucked up by the roots; and behold, in this horn were eyes like the eyes of a man, and a mouth speaking great things. 9As I looked, thrones were placed and one that was ancient of days took his seat; his raiment was white as snow, and the hair of his head like pure wool; his throne was fiery flames, its wheels were burning fire. 10A stream of fire issued and came forth from before him; a thousand thousands served him, and ten thousand times ten thousand stood before him; the court sat in judgment, and the books were opened. 11I looked then because of the sound of the great words which the horn was speaking. And as I looked, the beast was slain, and its body destroyed and given over to be burned with fire. 12As for the rest of the beasts, their dominion was taken away, but their lives were prolonged for a season and a time. 13I saw in the night visions, and behold, with the clouds of heaven there came one like a son of man, and he came to the Ancient of Days and was presented	10It grew great, even to the host of heaven; and some of the host of the stars it cast down to the ground, and trampled upon them. 11It magnified itself, even up to the Prince of the host; and the continual burnt offering was taken away from him, and the place of his sanctuary was overthrown. 12And the host was given over to it together with the continual burnt offering through transgression; and truth was cast down to the ground, and the horn acted and prospered. 13Then I heard a holy one speaking; and another holy one said to the one that spoke, "For how long is the vision concerning the continual burnt offering, the transgression that makes desolate, and the giving over of the sanctuary and host to be trampled under foot?" 14And he said to him, "For two thousand and three hundred evenings and mornings; then the sanctuary shall be restored to its rightful state."		

	E.U. (Continued)	VISION INTERPRETATION	MEDO-PERSIA
The Kings of the South and the North		*The Interpretation of the Vision* 2"And now I will show you the truth. Behold,	three more kings shall arise in Persia; and a fourth shall be far richer than all of them; and when he has become strong through his riches, he shall stir up all against the kingdom of Greece.
Daniel's Seventy 'Weeks'		*Answer to Daniel's Prayer* 20While I was speaking and praying, confessing my sin and the sin of my people Israel, and presenting my supplication before the LORD my God for the holy hill of my God; 21while I was speaking in prayer, the man Gabriel, whom I had seen in the vision at the first, came to me in swift flight at the time of the evening sacrifice. 22He came and he said to me, "O Daniel, I have now come out to give you wisdom and understanding. 23At the beginning of your supplications a word went forth, and I have come to tell it to you, for you are greatly beloved; therefore consider the word and understand the vision.	24"Seventy weeks of years are decreed concerning your people and your holy city, to finish the transgression, to put an end to sin, and to atone for iniquity, to bring in everlasting righteousness, to seal both vision and prophet, and to anoint a most holy place.
Daniel's Vision of a Ram and a Goat		*The Interpretation of the Vision* 15When I, Daniel, had seen the vision, I sought to understand it; and behold, there stood before me one having the appearance of a man. 16And I heard a man's voice between the banks of the U'lai, and it called, "Gabriel, make this man understand the vision." 17So he came near where I stood; and when he came, I was frightened and fell upon my face. But he said to me, "Understand, O son of man, that the vision is for the time of the end." 18As he was speaking to me, I fell into a deep sleep with my face to the ground; but he touched me and set me on my feet. 19He said, "Behold, I will make known to you what shall be at the latter end of the indignation; for it pertains to the appointed time of the end.	20As for the ram which you saw with the two horns, these are the kings of Media and Persia.
Daniel's Dream of Four Beasts	before him. 14And to him was given dominion and glory and kingdom, that all peoples, nations, and languages should serve him; his dominion is an everlasting dominion, which shall not pass away, and his kingdom one that shall not be destroyed. *The Interpretation of the Dream* 15"As for me, Daniel, my spirit within me was anxious and the visions of my head alarmed me. 16I approached one of those who stood there and asked him the truth concerning all this. So he told me, and made known to me the interpretation of the things. 17These four great beasts are four kings who shall arise out of the earth.		
Nebuchadnezzar's Dream	*The Interpretation of the Dream* 36"This was the dream; now we will tell the king its interpretation. 37You, O king, the king of kings, to whom the God of heaven has given the kingdom, the power, and the might, and the glory, 38and into whose hand he has given, wherever they dwell, the sons of men, the beasts of the field, and the birds of the air, making you rule over them all -- you are the head of gold.		39After you shall arise another kingdom inferior to you,

Nebuchadnezzar's Dream	Daniel's Dream of Four Beasts	Daniel's Vision of a Ram and a Goat	Daniel's Seventy 'Weeks'	The Kings of the South and the North
			MEDO-PERSIA	**GREEK EMPIRE**
and yet a third kingdom of bronze, which shall rule over all the earth.		21And the he-goat is the king of Greece; and the great horn between his eyes is the first king. 22As for the horn that was broken, in place of which four others arose, four kingdoms shall arise from his nation, but not with his power.	25Know therefore and understand that from the going forth of the word to restore and build Jerusalem to the coming of an anointed one, a prince, there shall be seven weeks. Then for sixty-two weeks it shall be built again with squares and moat, but in a troubled time.	3Then a mighty king shall arise, who shall rule with great dominion and do according to his will. 4And when he has arisen, his kingdom shall be broken and divided toward the four winds of heaven, but not to his posterity, nor according to the dominion with which he ruled; for his kingdom shall be plucked up and go to others besides these. 5Then the king of the south shall be strong, but one of his princes shall be stronger than he and his dominion shall be a great dominion. 6After some years they shall make an alliance, and the daughter of the king of the south shall come to the king of the north to make peace; but she shall not retain the strength of her arm, and he and his offspring shall not endure; but she shall be given up, and her attendants, her child, and he who got possession of her. 7In those times a branch from her roots shall arise in his place; he shall come against the army and enter the fortress of the king of the north, and he shall deal with them and shall prevail. 8He shall also carry off to

	GREEK EMPIRE (Continued)	ROMAN EMPIRE
The Kings of the South and the North	Egypt their gods with their molten images and with their precious vessels of silver and of gold; and for some years he shall refrain from attacking the king of the north. 9Then the latter shall come into the realm of the king of the south but shall return into his own land. 10His sons shall wage war and assemble a multitude of great forces, which shall come on and overflow and pass through, and again shall carry the war as far as his fortress. 11Then the king of the south, moved with anger, shall come out and fight with the king of the north; and he shall raise a great multitude, but it shall be given into his hand. 12And when the multitude is taken, his heart shall be exalted, and he shall cast down tens of thousands, but he shall not prevail.	13For the king of the north shall again raise a multitude, greater than the former; and after some years he shall come on with a great army and abundant supplies. 14In those times many shall rise against the king of the south; and the men of violence among your own people shall lift themselves up in order to fulfill the vision; but they shall fail. 15Then the king of the north shall come and throw up siege works, and take a well-fortified city. And the forces of the south shall not stand, or even his picked troops, for there shall be no strength to stand. 16But he who comes against him shall do according to his own will, and none shall stand before him; and he shall stand in the glorious land, and all of it shall be in his power. 17He shall set his face to come with the strength of his
Daniel's Seventy 'Weeks'		26And after the sixty-two weeks, an anointed one shall be cut off, and shall have nothing; and the people of the prince who is to come shall destroy the city and the sanctuary.
Daniel's Vision of a Ram and a Goat		
Daniel's Dream of Four Beasts		
Nebuchadnezzar's Dream		40And there shall be a fourth kingdom, strong as iron, because iron breaks to pieces and shatters all things; and like iron which crushes, it shall break and crush all these.

	ROMAN EMPIRE (Continued)	EUROPEAN UNION
The Kings of the South and the North	whole kingdom, and he shall bring terms of peace and perform them. He shall give him the daughter of women to destroy the kingdom; but it shall not stand or be to his advantage. 18Afterward he shall turn his face to the coastlands, and shall take many of them; but a commander shall put an end to his insolence; indeed he shall turn his insolence back upon him. 19Then he shall turn his face back toward the fortresses of his own land; but he shall stumble and fall, and shall not be found. 20Then shall arise in his place one who shall send an exactor of tribute through the glory of the kingdom; but within a few days he shall be broken, neither in anger nor in battle.	21In his place shall arise a contemptible person to whom royal majesty has not been given; he shall come in without warning and obtain the kingdom by flatteries. 22Armies shall be utterly swept away before him and broken, and the prince of the covenant also. 23And from the time that an alliance is made with him he shall act deceitfully; and he shall become strong with a small people. 24Without warning he shall come into the richest parts of the province; and he shall do what neither his fathers nor his fathers' fathers have done, scattering among them plunder, spoil, and goods.
Daniel's Seventy 'Weeks'		Its end shall come with a flood, and to the end there shall be war; desolations are decreed. 27And he shall make a strong covenant with many for one week; and for half of the week he shall cause sacrifice and offering to cease; and upon the wing of abominations shall come one who makes desolate, until the decreed end is poured out on the desolator."
Daniel's Vision of a Ram and a Goat		23And at the latter end of their rule, when the transgressors have reached their full measure, a king of bold countenance, one who understands riddles, shall arise. 24His power shall be great, and he shall cause fearful destruction, and shall succeed in what he does, and destroy mighty men and the people of the saints. 25By his cunning he shall make deceit prosper under his hand, and in his own mind he shall magnify himself. Without warning he shall destroy many; and he shall even rise up against the Prince of princes; but, by no human hand, he shall be broken. 26The vision of the evenings
Daniel's Dream of Four Beasts	18But the saints of the Most High shall receive the kingdom, and possess the kingdom for ever, for ever and ever.' 19Then I desired to know the truth concerning the fourth beast, which was different from all the rest, exceedingly terrible, with its teeth of iron and claws of bronze; and which devoured and broke in pieces, and which stamped the residue with its feet; 20and concerning the ten horns that were on its head, and the other horn which came up and before which three of them fell, the horn which had eyes and a mouth that spoke great things, and which seemed greater than its fellows.	
Nebuchadnezzar's Dream	41And as you saw the feet and toes partly of potter's clay and partly of iron, it shall be a divided kingdom; but some of the firmness of iron shall be in it, just as you saw iron mixed with the miry clay. 42And as the toes of the feet were partly iron and partly clay, so the kingdom shall be partly strong and partly brittle. 43As you saw the iron mixed with miry clay, so they will mix with one another in marriage, but they will not hold together, just as iron does not mix with clay. 44And in the days of those kings the God of heaven will set up a kingdom which shall never be destroyed, nor shall its sovereignty be left	

Nebuchadnezzar's Dream	Daniel's Dream of Four Beasts	Daniel's Vision of a Ram and a Goat	Daniel's Seventy 'Weeks'	The Kings of the South and the North
EUROPEAN UNION, THE ANTICHRIST, & THE END-TIMES (Continued)				
to another people. It shall break in pieces all these kingdoms and bring them to an end, and it shall stand for ever; 45just as you saw that a stone was cut from a mountain by no human hand, and that it broke in pieces the iron, the bronze, the clay, the silver, and the gold.	21As I looked, this horn made war with the saints, and prevailed over them, 22until the Ancient of Days came, and judgment was given for the saints of the Most High, and the time came when the saints received the kingdom. 23"Thus he said: 'As for the fourth beast, there shall be a fourth kingdom on earth, which shall be different from all the kingdoms, and it shall devour the whole earth, and trample it down, and break it to pieces. 24As for the ten horns, out of this kingdom ten kings shall arise, and another shall arise after them; he shall be different from the former ones, and shall put down three kings. 25He shall speak words against the Most High, and shall wear out the saints of the Most High, and shall think to change the times and the law; and they shall be given into his hand for a time, two times, and half a time. 26But the court shall sit in judgment, and his dominion shall be taken away, to be consumed and destroyed to the end. 27And the kingdom and the dominion and the greatness of the kingdoms under the whole heaven shall be given to the people of the saints of the Most High; their kingdom shall be an everlasting kingdom, and all dominions shall serve and obey them.	and the mornings which has been told is true; but seal up the vision, for it pertains to many days hence."		He shall devise plans against strongholds, but only for a time. 25And he shall stir up his power and his courage against the king of the south with a great army; and the king of the south shall wage war with an exceedingly great and mighty army; but he shall not stand, for plots shall be devised against him. 26Even those who eat his rich food shall be his undoing; his army shall be swept away, and many shall fall down slain. 27And as for the two kings, their minds shall be bent on mischief; they shall speak lies at the same table, but to no avail; for the end is yet to be at the time appointed. 28And he shall return to his land with great substance, but his heart shall be set against the holy covenant. And he shall work his will, and return to his own land. 29At the time appointed he shall return and come into the south; but it shall not be this time as it was before.30For ships of Kittim shall come against him, and he shall be afraid and withdraw, and shall turn back and be enraged and take action against the holy covenant. He shall turn back and give heed to those who forsake the holy covenant. 31Forces from him shall appear and profane the temple and fortress, and shall take away the continual burnt offering. And they shall set up the abomination that makes desolate. 32He shall seduce with flattery those who violate the covenant; but the people who know their God shall stand firm and take

Nebuchadnezzar's Dream	Daniel's Dream of Four Beasts	Daniel's Vision of a Ram and a Goat	Daniel's Seventy 'Weeks'	The Kings of the South and the North
EUROPEAN UNION, THE ANTICHRIST, & THE END-TIMES (Continued)				
				action. 33And those among the people who are wise shall make many understand, though they shall fall by sword and flame, by captivity and plunder, for some days. 34When they fall, they shall receive a little help. And many shall join themselves to them with flattery; 35and some of those who are wise shall fall, to refine and to cleanse them and to make them white, until the time of the end, for it is yet for the time appointed. 36"And the king shall do according to his will; he shall exalt himself and magnify himself above every god, and shall speak astonishing things against the God of gods. He shall prosper till the indignation is accomplished; for what is determined shall be done. 37He shall give no heed to the gods of his fathers, or to the one beloved by women; he shall not give heed to any other god, for he shall magnify himself above all. 38He shall honor the god of fortresses instead of these; a god whom his fathers did not know he shall honor with gold and silver, with precious stones and costly gifts. 39He shall deal with the strongest fortresses by the help of a foreign god; those who acknowledge him he shall magnify with honor. He shall make them rulers over many and shall divide the land for a price. 40"At the time of the end the king of the south shall attack him; but the king of the north shall rush upon him like a whirlwind, with chariots and horsemen, and with many ships;

Nebuchadnezzar's Dream	Daniel's Dream of Four Beasts	Daniel's Vision of a Ram and a Goat	Daniel's Seventy 'Weeks'	The Kings of the South and the North
				EUROPEAN UNION, THE ANTICHRIST, & THE END-TIMES (Continued)
				and he shall come into countries and shall overflow and pass through. 41He shall come into the glorious land. And tens of thousands shall fall, but these shall be delivered out of his hand: Edom and Moab and the main part of the Ammonites. 42He shall stretch out his hand against the countries, and the land of Egypt shall not escape. 43He shall become ruler of the treasures of gold and of silver, and all the precious things of Egypt; and the Libyans and the Ethiopians shall follow in his train. 44But tidings from the east and the north shall alarm him, and he shall go forth with great fury to exterminate and utterly destroy many. 45And he shall pitch his palatial tents between the sea and the glorious holy mountain; yet he shall come to his end, with none to help him.
				12 1"At that time shall arise Michael, the great prince who has charge of your people. And there shall be a time of trouble, such as never has been since there was a nation till that time; but at that time your people shall be delivered, every one whose name shall be found written in the book. 2And many of those who sleep in the dust of the earth shall awake, some to everlasting life, and some to shame and everlasting contempt. 3And those who are wise shall shine like the brightness of the firmament; and those who turn many to righteousness, like the stars for ever and ever. 4But you, Daniel, shut up the words, and seal the book, until the time of the end. Many shall run to and fro, and knowledge

	EUROPEAN UNION, THE ANTICHRIST, & THE END-TIMES (Continued)			
Nebuchadnezzar's Dream	*Daniel's Dream of Four Beasts*	*Daniel's Vision of a Ram and a Goat*	*Daniel's Seventy 'Weeks'*	*The Kings of the South and the North*
				shall increase." 5Then I Daniel looked, and behold, two others stood, one on this bank of the stream and one on that bank of the stream. 6And I said to the man clothed in linen, who was above the waters of the stream, "How long shall it be till the end of these wonders?" 7The man clothed in linen, who was above the waters of the stream, raised his right hand and his left hand toward heaven; and I heard him swear by him who lives for ever that it would be for a time, two times, and half a time; and that when the shattering of the power of the holy people comes to an end and all these things would be accomplished. 8I heard, but I did not understand. Then I said, "O my lord, what shall be the issue of these things?" 9He said, "Go your way, Daniel, for the words are shut up and sealed until the time of the end. 10Many shall purify themselves, and make themselves white, and be refined; but the wicked shall do wickedly; and none of the wicked shall understand; but those who are wise shall understand. 11And from the time that the continual burnt offering is taken away, and the abomination that makes desolate is set up, there shall be a thousand two hundred and ninety days. 12Blessed is he who waits and comes to the thousand three hundred and thirty-five days.

	CLOSING			
Nebuchadnezzar's Dream	*Daniel's Dream of Four Beasts*	*Daniel's Vision of a Ram and a Goat*	*Daniel's Seventy 'Weeks'*	*The Kings of the South and the North*
A great God has made known to the king what shall be hereafter. The dream is certain, and its interpretation sure." ₄₆Then King Nebuchadnez'zar fell upon his face, and did homage to Daniel, and commanded that an offering and incense be offered up to him. ₄₇The king said to Daniel, "Truly, your God is God of gods and Lord of kings, and a revealer of mysteries, for you have been able to reveal this mystery."	₂₈"Here is the end of the matter. As for me, Daniel, my thoughts greatly alarmed me, and my color changed; but I kept the matter in my mind."	₂₇And I, Daniel, was overcome and lay sick for some days; then I rose and went about the king's business; but I was appalled by the vision and did not understand it.		₁₃But go your way till the end; and you shall rest, and shall stand in your allotted place at the end of the days."

Appendix II
Other Oddities

During the time I wrote this book, other thoughts came to mind that seemed odd to me, but appeared not to fit with the methods God uses, as I understood them. I have recorded them in this Appendix, in the chance I may be mistaken, so that other Christians who study, in depth, the things I write about might see how or if they apply.

1.) There are 29.53059 days in one lunar cycle. Another way to say this is the average number of days between full moons is 29.53059 days. The Jews use a calendar based on the phases of the moon whereas the Gregorian calendar of today is based on the sun. This cycle is used to set the timing for the Jewish months of the year. Therefore, the number of days in a normal Jewish year is twelve months times 29.53059 days, which equals 354.367 days.

Rounding gives us 354 days for a regular Jewish year and every so often, as we have discussed, a "thirteenth" month is added to keep the calendar in alignment with the seasons of the year. Now we know that the highest form of the number of completion is one thousand and if you add this number to 354.367 days for the Jewish year, we get 1,354.367 days. This time is extremely close to Daniel's 1,335 prophetic days, which are 1,354.4688 real days. This is a difference of .10175 days or two hours, twenty-six and a half minutes. Is this the real reason why Daniel was told 1,335 days or was it because the time was based on using God's number times forty and adding the time Christ was dead?

2.) Converting twenty-three hundred prophetic years into real time yields an extra 33.541666 years.[21] This number is identical to the number of years for Christ's first coming determined in chapter 17. Hence, 33.541666 years is 12,175 days plus 75 P-days. Why are these numbers the same? Also, if Daniel's time is really twenty-four hundred days instead of twenty-three hundred days, then the antichrist will begin on July 22, 2018 instead of September 21. This was the day determined for the baptism of Christ.

3.) Christ was resurrected on Nisan 16 and, after forty days, ascended into heaven. This would have been on the twenty-seventh day of the second Jewish month Iyar (Ziv in the Bible) or Saturday, May 16, 33 A.D. In my first book, I determined the flood began in the year 2403 B.C. and ended one year later. The Bible says in Genesis that Noah entered the ark on the forty-seventh day of the year and left the ark on the fifty-seventh day of the year. Iyar 27 is the fifty-seventh day of the year! Furthermore, in the year 2403 B.C. the seventeenth day of Iyar (47th day of the year when Noah entered the ark) was May 17 on the Julian calendar.

This time differential is strange because it is nearly 2,435 years exactly (within a day) and suggests two things of further interest. First, that Christ may have ascended into heaven on the same day of the year as Noah entered and left the ark (separated by one year). Secondly, that twenty-four hundred prophetic days (2,435 years) may be the right choice instead of twenty-three hundred P-days. Let us look at a timeline for God's overall plan for mankind that has this information on it, as well as other milestones that are different from the symmetrical timelines that were provided in chapter one.

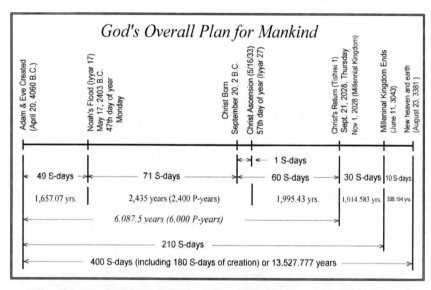

Adding the 180 S-days from creation to the 210 S-days yields 390 S-days total time from creation to the end of the Millennial Kingdom. Based on our studies, this would imply we are missing ten S-days and so I showed them on this chart so that we could get to four hundred S-days—a number God uses. These ten S-days could easily have been before creation, but I put them here which suggests this is the amount of time needed to remake heaven and earth. Ten S-days is 333.333 P-years or 338.194 real years... God's number in one of its highest forms. If we were to add up the total times, we would get 13,527.777 years (12,175 years + 1,014.5833 + 338.194 years). The same number as is in the plan for the Messiah applying a day for a year! And if we add the time for the Kingdom with these unknown ten S-days, we get 1,352.777 years. The numbers suggest it will take 333.333 P-years to remake heaven and earth and that they will be officially reborn in the year 3381 (God's converted number).

Calculations

Calculations are shown in the chapters where they are first discussed. Occasionally some equations are revisited in later chapters. When this occurs, those equations reference back to the original chapter they were first examined to eliminate redundancy.

CHAPTER 1

1) $2,000 \text{ P-yrs.} \times \dfrac{365.25 \text{ days}}{1 \text{ year}} \times \dfrac{1 \text{ year}}{360 \text{ P-days}} = 2,029.1666 \text{ yrs.}$

2) $1,000 \text{ P-yrs.} \times \dfrac{365.25 \text{ days}}{1 \text{ year}} \times \dfrac{1 \text{ year}}{360 \text{ P-days}} = 1,014.58333 \text{ yrs.}$

   $\& \ .5833 \text{ yrs.} \times \dfrac{12 \text{ months}}{1 \text{ year}} = 7 \text{ months}$

   So 1,014.5833 years equals 1,000 years plus 14 years plus 7 months exactly!

CHAPTER 2

3) $2,450 \text{ P-yrs.} \times \dfrac{365.2425 \text{ days}}{1 \text{ year}} \times \dfrac{1 \text{ year}}{360 \text{ P-days}} = 2,485.68 \text{ yrs.}$

4) -457.78 (March 21, 458 B.C.; B.C. numbers are negative) + 2,485.68 years = 2,027.898 years.

5) $69 \times 7 \text{ P-yrs.} = 483 \text{ P-years}$

6) $483 \text{ P-yrs.} \times \dfrac{365.2425 \text{ days}}{1 \text{ year}} \times \dfrac{1 \text{ year}}{360 \text{ P-days}} = 490.03369 \text{ yrs.}$

   $\& \ .03369 \text{ yrs.} \times \dfrac{365.2425 \text{ days}}{1 \text{ year}} = 12.3041 \text{ days or 13 days}$

   when rounding. So 483 P-yrs. equals 490 years plus 13 days!

7) 490 P-yrs. x 5 = 2,450 P-years.

8) 7 x 7 (weeks) = 49 weeks and since weeks are substituted with Jubilee years we get 49 x 50 years = 2,450 years.

9) -2 B.C. + 2,029.1666 years + 1 year for the B.C./A.D. transition = 2028 A.D. Negative numbers are used for B.C. dates.

10) 49 P-yrs. x $\frac{365.25 \text{ days}}{1 \text{ year}}$ x $\frac{1 \text{ year}}{360 \text{ P-days}}$ = 49.71 yrs. So

$$-458 \text{ B.C.} + 49.71 \text{years} = 408 \text{ B.C.}$$

11) 2,400 P-yrs. x $\frac{365.25 \text{ days}}{1 \text{ year}}$ x $\frac{1 \text{ year}}{360 \text{ P-days}}$ = 2,435 yrs. exactly

12) -408 B.C. + 2,435 years + 1 year B.C./A.D. adj. = 2028 A.D.

13) 2,300 P-yrs. x $\frac{365.25 \text{ days}}{1 \text{ year}}$ x $\frac{1 \text{ year}}{360 \text{ P-days}}$ = 2,333.54 yrs.

& -356 B.C. + 2,333.5 years + 1 year B.C./A.D. adj. = 1979 A.D.

14) 3043 A.D. - 1,014.58333 years = 2028 A.D.

15) -2 B.C. + 2,029.1666 + 1 yr. B.C./A.D. adj. = 2028 A.D.

16) 2,029.1666 years / 42 generations = $\frac{48.313 \text{ yrs.}}{1 \text{ generation}}$

CHAPTER 3   ## CHAPTER 4

None   None

CHAPTER 5

17) 70 'weeks' = 70 Jubilees = 70 x 50 years = 3,500 years

18) 2,000 P-yrs. x $\dfrac{360 \text{ P-days}}{1 \text{ year}}$ x $\dfrac{1 \text{ year}}{365.25 \text{ days}}$ = 1,971.25 P²-yrs.

19) -458 B.C. + 2,300 yrs. + 1 yr. B.C./A.D. adj. = 1843 A.D.

20) -306 B.C. + 2,333.5 yrs. = 2028 A.D.

21) 2,300 P-days x $\dfrac{1 \text{ year}}{360 \text{ P-days}}$ = 6.388 yrs. Translating these back into real

days: 6.388 years x $\dfrac{365.25 \text{ days}}{1 \text{ year}}$ = 2,333.541666 days.

22) $\dfrac{12{,}000 \text{ P-yrs.}}{1 \text{ Season}}$ x $\dfrac{1 \text{ Season}}{360 \text{ S-days}}$ = $\dfrac{33.333 \text{ P-yrs.}}{1 \text{ S-day}}$

23) 2,333.5 yrs. x $\dfrac{1 \text{ S-day}}{33.333 \text{ P-yrs.}}$ = 70 S-days

24) 2,435 yrs. x $\dfrac{1 \text{ year}}{365.25 \text{ days}}$ = 6.666 years

25) $\dfrac{365.25 \text{ days}}{1 \text{ year}}$ x $\dfrac{1 \text{ year}}{360 \text{ P-days}}$ = $\dfrac{1.01458333}{\text{P-}}$ conversion factor

26) 1 'week' = 7 P-yrs. &

7 P-yrs. x $\dfrac{1.01458333}{\text{P-}}$ = 7.1020833 years

27) 3.5 P-yrs. x $\dfrac{365 \text{ days}}{1 \text{ year}}$ x $\dfrac{1 \text{ year}}{360 \text{ P-days}}$ = 3.5486111 years

28) $\underline{24 \text{ hours}}$ x .5486111 days = 13.1666 hours
    1 day

29) 1 day = 24 hours so 6 days = 6 x 24 hours or 144 hours

30) Substituting a year for a day in equation 28 above we get:

6 years = 144 hours and multiplying by 1,000,

6 yrs. x 1,000 = 144 hrs. x 1,000 so 6,000 yrs. = 144,000 hrs.

6,000 P-years = 144,000 P-hours or $\underline{144,000 \text{ P-hours}}$
                                        6,000 P-years

31) From equation 22: $\underline{33.333 \text{ P-yrs.}}$ x $\underline{\text{ 1 S-day }}$ = 1.3888 P-yrs.
                          1 S-day   24 S-hrs.   1 S-hour

$\underline{1.3888 \text{ P-yrs.}}$ x $\underline{1.01458333}$ = $\underline{1.40914 \text{ yrs.}}$ Therefore:
    1 S-hour   P-   1 S-hour

$\underline{1.40914 \text{ yrs.}}$ x 0.5 = $\underline{0.7045717 \text{ yrs.}}$ Or: $\underline{8 \text{ months \& 14 days}}$
    1 S-hour   ½ S-hour   ½ S-hour

32) $\underline{\text{ 6,000 P-yrs. }}$ = $\underline{\text{ 1 year }}$ x $\underline{365.25 \text{ days}}$ = $\underline{15.21875 \text{ days}}$ &
    144,000 P-hrs.   24 hrs.   1 year   1 hour

$\underline{15.2188 \text{ days}}$ x .5 = 7.609375 days or 7 days 14 hrs. 37.5 mins.
    1 hour

$\underline{\text{ 1 year }}$ x $\underline{365 \text{ days}}$ = $\underline{15.20833 \text{ days}}$ &
    24 hrs.   1 year   1 hour

$\underline{15.20833 \text{ days}}$ x .5 = 7.604166 days or 7 days 14 hrs. 30 mins.
    1 hour

$\underline{\text{ 1 year }}$ x $\underline{366 \text{ days}}$ = $\underline{15.25 \text{ days}}$ &
    24 hrs.   1 year   1 hour

$\underline{15.25 \text{ days}}$ x .5 = 7.625 days or 7 days 15 hrs. exactly
1 hour

CHAPTER 6

33) 12,000 P-yrs. x $\dfrac{365.25 \text{ days}}{1 \text{ year}}$ x $\dfrac{1 \text{ year}}{360 \text{ P-days}}$ = 12,175 yrs.

34)　　　　1 S-day = 33.333 P-yrs. & So:

33.333 P-yrs. x $\dfrac{360 \text{ P-days}}{1 \text{ P-year}}$ = 12,000 P-days

35)　　　$\dfrac{365.25 \text{ days}}{1 \text{ year}}$ x $\dfrac{1 \text{ year}}{360 \text{ P-days}}$ = $\dfrac{1.01458333}{\text{P-}}$

12,000 P-days x $\dfrac{1.01458333}{\text{P-}}$ = 12,175 days

36) 6,087.5 years x $\dfrac{1 \text{ Jubilee}}{50 \text{ years}}$ = 121.75 Jubilees

37) $\dfrac{365.25 \text{ days}}{1 \text{ year}}$ x $\dfrac{1 \text{ year}}{12 \text{ months}}$ = $\dfrac{30.4375 \text{ days}}{1 \text{ month}}$

38) 1,260 P-days x $\dfrac{1.01458333}{\text{P-}}$ = 1,278.375 days or 1,279 days

39) 12,175 years x $\dfrac{12 \text{ months}}{1 \text{ year}}$ = 146,100 months

40) 144,000 P- x $\dfrac{1.01458333}{\text{P-}}$ = 146,100

41) 2,108.771858 days x $\dfrac{\text{P-}}{1.01458333}$ = 2,078.460969 P-days

42) $\underline{\text{12,175 years}}$ x $\underline{\text{365.25 days}}$ = 4,446,918.75 days
     Overall Plan   1 year

43) $\underline{\text{12,000 P-years}}$ x $\underline{\text{360 P-days}}$ = 4,320,000 P-days
     Overall Plan   1 year

44) $\underline{\text{4,320,000 P-days}}$ x $\underline{\text{Overall Plan}}$ = 1,440,000 P-days
     Overall Plan   3 trimesters

45) 40 P-hours x $\underline{\text{1.01458333}}$ = 40.58333 hours
                  P-

46) 3,600 P-days x $\underline{\text{1 Gn-day}}$ = 108 Gn-days
                  33.333 P-days

47) 2,333.54166 days x $\underline{\text{1 Gn-day}}$ = $\underline{\text{70.00625 Gn-days}}$ =
                  33.333 P-days   P-

     $\underline{\text{70 Gn-days}}$ + $\underline{\text{.00625 Gn-days}}$ & "a day is like a year" $\therefore$
       P-   P-

$\underline{\text{.00625 Gn-yrs.}}$ x $\underline{\text{33.333 P-yrs.}}$ x $\underline{\text{P-}}$ x $\underline{\text{365.25 days}}$ = 75 P-days
    P-   1 Gn-yrs.   1.0145833   1 year

----

48) 7 years x $\underline{\text{360 P-days}}$ = 2,520 P-days
              1 year

49) 1,335 P-days x $\underline{\text{1 Gn-day}}$ = 40.05 Gn-days
                  33.333 P-days

50) 0.05 Gn-days x $\underline{\text{33.333 P-days}}$ = 1.666 P-days or 40 P-hrs.
                 1 Gn-day

51) 6 H-days = 6,000 P-years x $\underline{\text{1.01458333}}$ = 6,087.5 years
                                   P-

52) 12 hours x <u>60 minutes</u> x <u>60 seconds</u> = 43,200 seconds... or
      1 hour   1 minute

    3 blocks of 14,400 seconds!

53) <u>43,200 seconds</u> x <u>Master Clock</u> = 3.54825462 sec. per year
      Master Clock   12,175 years

54) 3.548255 sec. x <u>   P-   </u> = 3.497253 P-sec. or 3.5 P-sec.
                          1.01458333

55) 20 years x <u>3.5 P-seconds</u> = 70 P-seconds
                        1 year

56) .037037 yrs. x <u>365.25 days</u> = 13.527.77 days; 14 days rounded.
                        1 year

57) 37.037037037 years x <u>365.25 days</u> = 13,527.777 days
                                1 year

58) 1,333.333 P-days x <u>1.01458333</u> = 1,352.777 days
                                P-

59) 12,175 days x <u> 1 year </u> = 33.333 years
                        365.25 days

CHAPTER 7

None

CHAPTER 8

60) "time" = t , so "time (t), times (2t), and ½ time (½ t)" = 3½ t

3½ t = 1,260 P-days   Therefore: t = <u>1,260 P-days</u> = 360 P-days
                                              3.5

CHAPTER 9	CHAPTER 10
None	None

CHAPTER 11

61) 180 miles x $\frac{5{,}280 \text{ feet}}{1 \text{ mile}}$ = 950,400 feet = 6.6 x 144,000

CHAPTER 12

62) Jerusalem: Latitude: 31 degrees 47 minutes North

               Longitude: $\underline{35 \text{ degrees} \quad 13 \text{ minutes}}$ East (2 hrs)

                      66 degrees 60 minutes $\Rightarrow$ 66.6

CHAPTER 13	CHAPTER 14
None	None

CHAPTER 15

None

CHAPTER 16

63)   4 years x $\frac{365.25 \text{ days}}{1 \text{ year}}$ = 1,461 days

     1,440 P-days x $\frac{1.01458333}{\text{P-}}$ = 1,461 days

---

       40 years x $\frac{365.25 \text{ days}}{1 \text{ year}}$ = 14,610 days

   14,400 P-days x $\frac{1.01458333}{\text{P-}}$ = 14,610 days

---

     400 years x $\frac{365.25 \text{ days}}{1 \text{ year}}$ = 146,100 days

144,000 P-days x <u>1.01458333</u> = 146,100 days
P-

4,000 years x <u>365.25 days</u> = 1,461,000 days
1 year

1,440,000 P-days x <u>1.01458333</u> = 1,461,000 days
P-

----

64)   <u>12,175 years</u>   = 16.9097222 years per minute
720 minutes

65) 40 P-hours x <u>1 day</u> = 1.666 P-days
24 hours

66) 1.666 P-days x 1.01458333 = 1.69097222 days

67) 50 S-days x <u>33.819444 years</u> = 1,690.97222 years
1 S-day

68) 4,000 + 400 + 40 + 4 + .4 + .04 + .004 + etc. = 4,444.444

69) 7,000 + 700 + 70 + 7 + .7 + .07 + .007 + etc. = 7,777.777

70) $\sqrt{4,444.444}$ = 66.666

71) 1,352.777 days + 40 P-hrs. (1.69097222 days) = 1,354.469 days

72) 1,354.469 days x <u>P-</u> = 1,335 P-days
1.01458333

73) 13.54469 hours - 13.52777 hours = .0169097222 hours

74) .0169097222 hours x <u>60 minutes</u> = 1.01458333 minutes
1 hour

75) 3.7 years x $\underline{365.25 \text{ days}}$ = 1,351.425 days
                          1 year

76) 1,354.46875 days − 1,351.425 days = 3.04375 days

CHAPTER 17

77) 1,260 P-days x $\underline{365 \text{ days}^*}$ x $\underline{\quad 1 \text{ year} \quad}$ = 1,277.5 days
                          1 year   360 P-days

* no leap-days during this span (1,277 days when rounding down)

78) 12,175 days + 75 P-days. x $\underline{1.01458333}$ = 12,251.09375
                                              P-

CHAPTER 18

None

CHAPTER 19

None

References

CHAPTER 1

[a] Genesis 12:1-4
[b] Revelation 21:1
[c] Luke 20:34-36
[d] Acts 17:10-15

CHAPTER 2

[a] Daniel 12:7
[b] "2028"; Vargo, pgs.80-81
[c] Ezra 7:8
[d] Ezra 8:31
[e] Matthew 24:22
[f] Luke 9:22, 18:32
[g] Psalm 90:4
[h] Daniel 7:25
[I] Revelation 4:4
[j] Daniel 8:1-14
[k] "2028"; Vargo, pg.186
[l] "2028"; Vargo, pg.241
[m] Revelation 17:11
[n] Leviticus 25:8-12
[o] Revelation 21:1
[p] Revelation 20:2
[q] Matthew 13:36-43
[r] Isaiah 46:10-11
[s] Psalm 90:10
[t] Revelation 11:1-2
[u] http://ezinearticles.com/
?Pregnancy-And-
Childbirth---What-Are-
The-Three-Trimesters?
&id=476452
[v] Genesis 6:3
[w] Exodus 16:34
[x] Genesis 15:13-16
[y] Matthew 24:43-44

CHAPTER 3

[a] Ezra 7:8-9
[b] Matthew 12:40
[c] Exodus 12:21-25
[d] Matthew 27:55-61
[e] Luke 23:55-56

CHAPTER 4

None

CHAPTER 5

[a] Leviticus 25:8-12
[b] Ezekiel 4:4-5
[c] Daniel 11:32
[d] Genesis 7:11, 8:14
[e] Revelation 13:18
[f] Revelation 13:18
[g] Revelation 8:1

CHAPTER 6

[a] "2028", pg. 57, pub. 2008
[b] "2028"; Vargo, pgs. 34-38,

pub. 2008
(c) Genesis 6:3
(d) Revelation 11:15
(e) Revelation 13:4
(f) Revelation 11:15
(g) Luke 12:49-53
(h) Daniel 9:27
(I) Matthew 25:1-13

CHAPTER 7

(a) http://www.libertyforlife
.com/jail-police/us_
concentration_camps.htm
(b) http://www.handofhelp.com
/vision_1.php

CHAPTER 8

(a) Matthew 24:15
(b) Amos 3:1-7
(c) Matthew 12:38-39
(d) 2nd Thessalonians 2:9-10

CHAPTER 9

(a) Luke 12:46
(b) Revelation 19:15
(c) Revelation 7:9
(d) Revelation 17:15
(e) Revelation 18:10
(f) Daniel 8:25, 10:13, 12:1
(g) Revelation 12:4
(h) Revelation 12:4
(I) (http://volcano.und.edu/
vwdocs/frequent_questions/
grp7/asia/question879.html)
(j) Daniel 7:4, Revelation 4:7,
Hosea 8:1, Ezekiel 1:10
(k) Daniel 11:44
(l) Zechariah 13:8-9
(m) Matthew 5:33-37
(n) Revelation 5:1-7
(o) Zechariah 13:1-6
(p) Matthew 24:16-21,
Mark 13:14-19
(q) Revelation 12:12
(r) 2nd Thessalonians 2:10-12
(s) Revelation 4:5, 8:5, 10:3-4,
14:2, 16:18, 19:6
(t) Zechariah 14:4

CHAPTER 10

(a) Revelation 16:14-16
(b) Malachi 4:5, Luke 1:17
(c) Matthew 11:11-15
(d) Luke 1:15
(e) Hebrews 9:27
(f) 2nd Kings 2:11
(g) Genesis 5:24
(h) Mark 13:22
(I) Luke 4:1-13
(j) Zechariah 4:14
(k) Zechariah 4:6,8
(l) Daniel 11:33-35
(m) Revelation 11:5
(n) Daniel 3:10-11, 6:7
(o) Ephesians 6:10-19
(p) Zechariah 4:9
(q) Daniel 11:24, 32, 39

(r) Revelation 17:8,11. "2028", Chapter 24, pub. 2008

(s) Daniel 11:45

(t) Acts 5:12

(u) Daniel 7:26-27

(v) Matthew 24:22

(w) Ezekiel 5:1-2, 12

(x) Revelation 14:1-6

CHAPTER 11

(a) Matthew 24:25

(b) Psalm 2, Rev. 3:26-27, Revelation 19:15

(c) Daniel 9:27, 11:30-32

(d) 2nd Thessalonians 2:4

(e) Matthew 24:16-22

(f) Revelation 11:13

(g) Revelation 12:9

(h) Revelation 17:15

(I) Daniel 7:23

(j) Rev.17:12, Daniel 2:42-44, Daniel 7:8 & 24

(k) Revelation 17:3, 10

(l) Revelation 17:3

(m) Daniel 2:39, 7:6 & 17

(n) Daniel 2:44, 8:8-9

(o) Daniel 2:39, 7:5 & 17

(p) Daniel 2:36-38, 7:4 & 17

(q) Daniel 4:4, 22, 25, 30

(r) 2nd Thessalonians 2:3-4

(s) Revelation 13:2, 4, 5, 7 & 17:12-13

(t) Revelation 13:3, 12,14

(u) Zechariah 11:4-5 & 17

(v) Matthew 25:1-13

(w) Revelation 17:8, 11

(x) Exodus 13:9 & 16. Deuteronomy 6:4-9, & 11:18, Ezekiel 9:4, Revelation 7:3, 9:4, 14:1, & 22:4

(y) Daniel 11:34

(z) Leviticus 19:28

(aa) Zechariah 14:4

(ab) Acts 1:10-12

(ac) Revelation 7:9-17

ad) Matthew 24:35

ae) Revelation 19:15

CHAPTER 12

(a) Daniel 2, 7, & 8

(b) Zechariah 13:8, Ezekiel 4:2

(c) Revelation 7:9

(d) Rev. 19:20, Daniel 11:45

(e) Revelation 19:20

(f) Daniel 7:7, 24

(g) Revelation 17:17

(h) Daniel 7:8, 20, 24

(i) Zechariah 3:1-10, Rev. 19:8

(j) Revelation 18:8, 18

(k) Revelation 18:10, 14, 17

(l) Revelation 18:17-18

(m) Daniel 11:45

(n) Daniel 11:25,28

(o) Daniel 11:29

(p) Daniel 11:27

(q) Daniel 11:30

(r) Daniel 9:27, 11:23 & 31

(s) Matthew 24:15

(t) Revelation 20:4

(u) Revelation 7:13-14

(v) Revelation 13:5. 2nd Thessalonians 2:4-10

(w) Matthew 25:2-3

(x) Matthew 25:1-13

(y) Ephesians 6:17, Hebrews 4:12-13, Revelation 1:16, 2:16, 19:21

(z) Zechariah 14:1-21

(aa) Matthew 13:37-43

(ab) Matthew 21:18-22

(ac) Revelation 20:5-6

CHAPTER 13

(a) Revelation 19:11

(b) Revelation 2:12

(c) Matthew 18:22

(d) Revelation 19:15

(e) Matthew 25:1-13

(f) 2nd Thessalonians 2:9-10

(g) Revelation 19:17-18

(h) Zechariah 14:5-7

(I) 1Corinthians 15:50-52

CHAPTER 14

(a) Daniel 11:33

(b) "2028"; Vargo, pgs. 72, 241, pub. 2008

(c) "2028"; Vargo, pg. 254, published: 2008

(d) Leviticus 25:1-7

(e) "2028"; Vargo, pg.186

(f) 2nd Thessalonians 2:6-7

(g) Daniel 7:25

(h) Zechariah 14:16-17

(I) Daniel 11:45

(j) Daniel 11:24

(k) Revelation 13:4-5

(l) "2028"; Vargo, pgs.186-187

CHAPTER 15

(a) Revelation 11:2

(b) "2028"; Vargo, pg.186-187

(c) Ephesians 2:9-10

CHAPTER 16

(a) Rev.14:9-10, 19:20:21

(b) John 10:9

(c) "2028", pgs. 134-137, pub. 2008

(d) Genesis 15:13, Acts 7:6

CHAPTER 17

(a) Leviticus 16:29-31

(b) "2028"; Vargo, pg.135

(c) Luke 3:23

(d) Leviticus 23:33-44

(e) Zechariah 14:18-19

(f) Luke 4:1-13, Mark 1:9-13

(g) John 2:19-22

(h) Leviticus 12:1-3

(i) Leviticus 12:4-8

(j) Luke 2:21-24

CHAPTER 18

(a) 2nd Timothy 4:1-5
(b) Matthew 13:41
(c) Revelation 16:13-14
(d) Zech. 14:1-2, Daniel 11:45
(e) Genesis 8-4
(f) Luke 12:46-48
(g) Matthew 12:8

CHAPTER 19

None

Index

Index

R

S

About the Author

R. H. Vargo is retired and lives in the Midwest part of the United States. He is married and has three adult children. He has no formal theological training other than years of self-study and shepherding from great pastors. He is an engineer by training and is knowledgeable in the areas of financial investing, eschatology, dieting and health, mathematics, computer software, and many other subjects. A regular "jack-of-all trades" and master of none. He is an avid reader of science fiction, prophecy, and the Bible. You may contact him at http://www.christsreturn2028.com.

LaVergne, TN USA
17 December 2009
167329LV00003B/2/P